SEEING THROUGH THE VEIL
OPTICAL THEORY AND MEDIEVAL ALLEGORY

Suzanne Conklin Akbari

Seeing through the Veil

OPTICAL THEORY AND MEDIEVAL ALLEGORY

UNIVERSITY OF TORONTO PRESS
Toronto Buffalo London

ISBN 0-8020-3605-8

Printed on acid-free paper

National Library of Canada Cataloguing in Publication

Akbari, Suzanne Conklin
Seeing through the veil : optical theory and medieval
allegory / Suzanne Conklin Akbari.

Includes bibliographical references and index.
ISBN 0-8020-3605-8

1. Literature, Medieval – History and criticism.   2. Allegory.
I. Title.

PN682.A5A32 2004   809'.915   C2003-906291-0

University of Toronto Press acknowledges the financial assistance to its
publishing program of the Canada Council for the Arts and the
Ontario Arts Council.

This book has been published with the help of a grant from the
Canadian Federation for the Humanities and Social Sciences, through
the Aid to Scholarly Publications Programme, using funds provided by the
Social Sciences and Humanities Research Council of Canada.

University of Toronto Press acknowledges the financial support for its
publishing activities of the Government of Canada through the
Book Publishing Industry Development Program (BPIDP).

For Omama

CONTENTS

Preface   ix

1   **Illumination and Language**   3

2   **The Multiplication of Forms**   21

3   **Guillaume de Lorris's *Roman de la rose***   45

4   **Jean de Meun's *Roman de la rose***   78

5   **Dante's *Vita nuova* and *Convivio***   114

6   **Dante's *Commedia***   138

7   **Chaucer's Dream Visions**   178

8   **Chaucer's Personification and Vestigial Allegory in the *Canterbury Tales***   211

9   **Division and Darkness**   234

Notes   245

Bibliography   307

Index   339

PREFACE

This is a study of several allegories by authors including Dante and Chaucer in the context of medieval theories of optics. It differs from previous studies both in the depth of its engagement with the medieval scientific and philosophical background, and in the breadth of the literature surveyed. The book is intended to be useful not only to readers interested in the specific allegories discussed here, but also to those interested in the scientific context of other medieval literature: chapter 2 includes a survey of optical theories prevalent in medieval Europe, allowing the reader to learn about the intellectual milieu of works not explicitly discussed in this book. It also suggests how medieval theories of vision relate to modern psychoanalytic theories that also use optical terminology to describe the link between subject and object.

Vision is almost always important in allegory, but its role varies: some early medieval authors presume the identity of sight and knowledge, while late medieval authors are often sceptical of the possibility of any certain knowledge, whether mediated by vision or by language. Such varied responses offer a way to track changes not only in the genre of allegory, but also in the medieval understanding of the individual's relationship to the surrounding world. This study attempts to follow the models of David Lindberg and Katherine Tachau in seeking to understand broad cultural changes as responses to medieval innovations in optics; it differs, however, in illustrating these changes not within science or philosophy, but in some of the best known literature of the Middle Ages.

The transparent mediation between subject and object, between reader and meaning, is the purpose of allegory. Some medieval allegories seem to fulfil this purpose, acting as a perfect mirror which unites the earthly domain of sense perception with the divine realm of intelligible knowledge. Other allegories, particularly those of the later Middle Ages, pursue a different end, amplifying what should be the imperceptible mediation of the text into something which draws attention to itself, even at the expense of the allegory's other, deeper meaning. By exploring how writers of allegory both adapt themselves to and resist established forms of allegory, it is possible to understand better how the genre develops and what its limitations might be. In the present study, I attempt to chart the development of the genre by examining the changing role of vision in several medieval allegories, and considering in each case what this suggests about man's ability to know through language. The final chapter, 'Division and Darkness,' centres on the legacy of allegory in the fifteenth century, and shows how a belief in language's fecundity, expressed during the thirteenth and fourteenth centuries by means of the analogy of the multiplication of visible forms, came to be replaced by a notion of language characterized by division and fragmentation.

This book has been a long time in gestation, and I would like to thank those who read it and generously contributed to what is useful and interesting in it: Renate Blumenfeld-Kosinski, Kathryn Gravdal, Robert Hanning, Ruth Harvey, Amilcare Iannucci, Joel Kaye, Sandra Prior, and two anonymous readers for the University of Toronto Press. I am especially grateful to have had the opportunity to work at the Library of the Pontifical Institute of Mediaeval Studies, paradise on earth. Joan Ferrante and Caroline Bynum nurtured this work through its formative stages, and offered support as it developed into a more mature piece of scholarship, while Edward Tayler, in his role as *dator formarum*, provided the ideas. It goes without saying that the faults remaining are all my own. I must also thank my family: Eddie and Yasin for babysitting, Sara for help in preparing the manuscript, and Camilla for her kisses. Finally, this book is dedicated to the memory of my grandmother, Elisabeth Straub Hoeflinger, who taught me the value of hard work and the pleasure to be found in a job well done.

SEEING THROUGH THE VEIL

Chapter One

ILLUMINATION AND LANGUAGE

There are two visions, one of perception, one of thought.

Augustine[1]

All of the senses are false in many ways; vision above all of them is the falsest.

William of Conches[2]

Seeing and knowing are inextricably linked in western culture. From antiquity to the present, vision is both explicitly and implicitly acknowledged as the highest of all senses. Plato makes this claim overtly in the *Timaeus*, where he asserts that 'vision, in my view, is the cause of the greatest benefit to us.' Because vision, for Plato, is the highest sense, it serves as the model for the workings of the mind: 'God devised and bestowed upon us vision to the end that we might behold the revolutions of Reason in the Heaven and use them for the revolvings of the reasoning that is within us, these being akin to those.'[3] In this passage, Plato expresses two concepts that richly inform medieval thought, both directly through Calcidius's translation of the *Timaeus* and indirectly through the medieval Neoplatonism of Augustine and Boethius: first, that vision is the basis of all human knowledge and, second, that vision consequently serves as the link between microcosm and macrocosm, between man and the universe. The special link between vision and knowledge does not, of course, preclude the existence of other ways of knowing, not only through the other senses but also through intellectual reasoning and divine revelation. But each of these other means of knowing is based on vision: the other senses are conventionally described as variations on the paradigm of vision, as by Aristotle in the *De anima*,[4] while both intellectual reasoning and divine revelation are metaphorically described as visual experiences. Plato's linkage of vision and human reason is still evident today: we say 'I see' to indicate that we understand an argument, explain our own 'point of view,' try to 'look' at the problem differently, and finally, we hope, achieve an 'insight.'[5] The act of vision is used as a metaphor not only for the faculty of judgment, but also for the other powers of the mind, divided during the medieval period into the three faculties of imagination, reason, and memory.[6] Each faculty is connected with vision, for the imagination is a repository of images, impressions received from the *sensus communis*, or common sense; the operation of reason or judgment is metaphorically described in terms of vision, as in the examples above; and memory, particularly purposeful memorization, is almost invariably constructed by means of visual images.[7]

Vision's role as a metaphor for divine revelation is in part based on Paul's description of the beatific vision experienced by the soul after the resurrection (I Corinthians 13:12).[8] It might be argued that New Testament writings, particularly those attributed to Paul, are informed by Platonic epistemology and therefore do not form a separate strand in western thought. But even Old Testament accounts of the experience of God centre on the visual sense: Moses can tolerate only an indirect glimpse of God, whose direct appearance is too blinding for the bodily eye to endure (Exodus 33:18–23). Clearly, then, the connection of seeing and knowing is not simply a legacy of Platonism but is instead the product of a more widespread belief in the primacy of vision.[9] Medieval Neoplatonism, however, especially as mediated through the writings of Augustine and pseudo-Dionysius, exercised a powerful influence on later theories of knowledge, culminating in the twelfth-century philosophical flowering associated with the School of Chartres. A fuller discussion of the relationship of vision and knowledge in the writings of these philosopher-theologians appears below in chapter 2. Yet it is not amiss here to sketch out the relationship between knowledge as it was understood in early medieval Christianity, and the Neoplatonism of Plotinus and Proclus that offered a framework for the development of Christian theology.

While Augustine emphasizes the primacy of vision in the experience of God, that sense of vision is associated as much with the passionate longing of the heart as the intellectual acuity of the mind. As Rowan Williams points out, in the *Confessions* Augustine describes his ecstatic foretaste of heavenly bliss 'as attained *ictu cordis*, "with the piercing glance of the heart.'"[10] Augustine's description of knowledge of the divine is further distinguished from the conventions of Neoplatonism by the presence of desire not only on the part of the soul for God, but on the part of God for the soul. In his seminal study of the relationship of Christian mystical theology to Neoplatonism, Andrew Louth makes it clear that, while the philosophical system of Plotinus offered Augustine a congenial model for the soul's movement toward God, a crucial difference remains. While 'Plotinus' One ... cares nothing for the soul,' Augustine's God offers enabling grace, 'so that God's own activity towards the soul vastly transcends Plotinus' notion of the soul's dependence on the One.'[11] Again, while the soul's ascent to God and the metaphor of the ladder are derived from Neoplatonism, they are quite differently developed by Augustine, as by other medieval Christian writers.[12] John Milbank puts it even more strongly when he characterizes the relationship of Platonic interiority to Augustinian 'inwardness' as not one of development and refinement, but rather 'something much more like its subversion.'[13] Like Augustine, pseudo-Dionysius drew upon the framework of Neoplatonism in his contribution to the emergence of a mystical Christianity; yet, again like Augustine, this development cannot be characterized simply in terms of assimilation. While Augustine drew mainly upon Plotinus, the Neoplatonism of Proclus was of more immediate use to pseudo-Dionysius, who in his *Celestial Hierarchy* and *Mystical Theology* constructed an elaborate system based on the three levels of reality delineated by Proclus.[14] Like Augustine, pseudo-Dionysius found it necessary to redefine the relationship between the soul and the One as defined in Neoplatonism to include the expressions of desire and grace, that is, the love of the created being for the Creator and vice versa. As Williams puts it, 'while the Neo-Platonic "One" proceeds into multiplicity and participation as a matter of course ... the Dionysian God *desires* to share himself.'[15] One further aspect of the Neoplatonic Christian mysticism of pseudo-Dionysius merits remark for its impact on subsequent understandings of vision's relationship to knowledge: that is, the apophatic understanding of God. Unlike cataphasis, knowledge of something by what is alike, apophasis is the knowledge of something by what is unlike, even contrary. In describing God, whose being surpasses every attempt to put it into words, apophasis (or negation) is better than cataphasis (or affirmation): for example, it is better to call God 'invisible' rather than 'most bright,' because he exceeds the capacity of vision. It is better to call him 'incomprehensible' rather than 'most perfect,' because he exceeds the human ability to assess perfection; and so on.[16] Both Augustine and pseudo-Dionysius preserve the Platonic emphasis on the primacy of vision, its status as the paradigm of how knowledge is acquired. Yet both also draw attention to the inadequacy of an epistemological model that does not make room for desire: the soul's desire for knowledge, and God's desire to be known, expressed through the exercise of grace. Further, pseudo-Dionysius in particular illustrates the potentially deceptive nature of vision, which even as it offers us enlightenment threatens to lead us astray. This potential danger is mirrored in language, for the cataphatic theology of pseudo-Dionysius's *On the Divine Names* affirmatively names the attributes of God, while never approaching His essence.[17] Paradoxically, that which offers us knowledge at the same time leads us away from it, both in the realm of sense perception and in the domain of language.

The belief that sense perception (particularly vision) and language are congruent is ubiquitous, found in both theological and rhetorical texts. For example, John of Damascus emphasizes the parallel of these two means of mediation when he states that 'just as words edify the ear, so also the image stimulates the eye. Just as words speak to the ear, so the image speaks to the sight; it brings us understanding.'[18] The mnemonic technique laid out in the classical *Rhetorica ad Herennium,* popular during the later Middle Ages,[19] depends upon a comparable analogy of letter and image: 'Those who know the letters of the alphabet can thereby write out what is dictated to them and read aloud what they have written. Likewise, those who have learned mnemonics can set in backgrounds what they have heard, and from these backgrounds deliver it by memory. For the backgrounds are very much like wax tablets or papyrus, the images like the letters, the arrangement and disposition of the images like the script, and the delivery is like the reading.'[20]

Language, like the senses, serves a mediating function, allowing the subject to know the object; but both language and the senses are always imperfect mediators, defective due to the Fall and to man's defiance of God's authority at the Tower of Babel. These two transgressions are linked not only by having both been caused by the sin of pride, as Augustine points out, but by being part of a repeated pattern of the early transgression of man against God, as Dante notes in his discussion of Babel in the *De vulgari eloquentia.*[21] Moreover, both transgressions are fundamentally intellectual sins, motivated by a desire for knowledge that goes too far: by desiring the sight of what should remain hidden, humanity denies the limitations placed on knowledge by God. Finally, both transgressions result in division: Eden ends with the division of humanity from God, Babel with the division of human beings from one another. In a world after the Fall, both vision and language are imperfect mediators, open to deception. Yet, paradoxically, as pseudo-Dionysius recognized, they represent the only possible approaches to knowledge. The persistence of this recognition can be seen in the second headnote to this chapter, which states that vision is both the truest and the falsest of the senses: it has the greatest capacity to reveal truth, and the greatest capacity to deceive. In medieval literature, the double nature of vision is often figured in the two opposing properties of mirrors, the good mirror which makes visible what could otherwise never be perceived, and the bad mirror which inverts the true image before it. In medieval iconography, these appear as the helpful mirror of foresight held by Prudence, and the dangerous mirror of narcissistic self-absorption held by Laziness.[22] In his *De planctu Naturae*, Alanus de Insulis calls Nature a mirror in order to emphasize her ability to convey divinity to mankind; but in his *Summa de arte praedicatoria*, he remarks that, in a mirror, 'the right parts appear to be on the left, and the left appear to be on the right.'[23] Similarly, the mirror of the text has the ability simultaneously to reveal meaning and to deceive: figurative language may be able to convey meanings that elude literal language, inadequate since the Fall; yet it too suffers from the postlapsarian defect that makes all language able to mislead the reader. The transparent mediation between subject and object, between reader and meaning, is the unreachable goal of language; in particular, it is the goal of allegory.

During the past half century, a number of critical studies of allegory have appeared, several of which are explicitly dedicated to the effort to produce a definition of the genre, based primarily on classical, medieval, or Renaissance uses of the term.[24] Efforts to define the genre generally take one of two directions: iconographic or rhetorical. Iconographic approaches focus particularly on the images represented alongside and (through ekphrasis) within the text; Tuve's *Allegorical Imagery* is exemplary of this approach. Such readings emphasize the symbolic, even transcendent nature of language as it is deployed in allegory. Rhetorical approaches, conversely, base their definition of allegory on Quintilian's classification of it as a figure of speech. The subtitle of Murrin's *Veil of Allegory* ('Some Notes Toward a Theory of Allegorical Rhetoric in the English Renaissance') signals reliance on the rhetorical tradition, while Copeland's work on translation as a rhetorical practice during the Middle Ages has produced some of the most illuminating discussions of allegory to date. The divergent foci of rhetoric and iconography remain evident in the modern theories of allegory associated with Paul de Man and Walter Benjamin: critical attention has shifted from de Man's emphasis on the status of allegorical language as rhetorical trope to Benjamin's notion of allegory as a preeminently visual, potentially transcendent mode.[25] It is striking that many of these studies begin with a paradigm of good or successful allegory in order to define the genre, even though establishing such an exemplar restricts the critic's definition to the qualities stressed in that particular allegory. For example, by calling Shakespeare's 'The Phoenix and the Turtle' 'the typical "pure" allegory,' Fletcher stresses personification above all other characteristics of the genre.[26] Similarly, Tuve devotes an entire chapter to Guillaume de Deguileville's *Pèlerinage de la vie humaine* because it concerns 'the basic allegorical theme ... the pilgrimage man takes through life to death and redemption.'[27] Even Quilligan, in her sensitive and insightful study of allegorical language, explicitly states that her study was motivated by her observation of punning in *Piers Plowman* and the *Faerie Queene*; therefore, her description of allegory focuses on polysemy, what she calls 'wordplay.'[28] In each case, the approach is flawed by the inevitable consequence that allegorical writings are made to fit the paradigm. Paradoxically, some of the most insightful studies of allegory end with what sounds like an admission of defeat. Honig concludes his dated but still perceptive study of allegory with a mystical view of the genre, one which asserts that the enigma it embodies extends to the term itself: 'And so the idea of allegory may only be ... a "name for something that never could be named" and actually should "bear no name" other than "inconceivable idea," existing "in the difficulty of what it is to be."'[29] Copeland and Melville aptly follow their definition of allegory, as 'a special province within a more general theory of communication that is capable of envisioning something like a complete fusion of or under-standing between speaker and hearer, reader and writer,' with the caveat 'So: we have a picture. It makes sense. And yet our every effort to take it seriously seems to turn us against it.'[30] Perhaps it is appropriate that a genre based on the ability of the word to signify more than one thing should itself resist being limited to a single definition.

I have referred above to 'the purpose' of allegory; but, of course, allegory has several purposes. In part, allegory functions to hide mean-ings that not everyone is worthy to understand, offering the full signifi-cance of the parable only to those who are fit by virtue of their faith. This aspect is foremost in Jesus' answer to his apostles when they asked why he spoke in parables: 'The reason I speak to them in parables is that "seeing they do not perceive, and hearing they do not listen, nor do they understand." ... But blessed are your eyes, for they see, and your ears, for they hear' (Matthew 13:13, 16). In addition, allegory acts as an aid to memory and, by increasing the pleasure of reading, facilitates learning. This didactic aspect of allegory is prominent in Martianus Capella's *De nuptiis Mercurii et Philologiae*, which employs allegory in order to teach students about the seven liberal arts in an engaging and, above all, memorable way. Dialectic, for example, is represented as a woman with intricately braided hair in order to recall the 'complex and knotty utter-ances' of logical argument.[31] The distinction I have drawn between rhetorical and iconographic approaches to allegory begins to break down when faced with allegory's mnemonic function, for it is precisely *as* image that allegory acts as an aid to memory even while it is being deployed as a rhetorical trope. It is nonetheless the case that critical studies of allegory tend to approach the genre either through iconogra-phy – allegorical image – or rhetoric – allegory as trope. The third, most crucial, purpose of allegory may help to explain why it is ultimately so difficult to separate its iconographic and rhetorical aspects: allegory conveys meaning that cannot be expressed directly through ordinary language. That is, by avoiding the limitations inherent in literal lan-guage, allegory creates meaning *within* the reader, bypassing the inevi-table degeneration of meaning as it passes through the obscuring veil of language. The paradox, of course, is that it is this veil which makes the transmission of meaning – the revelation – possible. Allegory is trope when it is expressed through language; but it is intelligible image at the moment of the reader's illumination.

At this point, it is helpful to turn to Aristotle's discussion of metaphor in order to understand more fully this third purpose of allegory, that is, the transmission of meaning that cannot be expressed directly through ordinary language. Aristotle's definition of metaphor anticipates the ambiguity inherent in many definitions of allegory: it is at the same time edifying and amusing, both enigma and riddle. Most importantly, meta-phor renders the abstract visible: it 'set[s] things before the eyes [πρὸ ὀμμάτων] ... [T]hings are set before the eyes by words that signify actuality.'[32] Through metaphor, it is possible to 'grasp the similarity in things that are apart,'[33] discerning the analogy that underlies the appar-ent enigma in a single perceptive glance: 'The right use of metaphor means an eye for resemblances [ὅμοιον θεωρεῖν].'[34] The enigma, at first obscure, becomes transparent through the reader's application of wit. At the same time, metaphor serves to amuse and divert the reader. Metaphors are 'smart sayings' that mislead the reader, so 'the mind seems to say, "How true it is! but I missed it."' A similar effect is produced when 'humorists make use of slight changes in words' or in 'jokes that turn on a change of letter; for they are deceptive.'[35] Aristotle's two senses of metaphor as philosophical enigma and amusing riddle reappear in the Middle Ages as two kinds of allegory. In the first, the reader 'sees' the similarity hidden in apparent dissimilarity; in the second, jokes turn on a change in letter. One seeks to make an otherwise inaccessible meaning available to human reason; the other is a self-referential game of words.[36] Unsurprisingly, Augustine values the philosophical enigma rather than the amusing riddle. He associates *aenigma* not with metaphor but with allegory, stating that 'an enigma is an obscure allegory.' He discusses the term at length in order to gloss Paul's famous statement that, as Augustine puts it, 'we see now through a mirror in an enigma.'[37] By declaring that enigma and mirror are identical, and that enigma is a variety of allegory, Augustine implies that at least some allegories are, figuratively, mirrors. They are not deceptive mirrors but revealing ones, which allow the viewer to glimpse things ordinarily hidden from human sight; to put it another way, these allegories allow the reader to apprehend meanings normally inaccessible through language. Augustine's equation of mirror and allegory prepares the way for twelfth-century Platonists such as Alanus de Insulis who identifies the mirror with the dream vision itself in his *De planctu Naturae*: in the poem's last lines, the narrator declares that 'when the mirror with these images and visions was withdrawn, I awoke from my dream.'[38] If the dream is a mirror, then the text that reproduces that dream is a mirror as well. Twelfth-century allegorists use personifications in keeping with the primary goal of metaphor as Aristotle had defined it: to make abstractions visible. Personification not only makes an abstraction visible; it makes it alive.[39] Throughout the twelfth century, the Augustinian legacy of allegory as specular enigma remained predominant; twelfth-century allegory was Neoplatonic and Augustinian, enriched by a concept of symbolism based on the writings of pseudo-Dionysius and developed more fully in the writings of the Victorines as well as in the works of William of Conches.[40] Not until the late Middle Ages did Aristotle's distinction between the two kinds of metaphor – philosophical enigma and amusing riddle – reappear.

Because allegories rely so frequently upon metaphors of vision, several studies have appeared which focus specifically on the significance of the act of seeing within allegory, such as Susan Hagen's study of the Middle English translation of Guillaume de Deguileville's *Pèlerinage de la vie humaine* and Patrick Boyde's work on Dante's *Commedia*. Others have cast their net more broadly, studying the phenomenon of sight in medieval works of other genres.[41] This is clearly a fruitful line of investigation, for the changes in allegory that have been noted by a number of critics correspond to changes in the understanding of how (and whether) vision leads to knowledge, both in the philosophical and the scientific contexts. Critics such as Delany, Bloch, and Sturges attribute the waning popularity of allegory during the later Middle Ages to the increasing currency of nominalist or modist theories of language during the fourteenth century.[42] It is not the case that there is a linear progression from one type of allegory to the other; there is, however, a gradual shift in emphasis regarding the status of the literal level of the text, the availability of meaning, and the mode of its transmission. Changes in the status of allegory must be understood in the context of changing views of allegoresis, that is, the reading of a text as an allegory. Allegoresis includes both the interpretation of scripture (biblical exegesis) and the interpretation of secular texts, which can be referred to (following Copeland) as the *enarratio poetarum*. This parallelism underlies Dante's influential description in the *Convivio* of the 'allegory of theologians' and the 'allegory of poets.' In a very crude sense, allegory and allegoresis can be understood as reciprocal acts of translation: allegory translates, as it were, plain truth into literary figures, while allegoresis translates the poetic language back into literal terms which can be readily understood. To understand allegoresis in this way is, of course, to play into the hands of the glossator: as Tuve suggests by referring to allegoresis as 'imposed allegory,' the glossator derives meaning from a text which often does violence, if not to authorial intention, then to the meaning that another reader might find in the text. Copeland has recently shown, however, that Tuve's account of 'imposed allegory' does not adequately describe the process of *enarratio poetarum*, the interpretation of the poets. She illustrates the extent to which the tools of rhetoric are brought to bear upon texts subjected to allegorical interpretation, both on the level of grammatical analysis and, more importantly, in the way in which the text is anatomized and its components set forth in the prologues to the gloss. Thus, Copeland argues, Tuve's notion of 'imposed allegory' does not reveal the extent to which allegoresis presents itself as, in a sense, 'anterior to the text,' supplying an authoritative intention (and hence meaning) that supercedes authorial intention, even rendering it irrelevant.[43] In her effort to bring out the complexity of the rhetorical strategies of allegoresis, Copeland draws a sharp distinction between the assumptions that govern biblical exegesis and those that apply to the *enarratio poetarum*. In so doing, she obscures the extent to which (as A.J. Minnis has shown) techniques of biblical exegesis affected allegoresis of secular texts, and vice versa.[44] As Copeland herself acknowledges, there was much interplay,[45] as can be seen in such texts as the early fourteenth-century *Ovide moralisé*, whose compiler includes biblical exegesis alongside the more literary exposition of Ovidian myth,[46] or Dante's *Commedia*, which uniquely sets out to straddle the gap separating the allegory composed by God from that composed by man.[47]

Tuve's description of 'imposed allegory' has exerted a strong effect on subsequent studies of allegory, sometimes with the effect of divorcing allegory from allegoresis rather than illustrating the extent to which they were presented (by the medieval glossator, at least) as reciprocal processes, both in the realm of biblical exegesis and that of the *enarratio poetarum*. Quilligan, for example, opposes allegoresis to allegory in a way that represents allegoresis as both prior to allegory and superseded by it. She resorts to spatial metaphors to do so: 'allegory works horizontally, rather than vertically, so that meaning accretes serially, interconnecting and criss-crossing the verbal surface long before one can accurately speak of moving to another level "beyond" the literal.' Quilligan goes on to explain the basis for this distinction: allegory's 'horizontal nature' is distinct from 'allegorical exegesis [which] has always stressed the opposite notion of vertical "layers" or "levels."'[48] This dichotomy is applied productively by Copeland and Melville in their distinction between the 'vertical textual archeology' characteristic of allegoresis and the 'horizontal polyvalence' of allegory.[49] Yet I would suggest that the 'horizontal' quality of allegory appears only gradually in the course of the thirteenth century, and that it develops in extraordinarily nuanced ways during the course of the later Middle Ages. I will continue to use this dichotomy of vertical and horizontal in order to distinguish between different kinds of allegory, arguing that when an allegory as a whole refers to a corresponding level of meaning, it can be characterized as vertical; conversely, when an allegory as a whole does not refer to some other level, but rather only sporadically signifies something other than what it literally says and even contradicts itself within the text, it can be characterized as horizontal. This approach is in keeping with other efforts to describe the shift in epistemological strategies which undoubtedly took place during the course of the later Middle Ages. Minnis has argued that, in the course of the thirteenth century, 'a new type of exegesis emerged, in which the focus had shifted from the divine author to the human author of Scripture.'[50] These changes made themselves felt in the vernacular translation and interpretation of pagan *auctores* as well, as Copeland has shown, so that 'the inventional power of the translator' comes to be dominant.[51] Not only interpretation but also literary invention partook in this shift in attitudes toward authorship, interpretation, and the status of language. Kelly argues that, during the later Middle Ages, love poetry moved away from 'abstraction' and toward 'a real, even contemporary, world of time and space,'[52] while Bloch sees such changes as part of a much larger movement. He suggests that a 'genealogically defined linguistic model,' characterized by the use of etymologies, was replaced by theories of language, generated by nominalist and modist philosophy, in which 'language, thought, and being exist parallel to but not contiguous with each other.'[53] Bloch's figure of the tree of etymology and genealogy implies a vertical relation of word and meaning. Sturges draws a comparable distinction between 'determinate' and 'indeterminate' modes of writing: a work written in the determinate mode specifies how the reader should interpret it, while a work written in the indeterminate mode requires the reader to fill in 'gaps' in the text. The former relies upon metaphor, which operates vertically; the latter upon metonymy, which operates horizontally.[54] In the chapters which follow, I too use the dichotomy of 'vertical' and 'horizontal' in order to discuss several of the best known and most influential medieval allegories. Like any dichotomy, this one is not an absolute definition of the way things are, but rather a tool for understanding how things work. It is easy to envision vertical and horizontal relationships: in the former, word and thing are seated within a hierarchy, one above the other; in the latter, they are on an equal footing, existing side by side. It is a spatial metaphor that helps the reader to understand a genre that, as noted above, refuses every attempt to name it. Perhaps it is appropriate to rely upon a metaphor in order to understand allegory.

Vertical allegory is characteristic of Neoplatonic ways of thinking in that it presumes that all things are linked to one another in a kind of chain or hierarchy. Each link in the chain reflects the preceding link just as a mirror reflects an object: as pseudo-Dionysius puts it, 'Hierarchy causes its members to be images of God in all respects, to be clear and spotless mirrors reflecting the glow of primordial light and indeed of God himself. It ensures that when its members have received this full and divine splendor they can then pass on this light generously and in accordance with God's will to beings further down the scale.'[55] Like any other link in the chain connecting humanity to the divine, the allegorical text mirrors a remote idea, making it accessible to human reason. Because it accords more importance to figurative than literal meaning, vertical allegory also places great emphasis on the primacy of form over matter. As a result, vertical allegory often includes what I will call 'structural allegory,' forms within the text that are generated by numerical patterns or symmetries and that contribute to the meaning of the work (see chapter 3). Structural allegory enables the form of the work to reflect its content, increasing the transparency of the text's mediation. Vertical allegory includes personifications that consistently conform to the abstraction they are supposed to embody; in horizontal allegory, personifications lose their fixed identity as embodied abstraction and behave in ways that suggest they are less personifications than personae, fictional characters with motivations and emotions. Vertical allegory points toward a hidden meaning that the reader must construct within his own mind, a transcendent truth that cannot be conveyed through literal language; horizontal allegory satisfies the reader here and now, exposing the double (or triple) meaning of language explicitly within the text as pun or euphemism. Vertical allegory aims to convey a transcendent truth that cannot be expressed through literal language, whether it concern God, creation, or the nature of identity; horizontal allegory celebrates the play of words and the unfixed nature of linguistic meaning. I want to stress once more that no allegory is a perfect example of either extreme of allegory; yet each can be characterized fruitfully by using the terms of this dichotomy, for between the two points stretches a broad spectrum of possible variations of the genre.

I have characterized vertical allegory at greater length because this is the mode of allegory I will discuss first in the chapters which follow; horizontal allegory, in contrast, will be defined in contradistinction to it. This approach particularly suits the first text I will discuss in depth, the *Roman de la rose*: it is composed by two different authors, each of whom presents a very different notion of the self and the possibility of self-knowledge and, hence, each uses a different model of allegory. For an articulate description of horizontal allegory, one need look no further than Quilligan's discussion of allegory's 'horizontal nature,' exemplified by the texts whose 'wordplay' she sees as typical of the genre, Spenser's *Faerie Queene* and Langland's *Piers Plowman*.[56] Copeland and Melville's reference to allegory as 'horizontal' is more tentative than is their assertion of allegoresis's 'vertical' nature; they are on firmer ground when bringing out the 'temporal complexity' of allegory, and the extent to which allegory and allegoresis are 'seamed always together' in the acts of encoding and translation.[57] Their discussion of allegoresis, however, is indirectly helpful to an understanding of how horizontal allegory works: it is significant that the rise of horizontal allegory corresponds chronologically to the increased practice of what Copeland, in her seminal study of medieval rhetoric, calls 'secondary' translation, a practice related to allegoresis. Primary translation is similar to modern notions of literal, word-for-word rendering of a text; in secondary translation, however, the creative capacity of the translator is brought out fully.[58] While the term *translatio* does appear in twelfth-century texts, it is used to mean 'metaphor' rather than anything like either primary or secondary translation, and is in keeping with Isidore of Seville's seventh-century definition of the term: 'Metaphor is a word drawn across ['translatio'] from another usage.'[59] (The relationship between horizontal allegory and secondary translation will be discussed in more detail in chapter 7.) Although I will allude to them frequently in this study, I do not suggest that works such as the *Cosmographia*, the *De planctu Naturae,* and the *Anticlaudianus* can be considered paradigmatic allegories. But it is certainly necessary to understand how allegory operates in these twelfth-century works in order to understand the changes that will be made by subsequent allegorists. The tradition of philosophical allegory written in Latin gives rise to a rich tradition of vernacular allegory in the thirteenth and fourteenth centuries. Each writer uses the method of allegory in conjunction with at least one other literary genre: Guillaume de Lorris mixes allegory with the genres of lyric and romance; Jean de Meun, with the encyclopedic tradition; Dante, with lyric, romance, and the encyclopedic tradition; Chaucer, with the fabliau. The fact that these authors can so easily merge allegory with other genres suggests that allegory is a genre identifiable less by its form or content than by how it works on the reader. The operation of allegory adds an energy to these vernacular works that is nothing other than the power of the word to generate multiple meanings.

The preliminary definitions of vertical and horizontal allegory offered above might suggest to the reader that horizontal allegory is fertile and dynamic in a way that vertical allegory is not. Yet it is not the case that vertical allegory is static, characterized by passivity on the part of the reader who receives knowledge from above. Both modes of allegory partake in an essential regenerative capacity embodied in the fundamental metaphor of meaning as seed: as Jesus puts it when he recounts the prototypical parable, 'The seed is the word of God' (Luke 8:11). The phrase is, of course, meaningful on a theological level, referring to the incarnation of the Word; yet it is also meaningful on a linguistic level, affirming the fertility inherent in language and conveying it through a powerful metaphor. To discover the generative power of the word, it is necessary to look backward in time, back to its origins. This philosophy of language is evident in the encyclopedia of Isidore of Seville:

> Etymology is the origin of ways of naming [*origo vocabulorum*], as when the meaning of the verb or noun is gathered by means of interpretation.
> (Isidore, *Etym.* I.xxix.1)

> Allegory is 'speaking other' [*alieniloquium*]. One thing is declared, and another thing understood.   (Isidore, *Etym.* I.xxxvii.22)

Isidore's *Etymologies* are perhaps the most elaborate and certainly the most widely circulated articulation of the early medieval belief that the supposed derivation of the word is intimately connected to its meaning, that finding the origin means finding truth.[60] Thus Isidore declares 'Etymologia est origo vocabulorum.' The twelfth-century writer Pierre Elie expands upon this definition of etymology, etymologizing the term itself: 'In fact, "etymology" is a compound noun composed from "etymos," meaning "true," and "logos," meaning "speech," and so it is said that "etymology" means, as it were, "speaking truly" [*veriloquium*], since whoever etymologizes attributes to a word its true (that is, its first) origin.'[61]

Pierre Elie's definition of etymology as *veriloquium* provides a striking counterpart to Isidore's definition of allegory as *alienoloquium*: the two modes of expression are opposite in the sense that, on a literal level, one is true and one false ('speaking truly' and 'speaking other'); but the two modes are at the same time complementary, for both seek to convey truth, one directly and the other indirectly, one literally and the other figuratively. While etymology reveals truth by revealing the origin of a word and, implicitly, of the thing it represents, allegory reveals truth by clothing an otherwise inexpressible meaning in words whose literal sense is at odds with the figurative sense: in this sense, etymology is opening up while allegory is closing, wrapping or clothing in a veil. Allegorical interpretation or allegoresis, on the other hand, is very similar to etymology, for both are interpretive endeavours dedicated to unveiling truth. Medieval writers commonly refer to the interpretive act as a cracking open of the shell to obtain the nourishing seed, in the context of both biblical exegesis and secular *enarratio*. For example, Dominicus Gundissalinus refers to the author's intention as the 'kernel'; to neglect the intention is to 'leav[e] the kernel intact and eat the poor shell.'[62] Both the writer who performs allegoresis on a classical text and the reader who interprets an allegorical fiction extract the kernel of truth from the husk, removing the veil or *integumentum* that conceals the meaning.

Meaning can be metaphorically represented as a seed, enfolded in the word just as the seed is wrapped in its husk. This metaphor suggests not only that meaning is found only through opening or unwrapping, but also that what is uncovered has the capacity for growth and regeneration. Through the metaphor of the seed, language is conceived of as a living thing, able to reproduce meaning by being passed from one text to the next just as the human form is reproduced by being passed from one generation to the next. Yet the metaphor of the seed can also be associated with the act of vision, as Lucretius does in the fourth book of his *De rerum natura*. At the beginning of that book, Lucretius states that objects continuously emit forms that are like 'membranae vel cortex' ['skins or bark']. Vision occurs when these forms strike the eyes of the viewer. Lucretius concludes the fourth book with another account of the transmission of forms, but here the forms are transmitted not in perception, but in conception. The impact of forms upon the body causes it to yearn to project forms from within itself, that is, to emit seed.[63] The *De rerum natura* was not directly influential during the twelfth century, when it seems to have been known only through Priscian.[64] But the link between the passage of forms in vision and the passage of forms in sexual reproduction continues to reappear throughout the Middle Ages. In the twelfth century, Bernardus Silvestris begins the last chapter of his *Cosmographia* with an account of vision, the highest of the five senses; he ends it with an account of the 'shining sperm' which 'reproduce[s] the forms of ancestors.' For Bernardus, vision and conception are reciprocal processes: just as forms from outside are received by the eye, forms from inside are projected by the genitals.[65] The correspondence between the intelligible forms present in the mind and the form imposed upon matter in the act of conception is reinforced in medieval medical theories of cerebral spermatogenesis, that is, the production of sperm in the brain.[66] Bernardus alludes to this theory in his description of how, in preparation for intercourse, 'blood, sent from the region of the brain, flows down to the loins. It has the appearance of white [or shining] sperm.' Like the sperm, the brain is formed of a 'soft liquid ... so that the

images of things might easily inhere in the liquid.'[67] In the thirteenth
century, Roger Bacon claims that the multiplication of forms causes
'every action in the world,' including both the propagation of visible
form and the regeneration of human beings.[68] Medieval allegorists similarly emphasize the link between the form in vision and the form in
sexual reproduction: this is particularly apparent in Jean de Meun's
continuation of the *Roman de la rose*, discussed in chapter 4. The multiplication of forms described by Bacon generates both visible forms and
living bodies; to put it more precisely, it produces both sensible species
and the human species. Yet neither one of these reproductive processes
takes place in a vacuum. The form conveyed by the sperm (according to
Aristotelian notions of biology) requires the passive matter of the female
in order for conception to take place; correspondingly, the visible form
requires some medium that it may pass through in order to be apprehended by the seeing subject. In each case, the medium – whether
female body or diaphanous air – has the potential to change the form
which passes through it, even to contaminate it. The allegories surveyed
in this study engage with precisely such epistemological problems in
their exploration of how knowledge can be mediated through language.
Like the female body, like the diaphanous air, allegory is a medium
which can degrade what it seeks to convey, but which is at the same time
essential to the effort.

In the *Anticlaudianus*, Alanus de Insulis describes the mirror Phronesis
uses to allow her to gaze upon what would otherwise be unbearable:
'The mirror acts as an intermediary to prevent a flood of fiery light from
beaming on her eyes and robbing them of sight.' This is the good,
revelatory mirror, for it makes the divine accessible to man. Yet the
image it reflects is undeniably a baser version of the original: 'In this
mirror ... the appearance of these things differs from the real objects.
Here one sees reality, here a shadow; here being, here appearance; here
light, there an image of light.'[69] Even in twelfth-century Neoplatonic
allegory, the dichotomy of the revelatory mirror and the deceptive mirror is not absolute. The revelatory mirror always necessarily degrades the
image in order to make it accessible to humanity; the deceptive mirror
always reflects its exemplar, however imperfectly. Similarly, the mirror of
the allegorical text may reflect a remote truth; but it can can never
transmit it as it really is. Yet Alanus claims that allegory can mediate
transparently between heaven and earth. For it to do so, the author must
abandon the role of the creator of the text: 'I appropriate a new speaking part, that of the prophet ... the language of earth will yield to and

wait on the language of heaven. ... I will be the pen in this poem, not the scribe or author.'[70] By becoming the instrument of God, Alanus minimizes the distance between human beings and the divine. Only if his text is authored by God can it be a perfect mirror, written in the language of heaven. The *Anticlaudianus* can be described as a vertical allegory because its text is sent down from heaven and it mirrors a divine reality. It is a link in the chain connecting God and man. The same is true of Alanus's *De planctu Naturae*: the text is the transcription of a divinely inspired dream. The images that the poet sees strike his imagination directly, for they rain down from above. Normally, forms are perceived by the imagination after being transmitted through the organs of sense and the *sensus communis*. In the case of an externally inspired dream, however, forms are transmitted directly from above to the imagination of the dreamer. Thus, if the allegorist suggests that the dream has an external cause, his dream vision is likely to be an example of vertical allegory; if he suggests that the dream has an internal cause, such as some bodily disturbance, his dream vision is likely to be an example of horizontal allegory. The former reflects a higher link in the chain of being; the latter reflects only the natural world. For predominantly vertical allegories such as the *De planctu Naturae* and the *Anticlaudianus*, dream, author, and text are all mirrors reflecting a divine idea. The reader's task is to interpret the allegory, that is, to see the original that is reflected in the mirror of the text.

By the early thirteenth century, when Guillaume de Lorris began the *Roman de la rose*, the Neoplatonic tradition that generated works such as the *De planctu Naturae* and the *Anticlaudianus* had begun to be challenged by new translations of scientific and philosophical texts appearing from the Islamic world. Guillaume's allegory is in some respects very much in the tradition of twelfth-century vertical allegory, and it is this tradition that provides the link between vision and knowledge (specifically self-knowledge) which is at the centre of the first *Rose*. Yet Guillaume is at the same time influenced by the scientific writings of William of Conches, whose discussion of optics betrays a dissatisfaction with the early medieval acceptance of vision's seamless mediation between subject and object. This leads Guillaume de Lorris to explore the three types of vision, direct, reflected, and refracted, in his allegory, and finally to leave his poem deliberately incomplete in imitation of man's ultimate incapacity to achieve knowledge of the self. Jean de Meun's continuation of the *Roman de la rose*, written in the latter half of the thirteenth century, demonstrates the writer's scepticism regarding allegory's capacity to reveal higher truths, a capacity which the twelfth-century allegorists had vigorously upheld. Jean subverts and obscures Guillaume's optical allegory of three types of vision, replacing several themes prominent in the earlier work in order to convert Guillaume's allegory of visually experienced knowledge of the self into an allegory of carnally experienced knowledge of the other. Later writers of allegory continue to explore the implications of these two divergent ways of knowing. In his *Commedia*, Dante makes what is arguably the most successful attempt to integrate the two forms of knowing, through the eye of the body and the eye of the mind. He incorporates many of the important themes and symbols of earlier allegories, such as the rose of Guillaume de Lorris's and Jean de Meun's poem, which reappears in Dante as the white rose made up of souls in heaven. By doing so, Dante implicitly comments on earlier examples of the allegorical genre, and offers a new way of accommodating contemporary theories of vision and of knowing. While Jean de Meun's rejection of the link of vision and knowledge took the form of new emphasis on knowing through the body, Chaucer rejects the link between vision and knowledge on different grounds, and finds a different solution. Jean's rejection of vision was based on the flaws implicit in the analogical way of knowing, the fact that each individual form necessarily reproduces the form immediately preceding it in the medium more closely than it reproduces the original object of vision. Chaucer's rejection of the link between seeing and knowing is based on the scepticism engendered by the claim that species do not exist, made by Ockham earlier in the century. Ockham's theory of direct realism and, especially, writings in logic and natural science influenced by his brand of nominalist philosophy, lie behind Chaucer's adoption of verisimilitude as the only viable way to convey reality in his *Canterbury Tales*.

Chapter Two

THE MULTIPLICATION OF FORMS

This chapter, a detailed account of optical theories particularly relevant to medieval literature, has two purposes. Its first purpose is to avoid repetition in the subsequent chapters on specific literary texts, so that, for example, 'intromission' need not be explained both in the chapters on the *Roman de la rose* and in those on Dante, and so on. Its second purpose is to provide a guide to the nexus of optical theory and medieval literature that goes beyond the specific works discussed here, so that a reader of Chrétien de Troyes's *Cligès* or Boccaccio's *Teseida*, for example, could easily determine which theories of vision were likely to provide a relevant context. Before offering this account, however, it is worth pausing to consider some of the difficulties that face modern readers of medieval texts, particularly with regard to the relationship of vision and knowledge, so consistently compared to one another from antiquity to the present. As Martin Jay has shown, twentieth-century uses of visual theory have a rich prehistory in classical, Cartesian, and Enlightenment formulations of the eye's relationship to the 'I,' and vision's role as mediator between what lies outside and what lies within. Although modern attitudes toward vision can be seen as a reaction against the Cartesian and Enlightenment apotheosis of vision (Jay's titular 'denigration'), vision, however debased, remained central to the philosophical project. As Jay acknowledges, even 'antiocularcentrism' displays a 'fascination with visual experience [which] often betrays a keen attraction to its pleasurable side.'[1] From Plato to Irigaray (if you like), the mechanism of vision is the underlying structure on which subjectivity is constructed. In response to the continued centrality of vision as a paradigm for knowledge, recent studies of vision in medieval literature show an eagerness to take into account modern theories of vision: while some focus exclusively on the medieval context, others more ambitiously attempt to integrate modern psychoanalytical theories of how subjectivity is constituted through the act of seeing, and how power is exercised through the gaze. Yet such efforts to reconcile medieval and modern theory inevitably end up by taking refuge in broad generalizations;[2] even the most successful of them produces moments where medieval and modern theories of vision seem to work at cross purposes.[3] How, then, to take both frames of reference into consideration? To resolve this dilemma it is useful to turn to a similar problem which has faced modern inquiries into somatic history. As Caroline Bynum has noted, it is possible to make rich and often fruitful comparisons between medieval and modern conceptualizations of body, particularly with regard to the way in which the body is related to personal identity.[4] It is necessary, however, to avoid superficial comparisons which elide crucial differences between medieval and modern cultures: instead, as Bynum points out elsewhere, 'if we situate our own categories in the context of our own politics, we must situate those of the Middle Ages in theirs. The relationship between then and now will thus be analogous and proportional, not direct.' She goes on to suggest that, '[b]y understanding the relationship of figures to contexts, and then the relationship of those relationships, we will often see that there is a large and developing issue with which both figures struggle, each in his or her own vocabulary and circumstances.'[5] With regard to theories of the body, for example, Bynum suggests that it is not very useful to make a direct comparison between the third-century theologian Origen and the twentieth-century feminist Judith Butler; instead, we must situate Origen in his context and Butler in hers, and then go on to compare the comparisons. By doing so, we arrive at an understanding through the process of analogy, reaching a conclusion which is sensitive to the substantial differences of time and place, the very different intellectual contexts experienced by each. What I wish to do here, then, is to unpack the first term in the analogy: that is, to offer a detailed and specific context drawn from contemporary works on optics in order to reveal the intellectual background of the vocabulary of vision as it appears in literature. After showing how the medieval subject is constituted through the act of vision, it becomes possible – *not* to make a definitive statement of what separates medieval from modern culture – but to compare medieval and modern subjectivity, each of which are constructed through the function of the gaze, albeit in very different ways.

The meeting of subject and object is the purpose of vision; the description of how this encounter takes place, and whether the encounter is direct or through a mediating third term, is the purpose of optics. In the wake of David Lindberg's masterful survey of the history of optical theory (*Theories of Vision*), scholars in a variety of disciplines have begun to take account of the broad influence of the perspectivist tradition. Yet even before Lindberg, scholars had begun to signal the broader implications of optical theory. Vescovini was among the first to point out the importance of medieval optics to an understanding of medieval epistemology, while Tachau has more recently shown that perspectivist theory underlies important developments in thirteenth- and fourteenth-century philosophy:[6] as Mark Smith puts it, what perspectivist optics offered was not simply a theory of light and vision, but a 'scientifically justified world view.'[7] The effect of Lindberg's work has come to be felt in studies of medieval literature, although the engagement with medieval optics is rather general and the use of specific models of vision rare. Here, I will argue that the perspectivist tradition had a strong effect on late medieval literature and that its effect was expressed in a variety of ways, by writers who drew upon a broad range of optical theories. Because earlier allegories, particularly twelfth-century allegory influenced by Neoplatonism, suggest that vision is the sense best able to mediate between subject and object, later allegories, informed by the innovations of perspectivist optics, modify vision's role in several ways: they introduce point of view into the work's visual imagery; they stress the fundamentally analogical nature of all knowledge, a consequence of the reproduction of the visible image in the medium; and they finally reject vision's claim to mediate between subject and object in favor of verisimilitude, creating an image that merely resembles its exemplar.[8]

In order to describe this phenomenon, it is useful to begin by offering a concise account of the history of medieval optics. This is a history which has been written a number of times, at greater and lesser length and within a variety of disciplines. Lindberg's survey is certainly the best account of the rise of perspectivist optics, but descriptions of vision from the perspective of art or medicine are useful supplements.[9] Here, I will focus particularly on sources which would have been available in Europe during the twelfth through fourteenth centuries and, in the following chapters, will refer often to the optical terminology and the theories of vision outlined in this chapter, particularly to the two opposing mechanisms of intromission and extramission. Although both mechanisms facilitate the purpose of vision – the meeting of subject and object – intromission stresses the primacy of the object, extramission the primacy of the subject. By showing whether an optical theory operates mainly through intromission or extramission, it is possible to draw some conclusions both about the status of the seeing subject and about the universe around him, the object of vision. The second-century physician Galen outlines these two models in his *De placitis Hippocratis et Platonis*: 'A body that is seen does one of two things: either it sends something from itself to us and thereby gives an indication of its peculiar character, or if it does not itself send something, it waits for some sensory power to come to it from us' (7.5). Intromission takes place when the visible form is literally 'sent into' the one who sees. The appearance of the object comes forth to meet the subject, creating a fundamentally passive viewer who receives the image of the object before him. Extramission, conversely, implies an active viewer, who radiates some sort of actualizing power that facilitates the transmission of the image of the object back to the subject. The extramitted visual beam, 'sent outward' from the viewer, reaches out (as it were) and apprehends the object of vision.

Both theories rely upon the basic distinction between form and matter. For the subject to know the object, the two must come into contact: the object must come to be, in some way, inside the subject. Since the material object cannot enter the subject, some immaterial representation of the object must take its place and enter the subject. That representation can be described as 'form'; in medieval optics, it is most commonly called the species. In the opening pages of his *De multiplicatione specierum*, the thirteenth-century philosopher-scientist Roger Bacon lists the various synonyms that can be employed to describe what he will normally call 'species': similitude, image, idol, simulacrum, phantasm, form, intention, passion, impression, and shadow (1.1). This use of the term species may seem at first strange to the modern reader, who is accustomed to understand 'species' as category, as in the phrases 'endangered species' or 'the evolution of species.' By the thirteenth century, as Michaud-Quantin has shown, the semantic field of the term species had become very broad indeed, referring not only to the universal category but also to the individual form apprehended by the senses, as in optical theory.[10]

Early Greek philosophers offered a variety of theories of vision: the atomists were the first to suggest that all objects constantly emit species (which they call *eidola*), numerous copies of the object's form, which penetrate the viewer's eye. Their intromission theory of vision was adopted and elaborated by the Epicureans; a version of it appears in the *De rerum natura* of Lucretius (d. 55 BCE), where the species are called *simulacra*

(4.54–61). The Pythagoreans objected that large objects would emit correspondingly large species, impossible for the eye to take in; therefore, they suggested, the viewer's eye must be the source of visual potency, emitting a fiery beam which enables vision. Like the Pythagoreans, the Stoics proposed an extramission theory, but they argued that the visual force emanating from the eye was not immaterial fire but rather a material, air-like substance which they called *pneuma*. The Stoics also added the detail that the visual force is emitted from the eye in the form of a cone, in the same way that light emanates from the sun.[11] Plato also followed the extramission theory, but allowed for some elements of the intromission theory as well. He suggested that the fiery beam emitted from the eye interacts with the force emitted by the object, the resulting beam returning to the eye with the impression of the object: 'whenever there is daylight round about, the visual current issues forth, like to like, and coalesces with it and is formed into a single homogeneous body in a direct line with the eyes, in whatever quarter the stream issuing from within strikes upon any object it encounters outside. So [this] ... causes the sensation we call seeing.'[12] Plato's assertion that the beam extramitted from the eye is made up of light was influential throughout the Middle Ages, mediated by the fourth-century translation of the *Timaeus* with commentary by Calcidius, as well as several later commentaries on Calcidius's text. It had particular impact, both directly and indirectly (through the Neoplatonists), on the development of a metaphysics of light during the thirteenth century.

Aristotle, as usual, takes a different line from Plato. In *De sensu* he rejects the extramission theory outright, stating that 'it is unreasonable to suppose that seeing occurs by something issuing from the eye.'[13] Aristotle agrees, however, that some mediating third term, situated between the seeing subject and the visible object, is necessary for the act of sight to be completed. He names this third term the 'diaphanous' and states that, when it is actualized by light, the diaphanous propagates the visible form from the object to the viewer (*De anima* 2.7.418–19). Aristotle's diaphanous differs from the visual fire proposed by the Pythagoreans, or the *pneuma* proposed by the Stoics, in that it does not emanate from the viewer, but rather exists separately. Aristotle explains neither the mechanism of the propagation of the form nor the physiology of the mind's apprehension of the form, merely stating that the senses receive 'the form of sensible objects without the matter, just as the wax receives the impression of the signet ring without the iron or the gold' (*De anima* 2.12,424a). This simile, frequently echoed by later commentators, made it possible to visualize the imposition of form upon the passive viewer according to the theory of intromission. In addition, Aristotle's use of a similar trope to describe the act of conception in *De generatione animalium* implies that the abstraction of form in the act of sense perception and the imposition of form in the act of creation must be reciprocal processes (1.21.729b). During the early Middle Ages, of course, Aristotle's theories were largely unknown; instead, various forms of Plato's extramission theory were current, mediated not only through Calcidius's translation of the *Timaeus* but also through the writings of Neoplatonists such as Plotinus, Porphyry, Proclus, and especially pseudo-Dionysius, who was translated into Latin during the ninth century by Eriugena and (significantly) during the thirteenth century by the scientist Robert Grosseteste.[14] In addition, Augustine, Macrobius, and Boethius all played important roles in the dissemination and propagation of a Platonic extramission model of vision.

Augustine (354–430) nowhere articulates a fully developed optical theory; indeed, in the *De Trinitate*, when a discussion of the nature of double vision threatens to become too focused on the material aspects of the process, Augustine withdraws, saying that the exact nature of binocular vision 'is a tedious subject, and not at all necessary for our subject to inquire and to discuss' (11.2.4). Nevertheless, Augustine's writings exerted a formidable influence on subsequent thought on the nature of vision, due both to his authoritative status in general, and to the central role of vision as a metaphor for knowledge in his works.[15] In the *De Trinitate*, Augustine distinguishes between corporeal vision and mental vision, and explains that knowledge of the divine Trinity is to be sought by examining the trinity within: that is, memory, understanding, and will (*De Trin.* 10.11.17–18; cf. 15.7.11–13). The mental gaze must be turned inward, not in a narcissistic act of self-contemplation, but in an effort to see beyond the created nature of the self to the transcendent nature of the Creator. This effort culminates in an ecstatic, transient glimpse of the divine light (*De Trin.* 8.2.3).[16] In the twelfth book of his *De Genesi ad litteram*, Augustine distinguishes between three kinds of vision: corporeal, spiritual, and intellectual. The lowest of these is, of course, corporeal vision, since it is the most prone to error due to its dependence upon the material world for its knowledge, both in the objects it apprehends and in the operation of the senses. Throughout his works, Augustine characterizes the mechanism of corporeal vision in terms of extramission. His most detailed account appears in the *De Genesi*, where he relates how the eyes emit 'a shaft of light' ['lucis ... jactus'] in the form of rays that can be 'pulled in' ['contrahi'] to examine near objects or 'sent forth' ['emitti'] to view objects farther away (1.16.31). Augustine's depiction of vision is thus far in keeping with Neoplatonic accounts. He differs from the Neoplatonic philosophers, however, in insisting that that shaft of light emitted from the eyes is material: 'Now this is certainly a ray of material light ['corporeae lucis'] that shines forth from our eyes' (4.34.54). In suggesting that the beam emitted by the eye is material, Augustine draws upon the Stoic notion of a material *pneuma* emitted by the eye which receives impressions of material forms. As Colish points out, however, Augustine incorporates the Stoic *pneuma* only in the context of extramission, and employs it in a way which 'neoplatonizes it quite appreciably.'[17] Like the Stoics, and unlike the Neoplatonists, Augustine states that the extramitted beam is material; unlike the Stoics, however, who had denied the Pythagorean theory of visual fire, Augustine unequivocally states that the light emitted by the eye is a fiery substance: 'the warm quality of fire' in the body rises to the brain, from whence flow 'the rays which go forth out of the eyes' (7.13.20). The emitted light is a material 'element' which is 'akin to soul' but, because it is material, not soul itself (12.16.32).

It is noteworthy that several of the examples cited by Augustine to illustrate the deceptive nature of corporeal vision reappear, in a much more fully elaborated form, in later medieval writings on optics. These include 'when people in a moving ship seem to see stationary objects on shore in motion,' and 'when the rays coming from the eyes are not focused, and two lamps seem to shine when there is only one, or when an oar in the water appears to be broken' (*De Gen.* 12.25.52). Augustine returns to the problem of double vision in the eleventh book of the *De Trinitate* in the context of a detailed analysis of the nature of the visible form. He explains that vision is possible only in the presence of three things: a visible object ['res quam videmus'], which exists even when it is not seen; 'the vision' ['visio'], which exists only when the object is actually seen; and 'the attention of the mind' ['animi intentio'] which is fixed on the object (*De Trin.* 11.2.2). According to Augustine, the production of the vision [*visio*] is a collaborative process: 'vision is produced both by the visible thing and the one who sees ['ex visibile et vidente'], but in such a way that the sense of sight ... come[s] from the one who sees, while that informing of the sense ['informatio sensus'], which is called vision, is imprinted by the body alone that is seen' (*De Trin.* 11.2.3). Both the seeing subject and the object seen participate actively in the production of vision, which takes place through the transfer, by means of reproduction, of the visible form. Augustine proves the existence of the transmitted form using an analogy which appears in Aristotle's *De anima.* We can clearly see, Augustine states, that an impression remains in wax after a signet ring is pressed there; similarly, if that ring were brought into contact with a fluid, our reason would tell us that an impression had been made there, even though no form remains. Just so, says Augustine, the sense of sight proceeding from the viewer receives the visible form impressed upon it by the object seen. Augustine also adduces as proof of the existence of visible form the persistence of an afterimage when one looks at a luminous body such as the sun (*De Trin.* 11.2.4). Like the examples of deceptive vision noted above (the broken stick, the double image of the lamp, and the apparent motion of stationary objects seen from a moving boat), this argument reappears in later medieval discussions of vision.

Macrobius (fl. ca. 400) offers an even less fully articulated account of the mechanism of vision than Augustine. Yet his description of the role of light in unifying creation with the Creator played a role in propagating a Neoplatonic metaphysics of light during the later Middle Ages.[18] Macrobius echoes Plotinus in stating that all created things are linked to God in a 'catena aurea,' or 'golden chain.' Each link resembles the other 'like a countenance reflected in many mirrors arranged in a row,' with 'one splendor lighting up everything and visible in all' (1.14.15). In the course of describing the nature of the cosmos, Macrobius explains how the light of the sun or moon is effaced in solar and lunar eclipses, and argues that the moon reflects the sun's light but not its heat just as the image of a bright object is reflected in a mirror (1.15.10–12; 1.19.12–13). Macrobius's commentary continued throughout the Middle Ages to be a widely influential Neoplatonic view of the cosmos; his *Saturnalia,* however, which includes a detailed account of sight, exerted an even more powerful (though short-lived) effect on twelfth-century science, as will be described below.

Throughout the *De consolatione Philosophiae* of Boethius (480–524), clear vision is a metaphor for understanding: the anguished narrator of the first book cannot recognize Philosophy because his 'sight was so dimmed with tears' (1.pr.1). Philosophy wipes his eyes with her garment, restoring his ability to see clearly with the eye of the mind (1.pr.2, 1.m.3, 1.pr.3). In the later books, too, Boethius repeatedly emphasizes that the intellectual gaze should be turned ever upward, toward the source of divine light: 'You who with upright face do seek the sky, and thrust your forehead out, / You should also bear your mind aloft' (5m.5; cf. 3.m.12).

Boethius goes on to describe the sense of vision more explicitly, comparing it to the sense of touch in order to explain how sense perception differs from higher thought: 'sight recognizes in one way and touch in another; the former ... looks at the whole at once by the light of its emitted rays, while the latter ... [perceives] by parts.' The distinction between the partial knowledge offered by touch and the whole, simultaneous knowledge offered by sight mirrors the distinction between sense perception and reason: 'the eye of intelligence ['oculus intellegentiae'] is set higher still; for passing beyond the process of going round the one whole, it looks with the pure sight of the mind at the simple Form itself' (5.pr.4; pp. 410–11). By stating that sight and touch operate differently, Boethius distinguishes his account of sense perception from that of the Stoics, who had described vision in tactile terms: the object seen actually touches the *pneuma* which extends outward from the eye, 'as if in contact with a stick.'[19]

In his commentaries on Aristotle's *Peri hermeneias*, Boethius incorporates some elements of the intromission theory of perception, drawn both from the Stoics and from Aristotle's *De anima*, in order to account for the method by which the form inherent in the object is transmitted as visible form to the mind of the viewer. Yet even there he criticizes the Stoic theory of sense perception on the grounds that, while they do appropriately use metaphors such as wax impressed by a seal or paper inscribed by a pen, the Stoics fail to explain how such material impressions can be abstracted into insubstantial mental images.[20] In the *De consolatione Philosophiae*, Boethius's last work and the one which exerted the greatest effect on subsequent generations of medieval readers through Latin commentaries and vernacular translations, the intromissive aspects of sense perception are finally rejected due to their incompatibility with the Neoplatonic philosophy which permeates the work.[21] In keeping with his emphasis on the active effort of the mind to discover what lies outside itself, Boethius couches his description of vision in the *De consolatione* in terms of extramission, the eye gazing outward toward a fundamentally passive world. He refers dismissively to the intromission theory of the Stoics who claimed that 'sensible images / From bodies outside themselves / Are impressed upon men's minds,' just as marks are imprinted upon a page with a stylus (5.m.4; pp.412–13). While he denies that the mind is a purely passive receiver of impressions, Boethius acknowledges that the forms of objects are received by the mind, which itself actively participates in the acquisition of knowledge: 'when light strikes the eyes, / Or a cry in the ears resounds[,] / Then the mind's wakened power, / Calling upon these forms [specie] it holds within / To similar motions, / Applies them to the marks received from without / And joins those images / To the forms hidden within' (5.m.4; pp. 414–15). Boethius's description of the act of vision as a collaborative process which requires both the visible object and the power of sight located in the viewer recalls Augustine's account in the *De Trinitate* cited above. In both cases, the physical phenomenon of vision is described in terms of extramission (as when Boethius relates how the visual ray is emitted by the eye [*De cons.* 5pr.4]) as part of a larger effort to locate the power to see – and, by extension, to know – predominantly in the subject. Subsequent commentaries on and translations of Boethius's *De consolatione* offer interesting evidence regarding how the effort to locate the power of knowledge in the seeing subject was perceived during the later Middle Ages.[22]

As can be seen from the preceding summary, writers of philosophy, theology, and literature drew upon predominantly philosophical sources (particularly the Platonic extramission model of vision) in order to account for how human beings acquire knowledge of the world around them. Such writers show little or no awareness of the advances in optics that had been made by figures such as Euclid, who in the fourth century BCE produced a theory of geometrical optics positing the existence of the visual ray; and, in the second century CE, Ptolemy, who improved upon Euclid's description of the visual ray, and Galen, who added a detailed anatomical description of the eye. Galen's synthesis of what he derived from hippocratic medicine with his own anatomical investigations exerted a powerful influence not only in his own time but also later, in a more fully elaborated form, following the reintroduction into western Europe of galenic theories via the medical writings of scientists in the Islamic world.[23] The translations of Constantinus Africanus (d. 1087) during the late eleventh century were perhaps the single most important conduit for the dissemination of such notions found in Galen's synthesis as the presence of bodily 'humours,' their balance (or imbalance) in various forms of 'complexion,' and the presence of three 'souls' centred in the heart, the liver, and the brain.[24] The brain, according to Galen, is the source of the rational power of the soul, which is emitted through the hollow tubes (the optic nerves) which lead to the eyes and thus enable sense perception. Vision takes place when the luminous visual *pneuma* (a concept Galen derives from the Stoics) emitted from the eye encounters a visible object.[25] Galen's work on vision remained available in Europe before the twelfth century in only a very limited form; in the

Islamic world, however, where a much greater number of Greek texts remained accessible, significant advances in the knowledge of ocular anatomy took place. The *De oculis* of Hunain ibn Ishaq (808–77), who was until the beginning of this century known in the West as 'Johannitius,' was the single most important text on ocular anatomy introduced into medieval Europe, although the survey of medical theory and practice found in the *Isagoge* and the *Pantegni*, both also translated by Constantinus Africanus, were additional sources for galenic theory as developed and elaborated in the Islamic world.[26] Though the sections of the *Pantegni* treating the eye are translated by Constantinus from Haly Abbas ('Ali ibn al-'Abbas; fl. 900s), while the *De oculis* is translated from Hunain ibn Ishaq, both source texts base their accounts of ocular anatomy and the mechanism of vision on Galen.[27] In his description of extramission, as throughout, Constantinus's *De oculis* differs only in minor detail from the source text by Hunain: both state that the visual *pneuma* must exit from and return to the eye in order for an image to reach the brain.[28] The same extramission theory is present in Haly Abbas, and thus in Constantinus's *Pantegni*, though the former does not describe the mechanism in much detail due to his eagerness to limit his discussion to the topic of ocular anatomy.[29]

The desire of Haly Abbas to confine himself to medical aspects of the eye, and leave to others the task of accounting for the mechanism of vision, is typical of western writers on the eye during the later Middle Ages as well. The translations by Constantinus Africanus, among other writings turned out by the medical community in and around Salerno,[30] spawned both abbreviated adaptions and vernacular translations which display a comparable unwillingness to venture beyond ocular anatomy and a list of potential cures. Such treatment of the subject is probably due to the pragmatic opinion that it is not necessary to know how vision takes place in order to treat the diseased eye. In his *De probatissima arte oculorum*, for example, Benvenutus Grassus alludes only very briefly to the mechanism of vision, and instead devotes the bulk of his treatise to a practical account of the various remedies to be applied to the eye in case of disease.[31] Though he cites as his source 'Johannicius' (that is, Constantinus Africanus's translation of Hunain ibn Ishaq), he does not reproduce the account of vision to be found in the *De oculis* and even in the *Isagoge*, referring only briefly (in the words of his fifteenth-century Middle English translator) to the 'vysyble spyryt sent from the fantastical celle by a synew clepid neruus obticus,' that is, the visual *pneuma* sent from the foremost ventricle of the brain via the optic nerve.[32] This emphasis on practical opthalmology rather than visual theory may perhaps explain the popularity of Benvenutus's little treatise, attested to by its translations into Provençal and French as well as Middle English.[33] A similar reluctance to venture beyond the bounds of practical medicine can be found in the fifteenth-century commentary on Avicenna written by the Parisian physician Jacques Despars. Though he alludes to the perspectivist account of vision, based on the propagation of visible forms from the object seen and their subsequent intromission into the eye, he also reproduces (and glosses approvingly) the plainly contradictory extramission theory of vision as put forth by Galen. Jacquart suggests that this omission reflects a decision on the part of Jacques Despars not to engage himself too deeply in the debate regarding theories of vision. She notes that while literary texts of the period show a willingness to take part in such debate, contemporary medical writings do not. Jacquart quotes Jacques Despars's statement that the debate regarding intromission and extramission 'concerns, not physicians, but only the learned and the natural philosophers who argue with the aid of syllogisms about causes, and not effects.'[34] In its refusal to view visual theory as part of his purview, Jacques Despars's statement recalls Augustine's dismissal, one thousand years earlier, of the problem of double vision with the remark that 'it is a tedious subject, and not at all necessary for our subject' (*De Trin.* 11.2.4).

Yet medieval physicians' unwillingness to debate the merits of rival theories of vision does not signal a general lack of interest in the topic. Far from it: optics, or *perspectiva* as it came to be known in the thirteenth century, was an area of intellectual inquiry which generated extraordinary developments in science, philosophy, and theology. Beginning in the twelfth century, translations of Arabic treatises on optics began to appear which exercised a profound effect on the development of optics in the West.[35] Yet even before the major influx of scientific manuscripts translated from Arabic, works of natural philosophy had begun to appear which signalled a renascent spirit of inquiry and a yearning for new sources of scientific knowledge, in the form of both experimental inquiry and translated treatises. During the early twelfth century, natural philosophers such as Adelard of Bath (fl. 1116–42) and William of Conches (ca. 1090–post 1154) had to rely upon just a few scientific texts that had been preserved from antiquity, primarily the *Quaestiones naturales* of Seneca (ca. 4 BCE–65 CE) and the *Historia naturalia* of Pliny the Elder (23/24–79 CE). The very uncertain dating of Adelard's *Quaestiones naturales* makes it difficult to know whether, as Elford has suggested,

William bases his model of vision on that of Adelard, or whether both used Seneca as a source independently.[36] Nonetheless, William's remarkable ability to synthesize a wide range of observations and analysis of natural phenomena made his *Philosophia* (ca. 1125) and, even more, his magisterial revision of that work in his *Dragmaticon* (1147–9) crucial sources for later generations of natural philosophers. William's scientific writings are significant not only for their extraordinary synthesis of current work in physiology, meteorology, and optics,[37] but also for his placement of these observations on natural phenomena in the context of Neoplatonic philosophy.[38] He is believed to have taught at Chartres or possibly Paris,[39] and spent his later years as tutor to the children of Geoffrey Plantagenet, the duke of Normandy, one of whom later became Henry II of England. The *Dragmaticon*, written during that period, is couched in the form of a dialogue between William himself ('Philosophus') and the duke ('Dux').[40] William's Neoplatonism is derived from a number of sources, but one in particular is worth examining at length due to its integration of scientific matters and Neoplatonism (albeit not on the scale William himself would later achieve), especially on the topic of sight. This work is Macrobius's *Saturnalia* (ca. 400), a work which was widely known throughout the twelfth century, though in the later Middle Ages its repute was largely eclipsed by Macrobius's *Commentariam in Somnium Scipionis*. In his *Policraticus* (ca. 1160), John of Salisbury refers approvingly to 'the author of the *Saturnalia*': he, Adelard of Bath, and William of Conches all draw upon the *Saturnalia*.

In the seventh book of the *Saturnalia*, Macrobius offers a detailed explanation of the relative merits of the extramission and intromission theories. He dismisses the intromission theory as formulated by the Epicureans on the basis that the pupil of the eye is too small to receive so many emitted forms simultaneously: how, for example, could one see an entire army (7.14.11)? Macrobius's refutation of the Epicureans employs the arguments used by Galen in his *De placitis Hippocrates et Platonis* (7.5) and, indeed, Macrobius's account of the anatomy of the brain and spinal cord earlier in book 7 is clearly galenic in origin; especially notable is Macrobius's use of the Greek word *syzygy* to describe dense networks of nerves, as Galen does in the *De placitis*.[41] Following Plato, Macrobius states that the eye emits a luminous ray (not, as Galen would have it, a material *pneuma*). If the air is illuminated by an external light, the ray, which is enlarged at the end, proceeds until it strikes a visible object (7.14.13). Macrobius's model of vision is fundamentally Platonic, though he draws upon Galen both for anatomical information and for arguments to be used to refute intromission theories of vision. Further, his theory is both enhanced by a discussion of the refraction of the visual ray drawn from the fifth book of Ptolemy's *Optica*,[42] and dramatically re-shaped by being integrated within an emphatically Neoplatonic view of the world and the place of the seeing subject within it. In his description of the visual ray, Macrobius explains that vision fails at a great distance because the power of the ray extends over a distance of only about 180 stades. The viewer is thus always at the centre of a circle, illuminated from the centre by the light emitted by the seeing subject: 'Thus he who sees is always at the center of a circle bounded by the horizon. And since we have established the extent of the visual ray from the center to the circumference, it is evident that the diameter of the circular horizon is 360 stades; and if the observer steps forward or steps back, he will always see around him a circle of the same size' (7.14.16). The geometry here is Ptolemaic; but the metaphysical import of Macrobius's image of the subject illuminating the circular space around him is derived from the Neoplatonism of Plotinus and Porphyry which has been shown to be essential to Macrobius's conceptualization of the soul and its place in the universe.[43] The synthesis of galenic and Ptolemaic science with Neoplatonic metaphysics found in the *Saturnalia*, enriched by the fuller account of galenic anatomy and physiology made available in the late eleventh century by the translations of Constantinus Africanus, was a vital source for twelfth-century natural philosophy.

William of Conches's knowledge of Macrobius is attested to not only by his allusions to the *Saturnalia* and the *Commentariam in Somnium Scipionis* in his own works of natural philosophy, but by his authorship of a set of glosses on the latter work, written early in his career.[44] William also wrote commentaries on Boethius, Priscian, Martianus Capella, and, most importantly in the context of optics, Plato's *Timaeus*.[45] In the twelfth century, of course, the *Timaeus* was known only in the Latin translation with commentary by Calcidius, who substantially developed the account of vision found in Plato's text. Calcidius schematizes all vision into three types: direct, reflected, and refracted.[46] This threefold schema is reproduced and elaborated by William in his commentary,[47] and became the norm for virtually all subsequent accounts of vision in the medieval West.[48] Though it has sometimes been assumed that William's theory of vision is consistent in the *Glosae super Platonem*, the *Philosophia*, and the *Dragmaticon*, Elford and, more recently, Ricklin have argued that his optical model develops significantly from one text to the next.[49] In his works of natural philosophy, William uses an extraordinarily diverse range of sources, ranging from Neoplatonic texts such as the *Saturnalia* to the most up-to-date translations of Arabic medical treatises.[50] His account of optical phenomena extends to the macrocosm of the universe as well as the microcosm of man: he describes meteorological phenomena such as the rainbow (*Drag.* 5.4), the emanation of light and heat from the sun, solar and lunar eclipses, and moon spots (4.12–15). His description of ocular anatomy is followed by an account of the mechanism of vision, and explanations of how a variety of optical illusions, especially those involving mirrors, are effected (6.19–20). William's account of the effects of light on the macrocosmic level are based on Aristotle's *Meteorologica* and Seneca's *Quaestiones naturales*. His account of the effects of light on the microcosmic level, however, in the act of vision, is more innovative. Like Adelard of Bath, William presents an extramission theory which is fundamentally Platonic; unlike Adelard, however, William explains how exterior light interacts with the visual power emitted from the eye to enable sight. The visual power, William explains, using galenic physiology, is generated in the brain and sent to the eye by way of a hollow conduit, the optic nerve. Upon passing through the pupil, the visual power merges with the external light and extends, in the shape of a cone, until it reaches a visible object which acts as an obstacle, impressing its form upon the visual beam.[51] The beam then returns, says William, to the foremost part of the brain, where it impresses the shape and colour of the object seen upon the waxy and receptive imagination (*Drag.* 6.19).

William of Conches's writings already show some evidence of the dramatic effect of the translations from Arabic being produced during the late eleventh and twelfth centuries. But the extramission theory of vision which appears in his works would soon be rendered outdated by the extraordinary advances in optical theory made possible by new translations, especially the works of al-Kindi (d. ca. 866) and Alhazen (ibn al-Haytham; ca. 965–1039). Al-Kindi produced the most articulate and sophisticated version of the extramission theory that had ever been invented. He would continue to be cited by later writers on optics such as Grosseteste, Albertus Magnus, and Bacon whenever they wished to present the extramission theory as a viable alternative.[52] The *Perspectiva* of Alhazen (ca. 965–1039), however, more than any other translated work stimulated the flowering of optics in the West after 1200. While both Avicenna (ibn Sina; 980–1037) and Averroës (ibn Rushd; 1126–98) treated the mechanism of vision in their works, neither of them incorporated the geometrical optical theories of Euclid and Ptolemy to any significant extent. Both subscribed to fundamentally Aristotelian versions of the intromission theory, enhanced by descriptions of ocular anatomy drawn from Galen. The most significant of these enhancements concerns the function of the 'crystalline humour' (the lens) and the 'aranea' (lit. 'spider web'; the retina): the crystalline humour is the primary sensitive organ, which transmits to the brain the image intromitted into the eye by the object, while the aranea serves a secondary function, assisting in the transmission of the image.[53] The Aristotelian intromission model as developed by Avicenna and Averroës continued to be drawn upon throughout the Middle Ages, often for comparative purposes but sometimes represented as the most accurate account, as in the *De homine* and the commentary on Aristotle's *De anima* written by Albertus Magnus (ca. 1193–1280).[54]

Unlike Avicenna and Averroës, Alhazen offered a sophisticated intromission theory of vision which combined the early Greek theories of the visual cone with Euclid and Ptolemy's geometrical approach to optics. His theory would exercise a dominant and continuing influence on the development of optics in the West, both in its Latin translation, produced around 1200, and vernacular translations.[55] Although his work shows no evidence that he was familiar with Alhazen's theories, Robert Grosseteste (ca. 1168–1253) is generally considered the first significant writer on optics in the medieval West. He is best known for what is sometimes referred to as his 'metaphysics of light,' the intertwining of the science of optics with theology.[56] Grosseteste held that light was the fundamental substance of the universe, and that it was the material manifestation of divine grace: light is 'the first corporeal form,' which 'diffuses itself in every direction ... by multiplying itself.' While light is 'the form [*species*] and perfection of all bodies ... in the higher bodies it is more spiritual and simple, whereas in the lower bodies it is more corporeal and multiplied.'[57] It is significant that Grosseteste also produced a new translation with commentary of the works of pseudo-Dionysius (written 1239–43), for Grosseteste's 'metaphysics of light' unifies creation through radiance in a way reminiscent of the celestial hierarchies of pseudo-Dionysius.[58] Grosseteste claimed that species were constantly emitted by both the object of vision and the subject, and that the species move between the object and the eye in accord with the laws of geometry. Grosseteste synthesized al-Kindi's extramission theory with a notion of the species drawn from intromission theory, bringing them together with a detailed application of geometrical optics and embedding the whole within a deeply Neoplatonic view of the spiritual value of light in unifying the cosmos.[59]

Grosseteste's works were read avidly by Roger Bacon (ca. 1220–92) who was, unlike Grosseteste, in a position to profit from the abundance of new translations from Arabic: most importantly, the *Perspectiva* of Alhazen.[60] Bacon elaborated on Grosseteste's work, producing a more fully developed theory of the 'multiplication of species.' According to this theory, each individual species emitted by the object stimulates the medium to create another, identical species, which in turn stimulates the medium to create another, and so on. The species of Grosseteste and Bacon differ from the *eidola* or *simulacra* of the Epicureans, however, in that they are emphatically not material. As Lindberg points out, 'The multiplication of species is more like the propagation of waves than the motion of projectiles.'[61] The theory of the multiplication of species proved to be very influential, not only within the field of optics, but also within the disciplines of philosophy and literature.[62] Bacon's work on optics was available both in an earlier treatise called *De multiplicatione specierum* (ca. 1262) and in the fifth book of his huge compendium, the *Opus maius*, written at the request of Pope Clement IV.[63]

Bacon's younger contemporary, John Pecham (ca. 1235–92), wrote a more succinct optical treatise, the *Perspectiva communis* (ca. 1265), which soon became the most widely read optical text of the later Middle Ages (no doubt due to its brevity).[64] Pecham incorporated the work of Grosseteste and especially Bacon, but he relied primarily on Alhazen's *Perspectiva*; for this reason, Pecham is sometimes seen as merely a popularizer of Alhazen. Pecham's contemporary, Witelo (d. post 1281) also based his *Perspectiva* (ca. 1273) on the optics of Alhazen.[65] Unlike Bacon and Pecham, however, Witelo avoids the term *species*, instead focusing primarily on geometrical descriptions of the motion of the light ray.[66] The European scientists who based their work on Alhazen, primarily Bacon, Pecham, and Witelo, are commonly refered to as 'perspectivists'; perspectivist optics continued to dominate university curricula until the time of Descartes.[67]

A significant challenge to the perspectivist account of perception was posed by the fourteenth-century philosopher William of Ockham (ca. 1285–ca. 1349). On the basis of the principle that any unnecessary step of a process must be eliminated (Ockham's famous 'razor'), the Venerable Inceptor declared that the visible species in the medium need not (and therefore did not) exist.[68] Even before Ockham, however, the thirteenth-century Franciscan theologian Peter Olivi had objected to the theory of vision by means of species on the basis that knowledge based on representations could never lead to knowledge of the original object.[69] Like Olivi, Ockham concluded that 'the representative will never lead to the cognition of what it represents.'[70] Intriguingly, however, as Tachau has shown, nominalist philosophers who followed in the footsteps of Ockham discounted their master's rejection of species in their own accounts of the relationship of perception to knowledge. Ockham's own secretary and disciple, Adam Wodeham (ca. 1298–1358), returned to the perspectivist account of species in the medium; Wodeham differs from the perspectivists, however, in his description of how the species in the medium comes to be apprehended as an intelligible species in the mind.[71] Developments in natural philosophy of the late fourteenth century, especially that produced by the 'Oxford Calculators,' a group of scholars associated with Merton College, show how Ockham's nominalism was adapted to a scientific model of nature that sought to take account of (and even measure) motion and change. Among these scholars, John Dumbleton (fl. 1331–49) is particularly significant, for he alone among the Calculators sought to integrate natural phenomena with a theory of language. In his *Summa logicae et philosophiae naturalis*, Dumbleton illustrates how motion and change can be observed and quantified both in the natural world created by God and in the semantic systems imposed by man.[72]

More than the abstruse, often highly technical analyses of scientists and philosophers, medieval encyclopedias would likely have been the source of general information on vision most useful and accessible to medieval readers, especially the voluminous and widely disseminated compilations of Isidore of Seville (d. 636), Bartholomaeus Anglicus (fl. 1230–50), and Vincent of Beauvais (ca. 1190–1264).[73] Isidore includes only a brief account of the anatomy of the eye in the eleventh book of his *Etymologiae*, which largely comprises a description of the parts of the human body. His account of how the act of vision takes place is both brief and ambiguous: he states that vision takes place 'either by an external ethereal light or an internal lucid spirit' ['aut externa aetherea luce, aut interno spiritu lucido'] extramitted from the eye.[74] This ambiguity reappears in Isidore's description of the eyes, which are called lamps [*lumina*] 'because light [*lumen*] shines forth from them ... or because they reflect light received from outside to produce vision' ['quod ex eis lumen manat ... aut extrinsecus acceptam visui proponendam refundant'] (11.1.36). Isidore's references to the vitreous humour and the tunics of the eye (11.1.2) reveal the galenic origins of his anatomy. Fuller accounts of ocular anatomy and the mechanism of vision appear in the thirteenth-century encyclopedias of Bartholomaeus Anglicus and

Vincent of Beauvais. The third book of Bartholomaeus's encyclopedia, like the eleventh book of Isidore's *Etymologiae*, concerns the human body; but it is only the first book of five (3–7) on anatomy and physiology to be found in the *De proprietatibus rerum*. Bartholomaeus devotes an entire chapter to the sense of vision, including not only the anatomy of the eye but also a relatively detailed account of the mechanism of vision. He ennumerates the three types of vision (direct, reflected, and refracted), citing as his source 'the first chapter of *Perspectiva*.' Bartholomaeus goes on to refer to the multiplication of the visible species [*multiplicantur species*], a formulation which could only be drawn from treatises on perspective.[75] He cites Aristotle as his source for the belief that sight is nothing other than the image issuing forth from the thing seen, and then cites the contrary opinion of Augustine that the power of vision resides in the eye of the viewer (3.17). Interestingly, Bartholomaeus elaborates on the extramission theory of vision which he attributes to Augustine by adding aspects of geometrical theories of vision which posit a visual pyramid, probably drawn from perspectivist writings, while continuing to attribute this modified theory to the saint. Overall, Bartholomaeus's account reflects to a remarkable extent contemporary developments in the science of optics, including the theory of the multiplication of species (to which he refers several times) first put forth by Grosseteste. Bartholomaeus also cites Constantinus Africanus as his source in his account of ocular anatomy and the function of various parts of the eye, both in the third book of the *De proprietatibus rerum* and in the more detailed description of the eye which appears in the fifth book (3.17; 5.5–7).

Vincent of Beauvais's *Speculum naturale*, one part of his vast *Speculum maius*, was not nearly as widely diffused as Bartholomaeus Anglicus's encyclopedia. Though much work has been done on the reception of Vincent's *Speculum historiale*, less is known about the reception of other parts of the *Speculum maius*, although Chaucer's use of them has been argued in a series of articles by Pauline Aiken. Vincent's account of the properties of the soul in the *Speculum naturale* is largely based on Albertus Magnus's commentary on Aristotle's *De anima*; it is therefore difficult to say how many references by other writers to Albertus Magnus or even to Aristotle, may in fact be references to their theories as found in Vincent's summary.[76] Discussions of vision recur several times in the *Speculum naturale*, in a variety of contexts. The very title of Vincent's work signals an awareness of the centrality of the metaphor of vision for the act of knowing, based on Neoplatonic and Augustinian developments of the theme.[77] His initial references to optical phenomena appear in the second book of the *Speculum naturale*, embedded within an account of the six days of creation. 'Fiat lux' gives rise to a sequence of no less than fifty-three chapters on the nature of light (2.32–84, cols. 99–133), in which Vincent frequently cites as his source Albertus Magnus's commentary on Aristotle's *De anima* (for example, 2.41, col. 105); when Vincent cites Aristotle, Avicenna, or Averroës, he does so via Albertus. His discussion of light addresses not only the question of whether light is substance or accident, but also the relationship of colour to light (2.56–71, cols. 114–25) and the properties of mirrors (2.72–81, cols. 126–31). Other optical phenomena addressed in the *Speculum naturale* include the rainbow (4.74–80, cols. 278–82), for which Vincent draws upon Aristotle's *Meteorologica* and William of Conches's *Dragmaticon*; the use of silver to make mirrors (7.20, cols. 436–7), and the emission of light by certain precious stones (8.51, col. 520; 8.62–3, cols. 525–6); the emanation of light from the sun (15.4, cols. 1095–6) and reflection of sunlight from the moon (15.8–9, cols. 1098–9); the light emitted by the eyes of certain animals (19.79, cols. 1425–6), and the anatomy of the eye (21.12–15, cols. 1567–9; 28.46–50, cols. 2023–6), on which he cites a variety of authors, most notably Constantinus Africanus's *De oculis* and the *Pantegni* (for example, 28.48, col. 2024) and al-Razi (28.50, col. 2026). He discusses at length the nature of vision (25.28–49, cols. 1793–1807), summarizing various theories of vision and addressing current topics in optical theory such as the status of the species in the medium, albeit in simple terms (25.42, cols. 1802–3). His explanation that the species exists in the eye as action, but in the medium only as potential (col. 1803), is based, like much of his treatment of optics, on Albertus Magnus's *De anima* commentary.

The encyclopedias also provide useful synopses of medieval thought regarding how the mind processes information received from the senses. The processes of sensation and thought could, of course, be described very elaborately and were the subject of much debate by philosophers and theologians, during the Middle Ages as in antiquity. As Vincent of Beauvais remarks, the cognitive faculties [*cognitivae*] can be subdivided in many different ways: in two, as the sensitive and the intellectual souls; in three, as sense perception, imagination, and intellect; in five, as sense perception, imagination, reason, intellect, and intelligence (*Spec. Nat.* 27.1, col. 1917). Vincent goes on to give a number of alternative ways to categorize the faculties of the mind, in keeping with the encyclopedist's task of summarizing what has been said by a host of authorities. He rehearses Augustine's influential division of the mind into the human trinity of memory, reason (which Vincent calls 'interna visione'), and will (27.6, col. 1921), as well as Avicenna's almost equally influential formulation of a mind in five parts, comprised of the three faculties responsible for the processing of images (the *sensus communis* or common sense; *imaginatio*, also called *phantasia*; and *imaginativa*), along with reason (*estimativa*) and memory (*memorativa*; 27.3, col. 1919).[78]

Avicenna's categories of the mind, coupled with galenic accounts of the anatomy of the brain, put the medieval encyclopedist on somewhat firmer ground in describing the parts of the mind. Vincent subsequently explains, following Constantinus Africanus, how the five faculties described by Avicenna are distributed among the chambers of the brain. While modern medicine teaches that the grey matter of the brain is where cognition takes place, medieval medicine taught that the ventricles of the brain were of primary importance: they were seen as the little rooms or *cellulae* where the cerebral spirit flowed, enabling the mind's cognition and its command of the body's motion.[79] Vincent explains that the imaginative faculties inhabit the foremost part of the brain, the reasoning faculty the middle, and the memory the rear (28.41, col. 2019). Like Vincent, Bartholomaeus Anglicus offers a variety of perspectives on the subdivisions of the mind; but he concludes, even more definitively than Vincent, that the mind or 'innere witte' (to be distinguished from the outer wits, the five senses) is divided into three *cellulae* containing, from front to back, the faculties of imagination, reason (*logica*) and memory (*De prop. rer.* 3.10, ed. 53, tr. 1:98). The three-cell theory simplifies Galen's (anatomically accurate) teaching that the brain actually contains four ventricles: a pair in front for perception and imagination, a central one for reason, and one in the rear for memory and voluntary movement of the body.[80] In the simplified form as found in the encyclopedias, the three-cell model of the mind is alluded to frequently in medieval literature.

Entire books could be (and have been) written on each of the mental faculties.[81] Here it suffices to outline very briefly the process by which the visible form inherent in the object makes its way into the mind of the observer, and the role of each of the faculties in assimilating the visual image. Whether by means of an extramitted ray or an intromitted species, the visible image makes its way through the pupil of the eye to the crystalline humour (the lens), which according to galenic theory is the principal sensitive organ of the eye. The image then is propagated along the optic nerve to the foremost *cellula* of the brain, where the *sensus communis* gathers impressions received from each of the 'outer wits' of sight, hearing, smell, taste, and touch. The imaginative faculty receives the sense impression just as soft wax receives a seal; reason judges the image; finally, memory stores the image for future reference.[82] Philosophers have endlessly speculated (and continue to do so) how the mind carries out the processes of thought, imagination, and memory. Imagination is in one sense a passive faculty, in that it receives images from without, as reflections appear in a mirror;[83] yet it has an active aspect as well, able to produce an image of something that has never been seen before by combining other images previously seen.[84] Reason, inhabiting the central *cellula*, is characterized most variously among the parts of the mind, as is evident from the many names assigned to the central faculty: *estimativa, cogitativa, ratio, logica* or *logistica*, judgment, and so on. Like imagination, reason is frequently characterized in terms of vision: imagination receives and sometimes recombines the images received from without, while reason is the exercise of the 'oculus intellegentiae,' according to Boethius, or 'interna visione,' as Vincent of Beauvais puts it.[85] Memory, too, has a visual component. This aspect is perhaps most obvious in mnemonic systems, such as that outlined in the pseudo-Ciceronian *Rhetorica ad Herennium*, where a series of images arranged in a particular order are used to commit to memory a sequence of some kind such as a speech or a list of items.[86] As Carruthers has pointed out, memory systems based on recall of the physical text (the open book) are at least as common in the Middle Ages as systems based on recall of images. Yet as Carruthers indicates by describing the two systems as 'visual' and 'pictorial,' memories are formed and retrieved through vision, regardless of whether they appear as words on the page of an open book or as symbolic images.[87]

Imagination, reason, and memory: all could be described using metaphors of vision. The developing science of optics generated a rich vocabulary that could be used to characterize such intangible concepts as the relationship of subject and object, the nature of mediation, and the process of understanding. The notion of the visible species, central to perspectivist optics, became a particularly useful means for medieval writers to register their ambiguity regarding the signifying properties of language. If the species in the medium reproduces its exemplar only imperfectly, how much more poorly must the word represent the idea? The very scientists who wrote treatises on perspectiva were aware of the poetic resonance of their specialized vocabulary. In his poetry, the perspectivist John Pecham describes the interrelation of the three parts of the Trinity in terms of the diffusion of light in the medium, and the generation of the Son from the Father in terms of reflection from a mirror: 'in the mirror of His light / An equal image shows itself.'[88] Even more poetically, Robert Grosseteste uses the phenomenon of refraction in his allegorical *Chastel de amur* to depict the diffusion of divine grace in the created world. In his *Opus maius*, Roger Bacon explains the figurative significance of refraction: 'vision is of three kinds: direct in those who are perfect, refracted in those who are imperfect, and reflected in evildoers and those who ignore God's commandments' (5.3.2). Similarly, in Grosseteste's *Chastel de amur*, refraction is used to show how grace permeates fallen mankind just as light is refracted through the medium. A rainbow, which is the main example used to illustrate the refraction of light in treatises on optics, shines above a castle which represents the body of the Virgin Mary. The refracted light of the rainbow, sign of God's covenant with man, corresponds to the colours of the castle: the foundation is green, the walls blue, the roof red. The inside of the castle is filled with pure white light ['clarté'], because 'God would only want to inhabit a beautiful and shining place' ['beau lu e ... cler'].[89] Here, the divine light, transparent and white within the castle, appears multicoloured on its exterior surface because it is refracted through the medium of the human body of Mary.

Developments in the history of optics are echoed in contemporary literature. At times, the literature appears to truly 'respond' to changes in scientific theory; at other times, new scientific theories are appropriated in the literature in order to provide the framework for a new, distinctive way of situating the subject in the world. The relationship between optical science and literature is particularly evident in the genre of allegory, simply because it relies so much on the figurative representation of knowledge in terms of vision, expressing spiritual illumination in terms of corporeal illumination. Changes in medieval theories of how words signify are accompanied and complemented by changes in optical theory. It is difficult to ascertain whether one generates the other, or whether both are part of a larger 'paradigm shift' which makes itself felt in the humanities as in the sciences.[90] Early medieval writers follow a theory of language which presumes that the word corresponds to the thing it represents; later medieval theories of language move away from this assumption and instead suggest that words are merely arbitrary signs assigned by man to stand in the place of things. The status of language as a mediator between man and the world is called into question: does language offer man access to the world, or does it interpose itself between him and the world? Medieval optical theory poses similar questions centered on the role of the mediator, whether the species or the diaphanous medium. The same questions persist: does the mediator provide access to knowledge, or does the imperfection of its transmission make it a barrier to clear and perfect knowledge?

The changes in language philosophy I have alluded to are sometimes expressed in terms of a shift from realism to nominalism. To do so is misleading, both in that it wrongly suggests that there was a linear progression from one way of thinking about signification to another, and in that it obscures the often dramatic differences that separate one 'nominalist' from another. This latter point applies not only to 'nominalist' philosophers widely separated in time, such as Abelard and Ockham, but also to those most intimately connected with each other, such as Ockham and his secretary Wodeham. A nuanced understanding of these various forms of nominalism, especially in connection with changes in language theory during the fourteenth century, offer productive insights into some of the most sophisticated literature of the period. Just as philosophers such as Ockham and Wodeham repeatedly resort to optical models in order to explain phenomena ranging from human cognition to the omnipotence of God, so too poets use vision as the fundamental model of knowledge and power. Mediation is increasingly the focus of attention, whether in the visual act or the significative act – to be ruthlessly swept away by a philosopher like Ockham, or delicately probed by a poet like Chaucer. In a world after the Fall and after Babel, both sense perception and language are imperfect mediators. It is therefore necessary to know things indirectly or, as Paul puts it, 'in a mirror, dimly' (I Corinthians 13:12). Just as the mirror conveys the image of the object that cannot be seen directly, so allegory conveys meaning that cannot be expressed directly. This analogy displays both allegory's strength and its weakness: the mirror reproduces the image, but not perfectly, just as the allegorical text conveys meaning, but never perfect truth. The next two chapters will show how this is illustrated by Guillaume de Lorris and Jean de Meun in the *Roman de la rose.*

Chapter Three

GUILLAUME DE LORRIS'S *ROMAN DE LA ROSE*

Each of the two parts of the *Roman de la rose* marks a transitional point in allegory. Guillaume's *Rose* is one of the first vernacular allegories, and is the first to treat a secular matter – the narcissism inherent in courtly love[1] – rather than a strictly religious topic. Although Chrétien de Troyes, Guillaume's predecessor in writing about the psychology of courtly love, incorporates allegorical elements into his romances, Guillaume de Lorris is the first to express both the intricate game of courtly love and an increasingly complex sense of self within the terms of Latin philosophical allegory. Furthermore, by concerning himself with the things inside man's mind rather than in the universe around him, with the microcosm rather than the macrocosm, Guillaume transforms twelfth-century philosophical allegory into a more personal, self-reflective genre.

Rather than discuss the *Roman de la rose* as a whole, I will discuss the first part (lines 1–4028), written by Guillaume de Lorris about 1225, and the continuation (4029–21750), written by Jean de Meun about 1265, separately.[2] This approach is characteristic of recent studies of the *Roman de la rose*. For many years it was assumed, on the strength of Jean de Meun's statement made within the poem (10496–586), that Guillaume was unable to finish the poem because of his sudden death, and that Jean was merely completing and elaborating on Guillaume's original plan. As Lori Walters has shown, certain illuminations in medieval manuscripts of the *Rose* reinforce the notion of a unified authorship; one illustration of Guillaume handing over a manuscript to Jean illustrates 'the transfer of literary authority' and 'the continuity of the poetic tradition.'[3] Until recently, modern readers have retained a similar notion of the relation of the two parts, largely under the influence of Alan Gunn's widely quoted *Mirror of Love* and (even more importantly) Dahlberg's popular translation.[4]

In the last few decades, however, readers have begun to suggest that Guillaume's poem was intentionally left unfinished, and that its apparent incompletion is actually part of the work's meaning. Hult notes that Faral (1926) was probably the first to suggest 'that the taking of the castle might destroy the poem's delicacy'; but the proposal became more widely accepted only after the studies of Strohm (1968) and Lejeune and Poirion (both 1973). Hult's own book on Guillaume's *Rose* (1986) contains the fullest argument for the poem's completion, including an account of its medieval reception as seen in its continuations, especially those of Gui de Mori and Jean de Meun.[5] In addition, in his studies of the poem's midpoint passage and Jean's authorial strategies, Kevin Brownlee has emphasized the ways in which Jean's poem differs from Guillaume's. By showing how the first *Rose* 'serves as a point of departure' for Jean, Brownlee implicitly argues that any reading of the whole poem must take into account the divergent aims of the authors.[6]

In keeping with the view that each part of the *Roman de la rose* must be read independently of the other, the following chapters will treat the work of Guillaume de Lorris and Jean de Meun separately. It may seem peculiar that Guillaume's portion of the poem is treated at such length, especially in view of the fact that it makes up only about one-fifth of the entire poem. This is because this study focuses particularly on the use of optical vocabulary, and while Jean's interest in optics is quite apparent from his emphasis on the subject in Nature's discourse, Guillaume's use of optical imagery and the extent to which he relies upon medieval writings on the mechanism of vision have not been established previously.

Structural Allegory

All narrative, by definition, is linear. Some narrative, however, is self-reflexive, generating a circular structure by, after the midpoint, repeating motifs from the first half in reverse order. Guillaume's *Rose* is one such text. Several critics have drawn attention to the peculiarly intricate structure of Guillaume's allegory. Verhuyck has suggested that the work is made up of a series of consecutive enclosures, like 'poupées russes' and 'boîtes chinoises.'[7] Barney uses the same metaphor to refer to Guillaume's '*loci amoeni*' which progressively 'enclose each other like Chinese boxes,' noting Guillaume's 'almost mathematical orderliness of presentation.'[8] By choosing these metaphors, Verhuyck and Barney stress the sense of repeated enclosure in the first *Rose*: these Chinese boxes and Russian dolls act as containers, enclosing layer upon layer of space within them. But they also implicitly extend the promise of ultimately containing more than mere space, for the *matrushka* doll finally yields the miniature doll that cannot be opened, while the nested Chinese boxes, by repeatedly deferring the gratification of penetrating the box and seeing what is inside, enhance the opener's curiosity and consequently augment the value placed upon the contents of the innermost box, that carefully hidden treasure.

Verhuyck suggests that, in the case of the nested enclosures of the first *Rose*, there is in fact no centre: Guillaume's 'multiplication des cadres' is perpetual, making the poem 'par essence, une oeuvre inachevée, une oeuvre sans dénouement.'[9] Despite the appeal of Verhuyck's notion of infinite regress, many readers have very definitely fixed on a centre of Guillaume's *Rose*: that is, the episode that takes place at the fountain of Narcissus. In a reading that in some ways evokes Verhuyck's, Kelly describes the circular structure of the poem and simultaneously stresses both the existence of a centre (that is, the 'danse in carol') and its inherent instability: 'the center itself seems constantly to recede, as changes in setting reform kaleidoscopically, bringing the parts together in new arrangements and images; features appear and disappear as a center begins to emerge, or rather open, like the rose itself.'[10] The perpetual centripetal motion emanating from the poem's centre described by Kelly, much like Verhuyck's 'multiplication des cadres,' conveys not only the presence of a central space in Guillaume's poem, but also its essential indefinability, the symmetry that surrounds the unattainable centre, and the infinite multiplication of images that emanate from it.

Both Verhuyck and Kelly draw attention to the crucial issue of symmetry in the poem's structure. What I wish to add to their analyses is an explication of the very specific optical terminology that appears in Guillaume's *Rose*. By 'optical terminology,' I mean not only those phrases that pertain to the eye and the act of seeing, but also those that refer to reflection and refraction, particularly as these phenomena are seen in bodies of water, in crystals, and in mirrors. Along with many readers of the poem, I would suggest that the centre of Guillaume's poem is located at the fountain of Narcissus and, more specifically, at the moment that the lover looks into the watery, mirroring surface of that fountain. But because that centre is a mirror, a surface which generates reflective images without any substantial reality, the centre remains undefinable. It is impossible to tell if the image emanates from substantial reality or from the mirror, making the search for the image's source repetitive and ultimately futile, as is implied by Verhuyck's 'multiplication des cadres' and Kelly's constantly receding centre.

I wish also to expand on the idea of a nested structure suggested by Verhuyck, Barney, and others. In their use of the analogy, the poem presents a series of enclosures which the lover (and the reader) penetrates consecutively. The final goal of the quest, what is contained within the innermost box – namely, the object of desire itself, represented by the rose – remains unattained and unattainable, eternally just beyond the lover's grasp in Guillaume's tantalizingly incomplete poem. The symmetry which so many readers have sensed in the poem is most evident in the series of enclosures that the lover must penetrate in the course of the narrative: he enters into the dream, passes through the door of the walled garden, and finally attempts to break through the hedge which encloses the rose. By the end of Guillaume's poem, the lover is even more firmly cut off from the object of his desire, looking up yearningly at the tower where Bel Acueil ('Fair Welcome') is imprisoned. On closer examination, these enclosures prove to be arranged in a symmetrical pattern. The dream at the poem's beginning and the tower at its end are, as it were, a mirror image of one another, related through a latinate pun drawn from Augustine's *De Trinitate*. The garden wall, whose entrance is attended by Oiseuse, and the hedge, which is guarded by Dangiers, form a comparable symmetrical set of enclosures. Appropriately, at the centre of the narrative, between dream and garden wall at the beginning and hedge and tower at the end, is the defining, central moment of Guillaume's poem: the lover's look into the reflective surface of the fountain of Narcissus.

Structurally, the dream serves as a framework within which the events of the narrative take place. Yet the dream performs the function of enclosure in another sense as well. In the poem's opening lines, Guillaume's narrator simultaneously points out the significance of the dream and its role as a container which encloses meaning: 'li plusor songent de nuiz / maintes choses covertement / que l'en voit puis apertement' ['most men dream at night of many things which are hidden that they afterwards see openly' (18–20)]. Here Guillaume simultaneously indicates the prophetic nature of dreams and shows how the dream functions as a receptacle of meaning. The meaning of dreams may at first be closed or hidden, and then be opened or revealed; the dream is, as it were, a covering which, when opened, exposes the meaning concealed within. This pattern of what is closed becoming open is repeated over and over in the poem as the narrator crosses boundaries into a variety of enclosed spaces, as each in turn changes from being 'coverte' to 'overte,' 'closed' to 'open.' The same terms are used to describe the changes in the rose, his object of desire, as it gradually blooms.[11]

Thematically, the dream carries with it implications regarding truth and falsehood, the veracity of the narrator's account and the accuracy of his perceptions as they are altered by the dream.[12] By opening the *Roman de la rose* with a dream, Guillaume invites the reader to question the truthfulness of the narrative or, alternatively, the accuracy of the narrator's perceptions in the course of experiencing the events of the dream. In the poem's opening lines Guillaume associates dreams with lies by using the rhyme pairs *songes/menconges, songier/mencongier*. While these words are not etymologically related, they have traditionally been linked in folk proverbs and in various literary sources.[13] But Guillaume does more than simply use a conventional rhyme: he first emphasizes the coupled terms by using the rhyme pair twice (1–2, 3–4), and then goes on to say that the dream's content is not lies, but merely what is hidden, and will later be 'bien aparant' ['quite apparent' (5)]. The revelation of what is hidden is the basis of the *songe* as *somnium* or enigmatic dream, and also sets up an opposition found repeatedly in the first *Roman de la rose*: that which is closed or hidden, opposed to that which is open or 'aparant.' It is for this reason that Guillaume points out that 'songes est senefiance' ['dreaming is meaningful' (16)].

The state of dreaming has long been described in terms of altered or veiled perception: in his *Commentariam in somnium Scipionis*, Macrobius quotes Porphyry as stating that, in dreams, 'when it gazes [the soul] does not see with clear and direct vision, but rather with a dark obstructing veil interposed.'[14] In this context, Macrobius cites Virgil's description in the sixth book of the *Aeneid* of the gates of horn and the gates of ivory, which for Macrobius represent two different states of dreaming: one that allows meaning to be perceived, though dimly, and another that is opaque and permits no perception. A similar optical metaphor, also based on the Neoplatonic equation of knowledge and vision, is famously used by Paul to describe how man knows now, in his imperfect, post-lapsarian state, in contrast with how he will know when he is in heaven: 'Videmus nunc per speculum in aenigmate, tunc autem facie ad faciem' ['For now we see in a mirror, dimly, but then we will see face to face' (I Corinthians 13:12)]. The images we see reflected in a mirror only imperfectly reproduce the object beyond. The mirror frequently appears in connection with the dream vision, both of them signifying the distorted perception characteristic of revelatory allegory.

Dreams are explicitly associated with mirrors in Alanus de Insulis's *De planctu Naturae*, a work that Guillaume de Lorris drew upon for both elements of imagery and the model of its dream structure. The work concludes, 'Accordingly, when the mirror with these images and visions was withdrawn, I awoke from my dream and ecstasy and the previous vision of the mystic apparition left me.'[15] In the *De planctu*, the mirror is both the medium of the dream and the mystical point of entry that permits access to knowledge not available by means of sense perception. Similarly, in the *Roman de la rose*, both the dream described at the opening of the poem and the mirror of Narcissus are liminal points where the narrator's perception is most acutely altered.

Alanus's dream in the *De planctu* is called a *somnium*, one of the five kinds of dreams described by Macrobius in his *Commentariam in somnium Scipionis*. Guillaume explicitly invites the reader to remember Macrobius's categorization of dreams in the opening lines of his poem:

> Aucunes genz dient qu'en songes
> n'a se fables non et menconges;
> mes l'en puet tex songes songier
> qui ne sont mie mencongier,
> ainz sont apres bien aparant,
> si en puis bien traire a garant
> un auctor qui ot non Macrobes,
> qui ne tint pas songes a lobes,
> ancois escrit l'avision
> qui avint au roi Scypion.

Many men say that there is nothing in dreams but fables and lies, but one may dream such dreams that are not at all lies, but rather[16] are afterward quite apparent. So may be taken as witness an author named Macrobius, who did not take dreams to be deceits, but rather wrote of the vision that came to the king, Scipio. (1–10)

Macrobius defines five types of dream: *somnium, visio, oraculum, insomnium,* and *visum.* The last two, Macrobius says, 'are not worth interpreting since they have no prophetic significance,'[17] while the *somnium, visio,* and *oraculum* are prophetic dreams that reveal a hidden truth to the dreamer. According to Macrobius's categories, the narrator's dream in the *Roman de la rose* is a *somnium,* for it is a prophetic dream:[18] as the narrator relates, looking back on his dream, 'en ce songe onques riens n'ot / qui tretot avenu ne soit / si con li songes recensoit' ['in this dream was nothing which did not happen just as the dream recounted it' (28–30)].

Within the *somnium* recounted by the narrator, Guillaume de Lorris also describes another sort of dream: the imaginary dream of a typical lover described by Amors, from which the dreamer awakes in a panic. This dream is an *insomnium*, 'ou il n'a que menconge et fable' ['where there is only lies and stories' (2434)], in contrast to the *somnium* experienced by the poem's narrator, which 'n'i a mot de menconge' ['has not a word of lies in it' (2074)]. This dream full of lies serves as a contrast to the dream containing a hidden truth which forms the narrative of the poem. Their conformity to the Macrobian definitions of *insomnium* and *somnium* shows not only that Guillaume de Lorris was aware of Macrobius's classification of dreams, but also that he wished to define the dream which is the poem's primary narrative device within that framework.

Macrobius states that, for a dream to be defined as a *somnium*, it must require an interpretation to elucidate its meaning: 'By an enigmatic dream we mean one that conceals with strange shapes and veils with ambiguity the true meaning of the information being offered, and requires an interpretation for its understanding.'[19] Guillaume's narrator gives an account of his dream several years after the fact, attempting to derive exegetically the truth hidden within his dream. It is debatable how successful his interpretation is;[20] it is at least clear that the narrator does not fully comprehend the deeper meaning of his *somnium* at the outset of the poem, only approaching a better understanding as the poem nears its conclusion. I will argue below that the narrator's increasingly accurate understanding of the meaning of his dream is precisely what causes the narrative to come to an abrupt end. The narrator gradually comes to realize that he, as the lover in the dream, and as the courtly lover in real life, reenacts the experience of Narcissus. His account of the dream stops just short of the recognition of the fact that the shifting object of his desire is actually himself.

The narrator first crosses over into the narrative enclosure of the dream; his second passage is into the enclosed garden, the walled *vergier* of Deduit. This garden is 'clos de haut mur' ['enclosed by high walls' (131)]. The narrator can gain access only by being allowed in, which occurs when he passes through the door which 'Oiseuse overt m'ot' ['that Laziness opened to me' (630)]. When Reson later attempts to dissuade the lover from his quest, she stresses the importance of Oiseuse's act (2986–94) using a homonymic rhyme that recalls the mirror Oiseuse holds in her hand (555): the doubled rhyme word evokes the false image within the mirror and the real object without. This mirror foreshadows the mirror of Narcissus found at the narrative centre of the poem. Oiseuse is Amant's partner in the dance in the garden (1249–50), affirming her pivotal role in permitting the lover to pass from the world outside the wall into the garden within.[21]

The narrator wanders through this enclosed garden, 'tant destre et senestre' ['to the right and left' (1415)], until he has explored the whole garden. At this point, he comes upon 'un trop biau leu ... en un destor' ['a very beautiful place, in a detour' (1423–4)], or a secluded area: this is the fountain of Narcissus, whose mirroring surface is the narrative centre of the poem. This is where the narrator's aimless wandering ends, and his single-minded quest begins, where he is transformed from dreamer into lover. But this transformation cannot take place until the narrator has fixed on an object of his potential love. He believes that, in 'li miroërs perilleus' ['the perilous mirror'], he could see 'sanz coverture' ['clearly' (1555)] and 'sanz decevoir' ['without deceit' (1558)]. But he soon discovers that 'cil miroërs m'a deceü' ['this mirror has deceived me' (1607)]. The mirror deceives him, because it inverts the actual order of perceived reality. As Alanus de Insulis writes in his *Summa de arte praedicatoria*, in a flat, man-made mirror, 'the right parts appear to be on the left, and the left appear to be on the right.'[22] Due to the mirror's inversion of the image, the lover can never attain in reality what he glimpsed in the mirror. From its inception, his love is doomed to be unsuccessful, as is illustrated by the increasing resistance of the barriers the narrator must cross. He progresses easily through the first enclosures of the poem, having only to ask Oiseuse in order to gain entry to the walled garden of Deduit. But in the second half of the poem, after the look into the fountain of Narcissus, the obstacles become painful and insurmountable: the hedge which surrounds the rosebushes is full of thorns, while the wall which is built around the roses and whose tower imprisons Bel Acueil poses a final, impenetrable barrier.

In the symmetrical structure of this poem, in which the events of the poem's second half mirror those of the first, the hedge serves as the emblematic counterpart of the walled garden: 'si vi un vergier grant et le, / tot clos de haut mur bataillie' ['I saw a garden, great and large, entirely enclosed by a high, fortified wall' (130–1)]; 'Li roser d'une haie furent / clos environ' ['The rosebushes were entirely surrounded by a hedge' (2763–4)]. The hedge, like the walls of the garden, encloses, keeping what is within hidden and protected. The walled garden, 'clos et barez' ['enclosed and barred'] is 'en leu de haies' ['in the place of a hedge' (466–7)]. The hedge is described as confining the rosebushes 'clos entor' ['entirely enclosed' (1616)], which can also be read 'clos en tor' ['enclosed in a tower'], thus foreshadowing the tower within which Bel Acueil and the roses will later be enclosed.

After the enclosure of the hedge has been penetrated by Amant, Jalousie decides to 'fere de noveil mur / clore les rosiers et les roses' ['build a new wall to enclose the rosebushes and the roses' (3592–3)]. The new wall is built to deter anyone who might wish to pick the roses, and contains a secondary enclosure designed to end what Male Bouche calls Amant's 'mauvés acointement' ['evil relationship' (3507)] with Bel Acueil. Jalousie plans to build 'une forterece / qui les rosiers clora entor. / El mileu avra une tor / por Bel Acueil mestre em prison' ['a fortress which will enclose the rosebushes entirely. In the centre it will have a tower to put Bel Acueil in prison' (3606–9)]. But this tower is more than just another wall separating the lover from his object of desire, another in the series of Russian dolls or Chinese boxes. It is the final, impenetrable enclosure, and it brings about the abrupt conclusion of the poem, not only because the tower (by association with Babel)[23] can be an emblem for the dissolution of language, but because the tower is the last enclosure needed to form a complete series of symmetrical barriers on either side of the look into the fountain of Narcissus.

Other signals also inform the reader that the narrative must draw to an end with the imprisonment of Bel Acueil. David Hult has pointed out grammatical changes in the poem that culminate in the concluding passage which begins 'Mes je, qui sui dehors le mur' ['But I, who am outside the wall' (3920)]. At this point, Hult argues, the voices of the omniscient narrator and the lover converge, so that the use of the personal pronoun 'becomes a coy way for the Narrator to insert himself into the dream story' (267).[24] The tower which appears at the end of the *Rose* similarly serves to demarcate the end of the structural allegory. The mirror at the centre of the work is, as it were, reflected in the poem's structure of symmetrical enclosures. The tower forms a structural parallel to the dream, for just as Bel Acueil is imprisoned within the tower, so the lover is enclosed within a dream whose meaning continues to be hidden from him.[25] Like the parallel images of the enclosed garden and the hedge surrounding the roses, so the dream and the tower mark the first and last barriers to the lover's self-knowledge. Each of these pairs – garden and hedge, dream and tower – has one part situated on each side of the central episode at the fountain of Narcissus.

Guillaume emphasizes the parallel of dream and tower by taking pains to describe how each performs the function of enclosure: in the poem's opening lines, Guillaume defines the dream as a container of meaning (18–20), while near the poem's end he repeatedly uses the verb *clorer* to describe the containment of the rosebushes and their attendant, Bel Acueil (3593; 3602; 3607). The dream's secondary function of enclosure, separating the lover within the dream from the real world outside, is echoed in the tower which separates Bel Acueil from Amant. Guillaume explicitly stresses the impenetrability of the latter enclosure: 'Bel Acueil est em prison / amont en la tor enserrez, / dont li huis est si bien barez / qu'il n'a pooir que il en isse' ['Bel Acueil is in prison, sealed up above within the tower, where the door is so well barred that there is no fear that he will come out' (3898–901)]. Correspondingly, the impenetrability of the former enclosure, the dream, is implicitly emphasized by the poem's abrupt end: the lover never awakens from the dream, just as Bel Acueil never emerges from the tower.

Dream and tower form a symmetrical pair, not only because both are described by Guillaume as loci of enclosure, but because each term is related to the word *speculum*, or mirror. The dream is figuratively called a mirror in the last lines of Alanus de Insulis's *De planctu Naturae*, a text that was Guillaume's source for conventional representations of homosexuality and for the image of the rose as an exemplar of beauty and desirability. Correspondingly, the tower is shown to be interchangable with the mirror in Augustine's *De Trinitate*, where Augustine distinguishes between 'tower' and 'mirror' in order to rectify the confusion sometimes caused by their Latin equivalents, *specula* and *speculum*:

> He uses the word *speculantes*, that is, beholding through a mirror [*speculum*], not looking out from a watchtower [*specula*]. There is no ambiguity here in the Greek language, from which the Epistles of the Apostle were translated into Latin. For there the word for mirror [*speculum*], in which the images of things appear, and the word for watch-tower [*specula*], from the height of which we see something at a greater distance, are entirely different even in sound; and it is quite clear that the Apostle was referring to a mirror [*speculo*] and not to a watchtower [*specula*] when he said 'beholding the glory of the Lord' [*gloriam domini speculantes*].[26]

In the *De Trinitate*, Augustine shows how introspection leads to knowledge of the divine, and consequently explores at length the relationship between vision and knowledge. As noted in the previous chapter, Augustine's descriptions of corporeal, spiritual, and intellectual vision in the *De Trinitate* and *De Genesi ad litteram* were frequently cited by medieval writers. The pun latent in the terms *specula* and *speculum*, which Augustine takes such pains to distinguish, may have provided the basis for Guillaume's use of the tower as the concluding emblem in the mirroring structure of the first *Rose*. While the dream which begins the structure is like a mirror in a symbolic sense, the tower which ends it is like a mirror in a linguistic sense. Playing on the Latin terms reinforces the central theme of the mirror, seen emblematically in the fountain of Narcissus and structurally in the symmetrical allegory of the poem, and brings out the more general importance of vision in the poem, for Augustine's cognate terms *specula* and *speculum* share the root *specio*, to see.

In the preceding pages, I have described the symmetrical structure of the narrative in considerable detail in order to fulfil two purposes: first, to show that the structural allegory strengthens the belief that Guillaume's poem is complete; second, to illustrate the extent to which reflection and the redoubling of images are central to the *Rose*. The narrator crosses barrier after barrier, so that each changes from being 'coverte' to 'overte,' 'closed' to 'open.' But when the narrator reaches the emblematic centre of the poem, the mirror of Narcissus, he believes that he has found something that is not closed, that he sees 'sanz coverture' (1555). But, in actuality, the mirror is the one thing which cannot be opened or 'overte': it has no inside. Because the barrier of the mirror can never really be penetrated, the obstacles that the narrator encounters in the second half of the poem become increasingly difficult to surmount, and ultimately impossible.

The Fountain of Narcissus

The narrator relates how, within the dream, he wandered throughout the garden until he came upon 'un trop biau leu ... en un destor' (1423–4) where he found a fountain bearing an inscription indicating that Narcissus had died in this place. He looked into the fountain, where he saw how the flowing water is supplied by two deep channels. At the base of the fountain, the narrator says, 'avoit .II. pierres de cristal / qu'a grant entente remirai' ['were two crystal stones, which I looked at intently' (1536–7)]. Then, he says, he saw something amazing:

Quant li solaus, qui tot aguiete,
ses rais en la fontaine giete
et la clarte aval descent,
lors perent colors plus de cent
ou cristal, qui par le soleil
devient inde, jaune et vermeil.

When the sun, that stimulates all things, casts its rays into the fountain, and
when its light descends to the bottom, then more than a hundred colours
appear in the crystal which, on account of the sun, becomes yellow, blue,
and red. (1541–6)

Many readers have stressed the importance of the allegorical meaning of
these '.II. pierres de cristal' to a satisfactory reading of the poem.[27] Lewis
seems to have been the first to suggest that the crystals allegorically
represent the eyes of the lady, an interpretation repeated by Frappier
and Köhler. Other readers have opted for a more strictly negative interpretation of the lover's motivation for love and its possibility of success.
Robertson and Fleming both assert that the crystals are in fact the eyes of
the lover, reflected in the fountain of Narcissus, and conclude that,
because his desire is fundamentally narcissistic, it can never be fulfilled.[28]

One obstacle to interpreting the crystals as the eyes of either the lady
or the lover has been the apparent inconsistency in Guillaume's reference to both a singular 'cristal' and plural 'cristaus.' In his elucidation of
this textual problem, Hult suggests that the switch from singular to
plural is intentional: 'Far from a philological oversight committed by
some early scribe ... the singular and plural variants of the word "cristal"
denote two entirely different concepts.' Hult argues that 'the transfer-
ence from two crystals to a single one presents a physical parallel to the
two types of perception that are being highlighted in the poem.'[29] Hult
rightly highlights the importance of perception, not only at the episode
centered on the fountain of Narcissus, but within Guillaume's poem as a
whole. He does not, however, seek out a contemporary scientific or
philosophical context for Guillaume's interest in the causes of distorted
perception.

The one reader who has done so is Kenneth Knoespel, who acutely
suggests that the episode at the fountain of Narcissus is 'an allegory of
vision itself'; the specific optical context he suggests is unfortunately less
persuasive. Knoespel argues that 'the fountain['s] ... allegorical meaning

is intimately related to the lover's own faculty of perception, for as a receptacle of images the fountain is a representation of the human eye.'[30] Knoespel is absolutely right to look to the medieval optical tradition; he errs, however, in his choice of text. Knoespel suggests that the crystals at the fountain's base represent the crystalline humour of the eye (the lens), as described in the account of ocular anatomy found in Constantinus Africanus's translation of Hunain ibn Ishaq's *De oculis*.[31] A few problems immediately arise: first, Guillaume emphasizes that the crystals are at the base of the fountain ['El fonz de la fontaine aval / avoit .II. pierres de cristal' (1535–6)], while Hunain states that the crystalline humour is located at the geometrical centre of the sphere of the eye.[32] In addition, Knoespel argues that the two channels (*doiz*) leading to the eye represent the optic nerves, which in the galenic anatomy of Hunain were believed to carry the optical *pneuma* from the brain to the eye. If, however, the fountain represents a single eye, there would hardly be two optic nerves leading to it. Knoespel tacitly admits this problem by shifting unobtrusively between discussion of one eye and both eyes.

I only belabour these difficulties in Knoespel's reading because I think he is absolutely right to identify the importance of vision and, more generally, the science of optics, in Guillaume's *Rose*. A more useful place to look, however, for the scientific context of Guillaume's poetic presentation of optical phenomena is the oeuvre of the great Chartrean master of natural philosophy, who also articulated an influential theory of allegorical interpretation: William of Conches. The importance of William's work for twelfth-century use of allegory has been ably demonstrated by Edouard Jeauneau and Peter Dronke and, as I will show, continued to influence conceptions of allegory into the thirteenth century. Jeauneau has surveyed the rich variety of contexts in which William employs the word *integumentum*, using not only William's glosses on Plato, Boethius, Juvenal, and Macrobius but also his writings on natural philosophy to illustrate his theory of allegoresis.[33] William uses the term *integumentum* (literally 'covering' or 'wrapping') to refer both to the cover itself and to that which it conceals: it is both the veil of allegorical language and the figurative meaning half-visible behind the veil.[34] In its primary use, the *integumentum* is used with regard to myth. For example, the four horses of Phoebus which draw the chariot of the Sun allegorically represent the four periods of the day, from dawn to sunset.[35] Such exegetical understanding of myth is ubiquitous; specific to the Chartrean context of the twelfth century, however, is the allegoresis of myth in the terms of natural philosophy. Such interpretations can be found in the commentaries on the *Aeneid* and on Martianus Capella attributed to Bernardus Silvestris.[36]

Integumentum, however, is not restricted to the exegesis of myth: as Jeauneau observes, it serves to unlock the hidden meanings of the philosophers as well. As Tullio Gregory showed in his pioneering study, the doctrine of the world spirit or *anima mundi* found in Plato's *Timaeus* is allegorically interpreted by William as an *integumentum* of the Holy Spirit.[37] This is only the most resonant of a number of instances in which William reconciles the divergence of Neoplatonism and Christianity through recourse to the *integumentum*. As Jeauneau puts it, only the interpretation of pre-Christian philosophy as *integumentum* allowed that philosophy to be assimilated to a Christian world view.[38] While both myth and philosophy can be interpreted as *integumenta*, there is nonetheless a difference of degree between them, if not of kind. The *integumentum* of myth is unfolded through examination of the letter, by penetrating the veil of allegorical language, which is why etymology plays an especially strong role in the process.[39] The *integumentum* of philosophy, by contrast, is unfolded through a recourse to geometry and mathematics: moving past the literal to the figurative, past the letter to the number. William thus begins his commentary on Plato's *Timaeus* (in the Latin version of Calcidius) with a meditation on the significance of number: '*One, two, three* ... One might ask why Plato, who does nothing without cause, begins his book with numbers ... Plato, as a good Pythagorean, knowing the great perfection that is in number (since no creature can exist without number, but number can exist without anything), in order to make manifest the perfection of his work, begins it with perfect numbers.'[40] The more elevated the hidden meaning, the more necessary it becomes to have recourse to the realm of pure form; that is, for the Platonist, number.

In keeping with this effort, William's engagement with the most profound *integumentum* of Plato's *Timaeus* is couched in terms of mathematics and geometry. The *integumentum* of the world spirit or *anima mundi* is expressed as a series of numbers arranged in a triangular array, which Jeauneau likens to a compass.[41] William uses the same figure in his discussion of the world soul in his glosses on Macrobius: there, however, he expands upon the figure of the triangular array, discussing the figurative significance concealed within the *integumentum* of the three-dimensional cone: 'The cone is an oblong geometrical form which, as it ascends, extends along its sides into a point ... The geometrical form of the sphere is said to be divine (that is, perfect) because it has no beginning or end. From this form, the soul descends, extended in the shape of a cone, because the cone is initially indivisible, but then can be divided. The soul is similar ... [for it] admits no division, but through its ability to be divided animates various things in various bodies.'[42] For William, the cone brings before the eye a hidden truth that would otherwise remain inaccessible. William makes clear the function of the visual image earlier in his commentary on the *Timaeus*, when he accompanies an account of the distribution of seas on the surface of the planet with a diagram: 'But because what is perceived by the eye is better received by the mind, we will display [*subiciamus*, literally, 'cast before the eyes'] a figure of the arrangement of the seas just described.'[43] The cone serves a similar function, making the meaning appear to the eye of the mind through recourse to the corporeal eye.

Further, the conical form extending downward from the fixed point of the heavens recalls the visual cone essential to William's essentially Neoplatonic model of vision, described above in chapter 2. The difference lies, first, in the source of the cone: on the microcosmic level, the cone emanates from the eye of man, while, on the macrocosmic level, the cone emanates from the Creator. A secondary difference concerns the interaction of the cone with what it encounters: the visual beam extends until it reaches the *obstaculum*, a sense object which deflects the visual power. As Ricklin points out, this model of vision has much in common with William's theory of how solar eclipses are occasioned, with the eye exerting a radiant force like that of the sun.[44] The divine Sun, however, does not have its rays deflected by any *obstaculum* in its path, but rather suffuses everything. As William puts it in his glosses on Macrobius, just as the corporeal sun simultaneously emits both light and heat, so the divine presence appears both in the Father and the Son, who are coeternal.[45]

William's glosses on Macrobius are, I suggest, essential to an understanding of Guillaume de Lorris's use of allegory. I will point out below specific passages on the nature of *somnia* to be found in William's glosses which shed light on the multiple personae who interact with Guillaume's narrator in the course of his dream. More generally, however, William of Conches's notion of *integumentum* as a way of figuratively understanding both myth and science plays a crucial part in Guillaume de Lorris's allegory of narcissistic love. Yet William's notion of *integumentum* is only one part of what the master offered to Guillaume de Lorris. His writings on natural philosophy, with their special focus on optical phenomena such as reflection, refraction, and deceptive images, illuminate the visual

experiences of Guillaume de Lorris's narrator, for they offered a vocabulary of vision which the writer of allegory could exploit. In his *Dragmaticon*,
William describes a number of optical phenomena. He describes the
deceptive nature of refracted vision, and the multiple colours generated
by refracted light in the rainbow. He offers a lengthy summary of the
various possible causes of the rainbow, concluding that the multiple
raindrops in a cloud reflect an image of the sun, having four colours
because the cloud is formed of the four elements.[46] In addition to his
practical illustration of the phenomena of reflection and refraction as
they appear in the rainbow, William also theoretically describes properties of refraction in his discussion of ice crystals. He states that the
refractive qualities of water can deceive the eye: 'Although each one of
our senses is deceptive in many ways, our sight is the most deceptive of
all. A staff, though in one piece, seems broken in the water. Two towers,
if viewed from afar, appear joined together, although they are at some
distance from each other.'[47] This passage is particularly suggestive for a
reading of Guillaume de Lorris's *Rose*: the deceptive quality of vision
described here resembles the deceptive nature of dreams alluded to in
the opening passage of the *Rose*. Additionally, the illustration of the two
towers which are seen as one may illuminate the final scene of the *Rose*,
where the lover is left lamenting his fate, shut out from the tower which
imprisons Bel Acueil.

William of Conches's *Glosae super Platonem*, however, is of greatest
importance in explicating Guillaume de Lorris's use of allegory. While
the *Dragmaticon* was a very popular work, surviving in some seventy
manuscripts, William's Timaean glosses were also extremely influential.[48] William's gloss on the *Timaeus* includes a discussion of dreams
especially pertinent to the allegorical dream vision of Guillaume de
Lorris, including an account of the causes of dreams, whether interior or ·
bodily (as in the case of the Macrobian *insomnium*), or exterior or
spiritually motivated (as in the *somnium*).[49] This discussion of dreams is
immediately followed by an analysis of the three forms of perception,
which William calls *contuitio, intuitio*, and *detuitio*. *Contuitio* is direct vision, without any reflected image; *intuitio* is reflective vision, 'when a
likeness appears to be on the surface, as in a mirror'; and *detuitio* is
refracted vision, 'when a likeness appears not superficially but profoundly,
as in liquids.'[50] The terminology of *contuitio, intuitio*, and *detuitio* is to
some extent familiar from Calcidius's fourth-century commentary on
the *Timaeus*, in which he describes three types of vision: '*tuitio*, which we
call "phasis," and *intuitio*, which we call "emphasis," and *detuitio*, which

we name "paraphasis."' While Calcidius uses the terms *tuitio* and *intuitio* repeatedly, he uses *detuitio* only once, substituting 'paraphasis' in his later discussion. Calcidius elaborates at length on the property of reflection, stating that it occurs 'in mirrors and in water,' and relating how different types of mirrors produce different effects.[51]

Calcidius's discussion of 'paraphasis' or refraction is brief and much less detailed. He states that it occurs 'not in the surface of mirrors, but more deeply.'[52] Instead of describing the two primary manifestations of refraction, the straight stick which appears to be bent in water and the band of colours seen in the rainbow, Calcidius mentions images which appear in smoke. William of Conches, however, repeatedly uses the term *detuitio*, adding to Calcidius's brief treatment of refraction his own knowledge of the phenomenon, both as derived from other authorities and, perhaps, from his own experience. William gives the etymology of *detuitio* as being from 'deorsum tuicio,' that is, 'looking downward,' indicating the refractive properties of deep water.[53] This etymology has less in common with modern philological efforts than with the perspective of the seventh-century encyclopedist Isidore, for whom the actual derivation of the word's source was relatively unimportant; instead, the purpose of etymology was both to demonstrate a link between the word's form and its meaning and to serve as an aid to memory.[54] The *Thesaurus Linguae Latinae* cites only one usage of the term, Calcidius's, and states that *detuitio* is derived from *de* and *tueri* (to look at). I will briefly compare the origin of William of Conches's term *detuitio* with that of the Old French term *deduit* for, although the terms are not etymologically related, Guillaume de Lorris uses the word *deduit* with a double reference, both to the poetic vernacular sense of joyous diversion and the technical Latin sense of refraction. He is able to do so because the words sound sufficiently alike and because their meanings are sufficiently congruent.

Deduit comes from the verb *deduire*, itself derived from Latin *ducere* or *deducere*, to lead away or from. This can be used in the literal sense, referring to the extension of some thing in a particular direction, or in the figurative sense, referring to a change in the state of the individual. The term is most frequently used in connection with one's entry into a state of joy or delight.[55] In his translation of the *Roman de la rose*, Dahlberg explains the poet's use of *deduit* by stating that 'Two senses of *diversion*, "having a good time" and "turning away from a (right) course," are implicit';[56] and, indeed, in the *Rose* that second meaning of *deduit* is highlighted by means of the optical allegory, as the changes in the narrator correspond to the changes in the refracted light ray as it is turned aside from its straight path. The similar meanings of *deduit* (turning away from one's usual course) and *detuitio* (the ray's turning away from its usual course), as well as their homophony, form the basis of a fictitious etymology exploited by Guillaume de Lorris. His use of the term *deduit* is a play on words: it refers not only to the pleasure or joy found in earlier uses of the word in such vernacular texts as the Old French *Eneas*, but also to very specific optical terminology that provides a scientific context with which Guillaume expected his reader to be familiar. Interestingly, in the *Eneas*, which is the earliest vernacular text to use the term *deduit*, the word is primarily associated with homoeroticism.[57] If the amorous desire of the narrator of the first *Rose* is also to be understood as homoerotic, as Harley and Uitti have argued in their analyses of the 'phallic rose,'[58] the *Eneas* may have stimulated Guillaume to anatomize precisely how self-love leads to same-sex love through the medium of the narcissistic gaze.

It is evident that the notion of the *integumentum* developed by writers associated with Chartres was the basis of Guillaume de Lorris's use of allegory in the *Roman de la rose*. The unfolding of the *integumenta* of myth and philosophy was illustrated in the commentaries on the *Aeneid* and on Martianus Capella attributed to Bernardus Silvestris, as well as in William of Conches's glosses on Plato's *Timaeus*, Boethius, and Macrobius. The *integumentum* itself could even be found, embodied in Nature's colorful gown in the *De planctu Naturae* of Alanus de Insulis – a literal *involucrum*, or veil. The scientific context brought to bear in these glosses provided fruitful ground for Guillaume de Lorris's own elison of myth and science in the *Rose* through the narrator's refracted look into the fountain of Narcissus. The figure of Narcissus is to be interpreted as an *integumentum*, like the figure of Orpheus in the *De consolatione Philosophiae* interpreted by William of Conches in his glosses on Boethius.[59] The optical phenomena of reflection and refraction are also to be interpreted as *integumenta*, like the cosmological phenomena found in the *Commentariam in somnium Scipionis* interpreted by William in his glosses on Macrobius. Yet William of Conches provided the first author of the *Rose* with more than the allegorical strategy of the *integumentum*: in his scientific writings and in the glosses on the *Timaeus*, he also supplied the terminology which provided the basis for Guillaume's polysemous use of the term *deduit*. The word for refraction used in William's Timaean glosses, *detuitio*, is employed in Guillaume's *Rose* to characterize the deceptive pleasures of the Garden of Deduit. It comes as no surprise that Guillaume explicitly associates *deduit*, the joy caused by love, with changes in vision: when one is in love, 'li oil sont en deduit' (2717). The eyes are in *deduit*, in a state of pleasure; but they are also in *detuitio*, refracted and thus partial, distorted vision.

William of Conches's passage on the three types of perception provides a useful context for reading, not only the pivotal scene at the fountain of Narcissus, but also the structure of Guillaume de Lorris's *Rose* as I have described it in the first section of this chapter. The poem's narrator experiences *contuitio*, or direct vision, as he looks at the wall paintings prior to his entry into the garden; he experiences *intuitio*, or reflected vision, when he passes the second barrier and enters the garden where he sees matched pairs of couples strolling about; and he experiences *detuitio*, or refracted vision, after he gazes into the depths of the pool at the heart of the garden and perceives the refractive nature of the crystals beneath the water.

Within the framework of Guillaume's allegory of vision, divided into the three types described by William of Conches, the wall paintings represent the most basic kind of perception, direct vision or *contuitio*. They appear on the outer surface of the garden, 'clos de haut mur bataillé' ['enclosed by a great reinforced wall' (131)], and depict such evils as Hatred, Villainy, Covetousness, and Avarice. They are placed there to serve as a warning and as an evidently effective aid to memory:[60] Guillaume's narrator prefaces his detailed description of the paintings with the promise to describe them 'si com moi vient a remenbrance' (138). The images are stylized, painted in blue and gold (463) rather than in the multiple colours that would more realistically represent people and objects. The narrator repeatedly refers to the paintings as 'ymages,' thus emphasizing that they are copies or simulacra of an original object.

After the narrator passes the imposing barrier of the garden wall, he experiences the second form of vision: reflected vision or *intuitio*. Hult illuminatingly contrasts the dancing couples in the garden to the wall paintings outside, noting that the 'elaborate figurative descriptions' of the dancers produces in the reader 'a sense of nostalgia for the figurative and intuitive simplicity of the initial static personifications.'[61] Hult is right to perceive that the pairs in the garden both correspond to the figures painted on the wall and are represented on a more complex level than the earlier images, for *intuitio* or reflected vision is a more complex optical phenomenon than *contuitio* or direct vision. Unlike *contuitio*, *intuitio* produces two apparently identical bodies: the original and its simulacrum. For this reason, Guillaume peoples the garden with couples who are perfect complementary pairs. The Old French 'pair' means both, as an adjective, 'similar, on a par' (Mod. Fr. *pareil*) and, as a noun, a 'pair' (Mod. Fr. *paire*).[62] Guillaume's couples in the garden are both these things to one another: each partner resembles his or her counterpart, and together they form a matched set. Guillaume's narrator begins by describing Deduit, who leads the carole, and his companion Leesce, who sings the song. These two show themselves to be related by their names: both 'deduit' and 'leesce' (from the Latin *laetitia*) are intimately connected with the experience of joy, both in their intrinsic meaning and in their use in the common formulaic phrases 'joie et deduit' and 'joie et lëece.'[63] (A variant of the former phrase is used earlier in the *Roman de la rose* [473].) Guillaume concludes his account of the pairs in the garden with a description of Joinece and her companion, emphasizing their similarity: 'Li valez fu joines et biaus, / si estoit bien d'autel aage / con s'amie, et d'autel courage' ['The youth was young and fair, and was of just the same age as his girlfriend, and of the same spirit' (1274–6)]. They kiss 'come .II. colombiaus' ['like two doves' (1273)], recalling the mirror image of two dancing ladies seen earlier in the carole: the narrator says that the two 'demoiselles' approached 'l'une ... contre l'autre, et quant eus estoient / pres a pres, si s'entregetoient / les bouches qu'i fust avis / qu'eus s'entrebessoient ou vis' ['one toward the other, and when they were right next to one another, they put their mouths close together so that it looked as though they kissed one another on the face' (764–8)].

Several homonymic rhymes appear in the course of the catalogue of dancers in carole. Amors, the god of love, is accompanied by Biauté, who is introduced with a homonymic rhyme (989–90). Similarly, the description of Largeice's companion, 'un chevalier dou lignage / le bon roi Artu de Bretaigne' (1175–6) includes the homonymic rhyme *contes*, both tales recounted and noblemen (1179–80). Cortoisie's friend is introduced with the homonymic rhyme *genz* (1245–6). These homonymic rhymes emphasize the similarity of each half of the pair, each a mirror image of the other. At the same time, each couple shares a link with the next couple in the carole, linking the pairs together in an allegorical dance literally represented by the dance in carole. The couple of Amors and Biauté is followed by Richece and her young man, 'un valet de grant biauté plain' ['a youth full of great beauty' (1108)]. Richece and the youth are followed by Largeice and her knight, who is so famed that 'l'en conte de li les contes / et devant rois et devant contes' ['one tells stories of him before both kings and noblemen' (1179–80)]. The punning on *conte* and *conter* here also recalls the other meaning of *conter*: to count the abundant wealth embodied by Richece. Largeice and her knight precede Franchise and her companion: 'Uns bachelers lez lui s'estoit / pris a Franchise lez a lez' ['a young man was at her side, side by side with Franchise' (1222–3)]. Here, the punning on *lez* evokes *lé*, meaning 'large,' the quality epitomized by Largeice. The carole is closed by the last two couples, the narrator accompanied by Oiseuse, who had let him into the walled garden, and Joinece, with her perfectly matched companion.

Guillaume's account of refracted vision or *detuitio* also alludes to the notion of pairs. This is appropriate, since medieval writers on optics emphasize that reflection and refraction are related phenomena, and some even fail to distinguish the two clearly. Early in the poem, Guillaume tells how in springtime the earth makes herself a 'novele robe ... de colors i a .C. peire' ['new robe of more than a hundred pairs of colours' (60–2)]. This description prefigures the phenomenon of refraction seen at the fountain of Narcissus, where the crystals beneath the water produce 'colors plus de cent ... inde, jaune, et vermeil' ['more than a hundred colors ... purple, yellow, and vermillion' (1544–6)]. The poet emphasizes that refraction is at work not only by describing crystals, but by placing them in deep water; crystals and deep water are two of the three conditions William of Conches states cause refraction. In the *Dragmaticon*, William describes the third of the refractive phenomena: he explains how a circular bow of many colours is produced by the refraction of sunlight through a cloud: 'Those who claim that the rainbow is not a substance say that it is an image of the sun. But because every image is similar to that of which it is an image, as the sun is round, so the rainbow appears round in shape. And to the extent that the rainbow is not a substance but the reflection of a substance, so there are no colors in it, but only reflections of color. ... There have been people who suggested that the rainbow was nothing but a cloud, neither too dark nor too bright but having four principal colors from the four elements: red from fire, purple from air, blue from the sea, and green from the earth because of the grass and trees.'[64] This manifestation of refraction resembles the marvel described by Guillaume de Lorris, where the sunlight streaming down generates 'more than a hundred colors in the crystal, which because of the sun becomes purple, yellow, and vermillion' (1544–6).

Elucidations of the rainbow's cause and meaning appear in both biblical and classical sources. In Genesis 9:13, the rainbow is taken as a sign of a new covenant between God and man while, in the third book of his *Meteorologica,* Aristotle offers a technical explanation of the phenomenon.[65] The most useful context, however, is provided by the commentary on Virgil's *Aeneid* attributed to Bernardus Silvestris, which includes a striking description of the bow's allegorical significance: 'Iris, who is multicolored and placed opposite to the sun, figures the senses, which are distinguished by their diverse species and powers and are contrary to reason. ... Thus we understand Iris, who is multicolored and set opposite the sun, to be the multiplex senses set contrary to reason.[66] In this commentary, the sun allegorically signifies reason, while its image, the rainbow, signifies the senses. Similarly, in Guillaume de Lorris's *Rose,* the rainbow produced by the crystals represents the sensuous pleasures of courtly love, the pleasures from which Reson unsuccessfully attempts to lure the lover. Guillaume represents Reson as the only force that might be able to persuade Amant from his destructive love, not only because she represents man's rational faculty as opposed to his deceptive senses, but because in twelfth-century allegory she is said to wield a tripartite mirror. In his *Anticlaudianus,* Alanus de Insulis writes that Ratio holds a mirror in her right hand: 'Her right hand is resplendent, aflame with the brightness of a threefold mirror [and there is] a triple reflection in the threefold mirror.'[67] The fact that she holds a mirror makes her a counterpart to Oiseuse, who 'en sa main tint un miroër' (555). Just as Amant was let into the garden by Oiseuse, so Reson could potentially let him out. Additionally, the tripartite nature of her mirror makes it possible for her (if she had the lover's cooperation) to counter the three modes of vision concealed in the optical allegory of the *Rose.*

The multiple colours produced by the crystals mark a liminal moment in the narrative: it is in the crystals that the lover spots the rose that becomes the object of his desire, his quest for the remainder of the poem. Simultaneously, the crystals mark a liminal moment in Guillaume's allegory of vision, for after the lover looks into them, he passes from the realm of reflected vision, *intuitio,* into that of refracted vision, *detuitio* or *deduit.* Just as the crystals, when struck by the sun, literally produce 'colors plus de cent' ['more than a hundred colours' (1544)], so they allegorically produce a multiplication of the self: after his look into the fountain, the lover begins to encounter multiple redoubled images of himself in Amors, Amis, Dangiers, and Bel Acueil. The crystals simultaneously produce multiplication of what should have been a single object of desire: instead of a single rosebud, the lover comes to desire the young man who attends the rosebushes, Bel Acueil.

Multiplication of the Self

In his glosses on the *Commentariam in somnium Scipionis*, William of Conches explicates each of the categories of dreams outlined by Macrobius. The final category, the *somnium*, is described by Macrobius simply as 'an enigmatic dream which requires an interpretation for its understanding.' William, however, amplifies this description: 'At other times, God intimates [the future] either by what is similar or by what is contrary. By what is similar occurs when I dream that my teeth come out or my blood is spilled, which signifies the death of friends. By what is contrary occurs when I dream that I smile, which signifies that I will weep, and the converse. And this kind of dream is called a *somnium*.'[68] In keeping with the two ways in which the *integumentum* of the *somnium* can appear – as what is similar or what is contrary – the redoubled images of the narrator that appear after the midpoint of the poem are both similar to him (Amis and Bel Acueil) and contrary (Dangiers). I will describe in turn how each of these figures reflects the narrator's own persona.

Guillaume de Lorris opens his poem by calling it the 'Romanz de la Rose, / ou l'art d'Amors est tote enclose' (37–8). The art of love is within the poem, not only implicitly in the example of the lover's quest for the rose, but explicitly, as a 740-line section of instructions from the god of love to the lover, enclosed at the midpoint of Guillaume's poem (2049–789). Amors offers his counsel to the lover only after he has first made him into one of his own men, striking him with the five arrows of love, which simultaneously give both pain and solace. But Amors can capture the lover only after the lover has fixed on an object of desire. After glimpsing rosebushes in the mirror of the fountain of Narcissus, the lover chooses a tightly closed bud instead of one of 'les roses overtes' (1643), because the buds last longer. After this eminently rational choice has been made, Amors attacks. He shoots his victim with five arrows – five, evoking the five petals of the rose. The arrows travel 'par l'ueil ou cuer' ['through the eye to the heart' (1741)], in imitation of the medieval conceptualization of the Platonic extramission theory of vision, in which the powerful visual beam radiating from the lady's eye penetrates the lover's eye and travels through the bodily humours to the heart, seat of the soul. After being struck by these arrows, Amant begins his vassalage to Amors, who lectures him on the art of love and locks the lover's heart with a key 'en leu d'outages' ['in lieu of hostages' (1991)].

There is apparently a great distance between the lover and Amors: one is the student, one the teacher; one is the narrator of the poem, one a god with a rich history of classical myth and fable behind him. But at several points in the poem, Amors and Amant appear to be peers, one almost a reflection of the other. Even their names are harmonious, both two syllables, starting with the same two letters. They share a kiss of fealty that recalls the mirror image of two embracing women seen earlier in the garden (757–68) as well as the kiss of Joinece and her perfectly matched companion (1273), and Amors describes a dream that simultaneously reflects and distorts the dream which the narrator experiences within the poem (2413ff.). The doubling implicit in Amors's relation to Amant is reinforced by Guillaume's allusion to the myth of Narcissus and Echo. While the lover reenacts the experience of Narcissus at the fountain, Amors appears in the role of Echo, stealthily tracking the lover through the garden of Deduit just as Echo stealthily follows Narcissus.[69] At the same time, however, the lover is unaware that he reenacts the role of Narcissus. After recounting the myth, he tells its moral: 'Dames, cest essample aprenez, / qui vers vos amis mesprenez; / car se vos les lessiez morir, / Dex le vos savra bien merir' ['Ladies, take this example, you who do ill against your sweethearts; for if you let them die, God will know how to well repay you' (1505–8)]. By likening the ladies he counsels to Narcissus, he likens their lovers (and, implicitly, himself) to Echo,[70] an impression reinforced by his description of her as 'li loial amant' (1463). The narrator believes that he reenacts the experience of Echo, but actually he relives the experience of Narcissus, blind to everything but his own image.[71]

In his *Dragmaticon*, William of Conches alludes to the myth of Narcissus and Echo when, following a comprehensive discussion of the mechanism of vision and the anatomy of the eye, the philosopher answers his interlocutor's question regarding the nature of echoes. After the duke draws an analogy between visible reflection and the echo, the philosopher asks, 'Do you not know, then, that this is performed by Echo the nymph?' The duke indignantly replies, 'I am not Narcissus to be pursued by her. I ask for a rational explanation of the phenomenon, not for a fable.'[72] In the *Rose* as in William of Conches's dialogue, to resemble Echo is to be like a mirror, for Echo reflects words, distorting their meaning, just as the mirror of Narcissus reflects images in a distorted or inverted fashion. Amors's reenactment of the nymph's pursuit of Narcissus makes him temporarily, as it were, Amant: Amors becomes (like Echo) the lover and Amant (like Narcissus) the beloved. At the same time, the passage quoted above from the *Dragmaticon* shows that the roles of Echo and Narcissus are not clearly separable, for in both cases the ultimate source of the visible or audible image remains elusive. This disrupts the stability of the dichotomy of Echo and Narcissus, lover and beloved, and further complicates the effort of self-recognition in the *Rose*.

After the lover's early efforts to approach the rose prove unsuccessful, he follows Amors's earlier counsel and seeks out Amis for help. Amis appears only briefly (3091–134), but his consolation of and advice to the lover show him to be Amors's surrogate in the quest for the rose. His name, Amis, echoes the name Amors, and also that of the lover himself, Amant. Just as Amors and Amant are twins, like the dancing pairs in the garden who represent *intuitio* or reflected vision, so Amis, Amors, and Amant are triplets, redoubled images of a single self representing *detuitio* or refracted vision. Amors describes the comfort offered by Amis as 'deduit' (2698), referring both to the pleasurable diversion his friendship can offer and his role as a projected image of the narrator himself, generated by the refractive properties of the crystals in the fountain of Narcissus.

One might argue that it is perfectly natural for Amors and Amis to share characteristics of Amant, for every allegory includes personae that are ultimately generated from within the protagonist's self. Amors is merely one of several voices within the narrator. He represents, as Fleming writes, a '"civilized" sentiment' within which is concealed 'an idolatrous passion'[73] – a passion which comes from within the narrator himself. Ferrante calls Amors 'the voice of love within the lover, the lust that is disguised in courtly behavior.'[74] But the male personifications in the first *Roman de la rose* are unusual, not simply because they deviate from the norm in personification allegory where a female character is used to embody an abstract quality (primarily because the noun's grammatical gender is feminine), but because each of the male personifications in the *Rose* reflects the lover's self. Amors is first Amant's complementary 'paire' and then, with Amis, one of three persons whose names affirm their identity and indistinguishability. Dangiers is both an inverted, opposite image of Amant's twin Amors as well as a figure of the dreaming self within the poem, while Bel Acueil gradually comes to take the place of the desired rose, a beloved object which is finally recognized as an image of the self.

The only other male personification in Guillaume's *Rose*, Male Bouche, is a particularly interesting case. He appears only briefly, but plays a pivotal role in denouncing Amant's association with Bel Acueil. It is not clear why he should be characterized as male, since the noun 'bouche' is grammatically feminine. In her discussion of the male personifications of Guillaume's *Rose*, Ferrante suggests that they are in some sense more closely related to Amant, noting that Bel Acueil and Dangiers are personified as male because they have the most direct contact with Amant: Bel Acueil is male 'because he is that aspect of the woman that plays the courtly game on the man's terms, the image of the man which he has projected onto the lady; Dangiers, because he represents the strength she possesses to repel the lover's attack.' Ferrante accounts for Male Bouche being personified as male by noting that 'he represents the *losenger* ... the men who tell what they see.'[75] But Guillaume's representation of Male Bouche's gender is deceptive and unstable: Male Bouche is at first 'la jangleor' (2819) and later, 'le jangleor' (3512). As soon as Amant kisses the rose, the narrator announces that it is time to tell how the new barrier of the wall and tower were erected to protect the roses and Bel Acueil, and immediately recounts the words of Male Bouche, who is now emphatically masculine, called 'il' five times in thirteen lines (3498–510). Earlier, Male Bouche had been called 'ele' (3020); Male Bouche takes on a masculine identity, however, as soon as the lover's object of desire changes from the rose, characterized as feminine, to Bel Acueil, the young man who attends the rosebushes. He becomes yet another redoubled image of the self, another indication of the self-referentiality of the lover's desire.

Charles Dahlberg has suggested that Dangiers is an element of the narrator's personality, his 'protector' or conscience.[76] Again, we can and should expect any personification in an allegory to reflect the protagonist to some extent. But Dangiers is characterized in terms that are also explicitly applied to Amant. When Bel Acueil rebuffs the lover, he calls him 'vilains' ['villainous'], thus implicitly likening him to Dangiers, who appears for the first time five lines later as 'Dangiers li vilains' (2904). The proximity of the two uses of the term 'vilains' serves not only to stress the connection but also to illustrate how Dangiers is, on one level, an embodiment of the lover's own 'villainy,' not the resistance offered by an external beloved. At the crucial moment when the lover steals a kiss from the rose, Dangiers is asleep (3651ff.), a reflection of the dreaming self who exists outside the narrative of the events taking place in the garden. Finally, the narrator calls Dangiers 'le peïsant' (3653), a term he applies to himself (3932) in the lamenting monologue with which he ends the poem.

Dangiers's position as guard at the newly fortified enclosure for the rosebushes and Bel Acueil is highly significant, as is shown by a comparison with the allegorical palace of love in the fifth dialogue of Andreas Capellanus's *De arte honeste amandi.*[77] Like this collection, Guillaume's *Rose* treats the subject of courtly love and contains an 'art of love,' recommendations for the deportment of a lover. The 'comandemenz' (2057) given by Amors in the *Rose* resemble the twelve commandments with which Andreas concludes his allegorical description of the palace. Guillaume's fortress is a revision of Andreas's palace:[78] instead of the marvellous palace of love inhabited only by ladies and the God of Love, Guillaume describes the stifling enclosure of love, which both makes the object of desire inaccessible and traps the lover himself. Andreas relates the significance of each of the four gates of the palace, stressing that 'the eastern gate the God of Love has reserved for himself alone.'[79] In Guillaume's description of the fortress, Dangiers appears in Amors's position, guarding the eastern gate: 'Dengier porte / la clef de la premiere porte / qui ovre devers Orïant' ['Dangiers holds the key of the first door which opens towards the east' (3851–3)]. Here Dangiers stands in the place of Amors; but, at the same time, he is diametrically opposed to Amors, both in being 'vilains' while Amors epitomizes courtliness, and in his position regarding the lover's quest for the rose. While Amors abets the lover's desire, Dangiers begins by being antagonistic and ends by being adamantly opposed: 'Des or est changiez mout li vers, / quar Dangier devient plus divers / et plus fel qu'il ne souloit estre' ['From here the song changes, for Dangiers becomes even more inconstant and more felonous than he had been before' (3743–5)]. Amors and Dangiers are both reflections of the lover, but at the same time they mirror one another, Dangiers an opposite, inverted image of Amors. As William of Conches writes, 'in a flat mirror, the right part is seen to be on the left and vice versa, so that if a man moves his right hand, the reflection moves its left and vice versa.'[80]

Finally, Bel Acueil, the young man who attends the rosebushes, is most obviously presented as a reflection of Amant. Bel Acueil's personification as male stresses the self-reflective nature of the narrator's desire since (as several of the *Rose*'s readers have noted), Guillaume must have made an effort to find a grammatically masculine abstraction he could personify as male.[81] In one manuscript of the *Rose*, Bel Acueil is represented as female, indicating that at least some medieval readers expected that the character should be female.[82] Such an expectation is fostered in Ovid's *Ars amatoria*, which counsels: 'Take care first to know the handmaid of the woman you would win; she will make your approach easy ... Corrupt her with promises, corrupt her with prayers; if she be willing, you will gain your end with ease.'[83] Since Guillaume overtly calls his work 'li *Romanz de la Rose*, / ou l'art d'Amors est tote enclose' ['the *Roman de la rose*, where the art of love is entirely enclosed' (37–8)], he undoubtedly expected his reader to be familiar with Ovid's *Ars amatoria*, perhaps in one of the Old French vernacular translations available early in the thirteenth century,[84] and therefore to note the peculiarity of the rose's 'handmaid' being a man.

There are, as one reader has put it, 'homoerotic overtones' in Amant's relation to Bel Acueil, just as there were in his relation to Amors.[85] Poirion writes that, in the context of the homosexuality evident in Ovid's version of the Narcissus myth, 'La substitution de Bel Accueil à la rose ... nous laisse sur une équivoque troublante.'[86] Amant's intimate ties to Amors reappear in his friendship with Bel Acueil: just as Amors takes Amant's heart in swearing him to vassalage (1994–5), so Amant desires the heart of Bel Acueil (3977). The implication that Amant and Bel Acueil's friendship is homoerotic is made explicit when Male Bouche accuses the two of having a 'mauvés acointement' ['evil relationship' (3507)].

Bel Acueil seems to be well aware of the dangerous nature of his relationship with Amant. He is at first friendly to the lover: 'Cil m'abandona le passage / de la haie mout doucement' ['he opened the passage of the hedge to me most sweetly' (2778–9)]. But when Bel Acueil realizes that Amant intends to take the rose, he reacts with astonishment and anger: 'Frere, vos beez / a ce qui ne puet avenir./ ... n'est pas droiture / que l'en l'oste de sa nature. / Vilains estes du demender!' ['Brother, you ask what can never come to pass ... It is not right to take it from its nature. You are villainous to ask it!' (2892–9)]. Bel Acueil refuses the lover's request on the basis that it 'n'est pas droiture,' a phrase Amors used earlier to describe homosexuality (2160–2). He also accuses Amant of attempting to denude the rose 'de sa nature.' Such a rejection of natural laws is the basis of Alanus de Insulis's scathing condemnation of homosexuality in the *De planctu Naturae*.

Because Amant and Bel Acueil are physically similar, both young men, their friendship is open to the accusation of being homoerotic. As the narrative progresses, Amant and Bel Acueil come to resemble each other even more. After Bel Acueil has shown himself to be responsive to the lover's entreaties, he is imprisoned with the rosebushes behind a fortified enclosure within the garden. Just as Bel Acueil is imprisoned within a physical enclosure, Amant is said to be enclosed by his untenable situation. He tells how he went for council to Amis and 'li desclos l'encloëure / dont je me sentoie encloé' ['disclosed to him the difficulties in which I felt myself enclosed' (3096–7)]. Amant is one of those 'qu'Amors a en prison' ['that Amors has in prison' (2605)], just as Bel Acueil, with the roses, is 'em prison' ['in prison' (3898)]. Bel Acueil comes to take the place of the rose in Amant's desire because he is, more than the rose itself, a reflected image of Amant. While the roses are concealed within the wall of the fortress, Bel Acueil is inside a double enclosure, imprisoned within a tower inside the fortress. Guillaume stresses the physical centrality of Bel Acueil:

> [La tor] est dehors environee
> d'un baille qui vet tot entor
> si qu'entre le baille et la tor
> sont li rosier espés planté
> ou il a roses a plenté.

The tower is entirely surrounded by a bailey that goes all around, so that between the bailey and the tower the rosebushes are densely planted, where there are numerous roses. (3830–4)

The rose, which was the lover's original object of desire, is actually physically displaced by Bel Acueil.

Near the poem's end, Amant complains of his loss of Bel Acueil; the rose has become merely an afterthought: 'ma joie ... / est toute en li et en la rose / qui est entre les murs enclose' ['my joy ... is entirely in him, and in the rose, that is enclosed within the walls' (3968–70)]. At this point, Amant has replaced not only his love for the rose but also his loyalty to Amors with his fidelity to Bel Acueil. While Amant now finds his joy in Bel Acueil (3968–9), earlier his joy was found only in Amors: 'J'atent par vos joie et santé' (1908). While formerly he declared that 'n'avoie en nului fiance / fors ou diex d'Amors' ['I have hope in no one except the god of Love' (2758–9)], he concludes the poem by placing his last hope in Bel Acueil: 'Ja mes n'iert rien qui me confort / se je pert vostre bienveillance, / car je n'ai mes aillors fiancé' ['There will never be anything that will comfort me if I lose your good will, for I have no other hope' (4026–8)]. The lover's desire shifts from Amors to Bel Acueil for the same reason that it shifts from the rose: Bel Acueil more closely resembles Amant himself. Both Amant and Bel Acueil are imprisoned, one by love and the other by stone walls: while Bel Aceuil is 'enmur[e]'

(3911), Amant is 'dehors le mur' (3920). The wall which separates them is no more than the impenetrable glassy surface of Narcissus's pool.

Just as the reader sees these points of identity, so does the narrator, so that as the poem nears its end, he reaches the penultimate moment of his reenactment of Ovid's tale of Narcissus. He stops just short of Narcissus's realization: 'I am he! I have felt it, I know now my own image. I burn with love of my own self; I both kindle the flames and suffer them.'[87] The narrative ends just before Amant discovers that Bel Acueil told the truth when he warned 'vos beez / a ce qui ne puet avenir' ['you desire what can never come to pass' (2892–3)], recalling the warning to Ovid's Narcissus: 'That which you behold is but the shadow of a reflected form and has no substance of its own. With you it comes, with you it stays, and it will go with you – if you can go.'[88] Amant cannot go, and does not go. Because the narrative does not end conclusively, he remains eternally 'dehors le mur' ['outside the wall'], on one side of the reflective glass.

It is significant that Bel Acueil begins to take the place of the rose after Amant has seen the rose again and found it more feminine and thus less an image of himself:

> [La rose] n'iere pas si overte
> que la graine fust descovierte;
> encois estoit encor enclose
> entre les fueilles de la rose
> qui amont droites se levoient
> et la place dedenz emploient,
> si ne pooit paroir la graine
> por la rose qui estoit pleine.

The rose was not so open that the seed was disclosed; but rather it was still enclosed in the petals of the rose which raised it upright and filled the space within, so that the seed, with which the rose was full, could not appear. (3345–52)

The rose has been most frequently interpreted as an idealized representation of the lady. But, more recently, critics have viewed Guillaume's description of the rose more literally, as 'a sexual metaphor,'[89] a 'pudendal emblem,'[90] or, more bluntly, 'the lady's genitalia.'[91] Yet this more physical metaphor is based on the image of the open, blooming rose featured in Jean's poem, not the upright bud found in Guillaume's poem (1663–5) which can be construed, as Harley puts it, as 'a phallic image.'[92] Poirion views the rose as 'le reflet du désir masculin,'[93] but the rose is more than generic masculine desire: it is the reflection of the lover's desire for the self, a desire which took shape as he gazed into the mirror of Narcissus.

Even on a linguistic level, physical consummation of the lover's desire is impossible: as Vitz puts it, 'What is a deflowered flower?'[94] Quilligan points out that the term 'deflorer,' in the carnal sense of deflowering a maiden, does not appear in the *Rose*;[95] but it does appear in contemporary Latin literature, including the *De planctu Naturae*,[96] which contains a description of the rose that may have contributed to Guillaume's choice of the rose to represent Amant's beloved. Among the changing pictures appearing on Natura's integumental gown, Alanus says that he saw 'the beautiful rose [*forma rosae*], faithfully reproduced, differing but little from its actual appearance.'[97] This description of the rose follows a catalogue of birds recalling that of the *Rose*'s opening, and is prominently placed in the *De planctu*, at the opening of a metrical section lauding the rose as first among the flowers of spring.

Guillaume chooses the rose to represent the object of desire, relying both on a courtly lyric tradition likening the beloved to a flower, often a rose, and on the classically influenced tradition of twelfth-century philosophical allegory.[98] Like Alanus's rose, but emphatically unlike Jean de Meun's, Guillaume's rose is not a real substance, but 'forma rosae,' the form of a rose. Guillaume's rose is similarly insubstantial, for its physical reality remains elusive. The rose first appears, not directly, but in a reflected image in the crystals of the pool of Narcissus. Even once the lover has set his sights on the rosebud itself, he can approach but never possess the object of desire. Once the rose ceases to resemble Amant, his desire shifts to Bel Acueil, a more accurate image of himself. Yet this desire can never be realized, for love of the self can never be fulfilled.[99] In the *De planctu Naturae*, Alanus describes the fate of Narcissus very much as Guillaume describes the plight of the lover in the *Rose*: 'Narcissus, when his shadow faked a second Narcissus, was reflected in a reflection, believed himself to be a second self, and was involved in the destruction arising from himself loving himself.'[100] The second self Amant pursues is Amors, Amis, Dangiers, the rose, and Bel Acueil. Once he comes close to the realization that it is himself he desires, the dream – and the narrator's account of the dream – has to end.

The interpretation of Guillaume's *Rose* I have just offered, coupled with the structural allegory outlined earlier, strongly suggests that the work is complete as it stands. One major question remains: should we view the work, as many critics have in the past, as a kind of *ars amatori*, a handbook or model of the right way to love? As Harley points out, those critics 'who see in Amant's experience at the fountain a transcendence of Narcissus's fate ... may well be viewing the text under the distorting light of an a priori perception – that of Guillaume de Lorris as "the conservative celebrant of courtly love."'[101] The poem may alternatively be viewed as a challenge to the existing standards, 'reveal[ing] the hypocrisy, the self-indulgence, and self-delusion inherent in the lover's stance.'[102] I would go so far as to suggest that Guillaume does not simply challenge the existing norms of courtly love, but rather contains them within his own poem. He explicitly describes this process in the opening lines of the *Rose*: 'je veil que li romanz / soit apelez que je comanz / ce est li *Romanz de la Rose*, / ou l'art d'Amors est tote enclose' ['I wish that the romance be called what I command: this is the Romance of the Rose, where the art of Love is entirely enclosed' (35–8)]. The art of love is enclosed, like so many images and characters in this narrative. The art of love referred to here is not the overall content of the poem: it is the art of Amors, his lecture to Amant on how to love correctly. This art of Amors is enclosed and thus ultimately inaccessible to the narrator: he is unable to successfully follow the commands of Amors, because his object of love was ill-chosen.[103] Viewed in this way, the rest of the poem becomes a frame to the 'art d'Amors,' just as the dream framework encloses the body of the narrative, as the walls enclose the garden, as the hedge encloses the rose, and so on.

The *Rose* is not an Art of Love; it simply encloses it. Enclosure is the thread which runs through the work, establishing barriers which turn out to be of a self-generated origin. Thus near the end of the poem, when the narrator calls himself a victim of Fortune's wheel, he says 'Et je sui cil qui est versez!' ['And I am he who is turned!' (3963)]. But he is not only 'versez' or 'turned' on the wheel; he is 'versez' in the sense of being put into poetic verse,[104] and he is also *vers*, opposite, looking across at himself. Here, a few lines from the end of the poem, the narrator is close to the revelation of Narcissus: 'iste ego sum.' Because this love is doomed, another level of love is similarly ill-fated; if we return to the opening lines of the poem, we recall that this story of love for a rose is itself a reflection of a 'real' love:

> or doint Dex qu'en gre le receve
> cele por qui je l'ai empris:
> c'est cele qui tant a de pris
> et tant est digne d'estre amee
> qu'el doit estre Rose clamee.

Now let God grant that she receive it with grace, she for whom I have undertaken it: it is she who is of such great price and is so worthy of being loved that she should be called Rose. (40–4)

'A tant de pris,' 'tant est digne d'estre amee,' 'doit estre Rose clamee': the lady is worthy of being loved and should be called Rose – but she is not. The conditional nature of the phrases meant to laud this lady to whom the poet dedicates his work undermine the declaration of devotion. He should love her but, implicitly, he cannot. He is still a victim of the self-love that rendered his quest within the dream futile.

Chapter Four

JEAN DE MEUN'S *ROMAN DE LA ROSE*

In this chapter, I will examine Jean de Meun's continuation of the *Roman de la rose* on two levels: first, as it illuminates the meaning of Guillaume's allegory as it was understood by Jean; second, as the completed poem represents a new formulation of the genre of allegory. It is possible to gain some insight into contemporary reception of Guillaume's poem by determining what aspects of the poem Jean attempts to obscure in enclosing the earlier work within his own continuation. Although Jean was not an exact contemporary of Guillaume, he was familiar with many of the sources, both literary and philosophical, that would have been available to Guillaume, and thus would have had the resources to understand the earlier poem within its intellectual context. By seeing precisely how Jean incorporates Guillaume's work into his own, we can know what aspects of the earlier poem Jean found worthy of being refuted. Modern criticism of the *Rose* has fallen for the most part into two camps: those who assume that Jean more or less carried out the original author's intention, and hence criticize the works as one,[1] and those who recognize the split between the two poems as significant, assuming that Jean did 'not really continue Guillaume's poem' because he did not grasp the subject matter or 'because he chose to disregard it.'[2] I will argue that Jean did understand the subject matter of Guillaume's poem, and that he proceeded, not merely to add to it, but to exercise control over the reader's understanding of the first part by altering the context surrounding the images and structures created previously by Guillaume.

Jean de Meun's subsumption of Guillaume's poem within his own is centered on two foci: sexuality and language.[3] Jean seems to have interpreted elements in Guillaume's poem as homoerotic – even though Guillaume implicitly demonstrates the futility of that kind of love – and thus chose for a main source of his continuation the *De planctu Naturae*, in which Alanus de Insulis vigorously condemns sexual perversion, particularly homosexuality. In addition, Jean makes extensive use of Alanus's work because he was aware of Guillaume's reliance on the *De planctu Naturae* as a model for his own allegory. Both Alanus and Guillaume use figurative language as a way of conveying truths that cannot be expressed literally; yet it is possible to express truth through language only if words and the things they signify are essentially linked. If the connection between words and things is only accidental, language cannot be a transparent medium of meaning. Jean's awareness of this philosophical position led him to write an allegory that is, in a sense, an anti-allegory: in his continuation of the *Rose*, he demonstrates that figurative language can never reveal truth.

The subsumption of Guillaume's poem takes place on a number of levels: authoritative, structural, and emblematic. I will discuss each of these levels in turn, beginning with Jean's use of texts alluded to by Guillaume in an effort to recontextualize them, retrospectively altering their significance in the first *Rose*. Next, I will show how Jean represents Guillaume's poem as a fragment and consequently is able to conceal its structural allegory. I will then point out how the emblem of Narcissus, central to Guillaume's depiction of the failure of courtly love, is explicitly replaced by Jean with the figure of Pygmalion. Instead of the garden of Deduit and the fountain of Amors, Jean offers another garden and fountain, these ones eternal. Instead of the marvellous crystals described by Guillaume, Jean produces a carbuncle which gives off its own light. Jean also takes up the emblem of the mirror, burying Guillaume's subtle optical allegory under a profusion of scientific details regarding various kinds of mirrors and lenses. He disguises the prophetic *somnium* of the Guillaume's *Rose* as an erotic *insomnium*.[4] The double meaning of *deduit*, both pleasurable diversion and optical refraction, is not lost on Jean: he counters Guillaume's multiplicity of images generated by refraction by gradually bringing together disparate loci of desire into one physically attainable object, the full-blown rose. I will conclude by explaining what a detailed understanding of the optical allegory of the continuation reveals about Jean's use of figurative language, particularly about his understanding of the link between words and the things they signify. To that end, I will discuss passages in which Jean relates sexual fertility to language, including his account of the castration of Saturn, in which Reson uses the indelicate term 'coilles,' and Genius's explication of the myth of Saturn, as well as Jean's translation of the letters of Heloïse and Abelard.

Jean often chooses the *auctores* he will cite on the basis of Guillaume's having used them in the first part of the *Roman de la rose*. By quoting their words in a different context within his continuation, Jean changes their meaning within the first part of the poem as well. Few readers have noted the importance of Alanus de Insulis's *De planctu Naturae* in Guillaume's *Rose*, probably because Jean's use of Alanus is so much more obvious, overshadowing the more subtle allusions made by Guillaume. For example, Wetherbee notes connections between Alanus's description of 'effeminate slackness' and the lover in Guillaume's poem, but refrains from making the connection explicit because he might 'be misrepresenting the spirit of Guillaume.'[5] Alanus's *De planctu Naturae* is the very first text quoted by Jean in his continuation (4249–328): as Lecoy puts it, 'Toute cette tirade est pratiquement traduite du *De planctu Naturae*.'[6] The description of love adapted from Alanus is followed by one from another source used by Guillaume, Andreas Capellanus's *De arte honeste amandi* (4347–58). In each case, Jean de Meun shows himself eager to show his command of both philosophical and courtly literary works alluded to by Guillaume.

Jean quotes not only Ovid's *Metamorphoses* but also his *Ars amatoria*, both important in the first *Rose*. But nowhere is Jean's desire to overshadow Guillaume more evident than in his citation of scientific texts. He explicitly names 'Euclidés' [Euclid (16141)], 'Tholomees' [Ptolomy (16141, 18542)], 'mestre Algus' [al-Khowarezmi (12760, 16141)], the *Meterologica* of 'Aristote' (18001), 'Alhacem, li nieps Huchaÿn' [Alhazen (18004)], and 'Albumasar' [Abu Ma'shar (19147, 19156)]. Jean seems to be primarily concerned to demonstrate his knowledge of writings on love and its relation to procreation (Ovid, Andreas Capellanus, Alanus de Insulis) and writings concerned with optical phenomena, both topics of central importance to Guillaume's *Rose*. By doing so, he causes Guillaume's exposition of the nature of courtly love and his optical allusions to become no longer merely subtle, but almost invisible.

Jean also manipulates the narrative structure of the *Roman de la rose* in order to subordinate the meaning of Guillaume's poem to the meaning of the completed work. Several readers have commented on how subtly and yet how skilfully Jean splices together his and Guillaume's narratives: as Leslie Brook puts it, Jean de Meun 'subverts the monologue from within. What better statement of intent could he make?'[7] Jean is careful to make the join between the two parts of the poem unobtrusive. Guillaume's narrator ends his complaint by lamenting his lack of hope, which he calls 'fiance' (4028). Jean immediately reiterates the concept of hope, but he calls it 'espoir' (4029). The difference between the two terms is subtle: 'fiance' denotes 'foi' ['faith'], especially in the sense of 'hommage, fidélité,' while 'espoir' refers to 'attente' ['expectation']. The first refers to a contract based in the rules of courtly love, the latter to an abstract philosophical and theological concept.

As soon as Jean fixes on the concept of hope, he immediately refers to its opposite, 'desespoir' (4030), only to reject despair and embark upon a description of 'Esperance,' hope personified. The rhetorical structure of his rejection of despair, with the pivotal word repeated twice, once positively, once negatively, reappears twice again, as the lover considers whether he should escape or repent of his love:

> [je] ne m'en desespoir.
> Desespoir! Las! je non feré.
>
> I am ready to despair of it.
> Despair! Alas, I will not do it.   (4030–1)
>
> il n'en istra, ce croi, ja vis.
> Istra! non voir.
>
> He [Bel Acueil] will not escape from there, so I believe, alive.
> Escape! Certainly not.   (4096–97)
>
> je m'en veill, ce croi, repentir.
> Repentir! Las, je que feroie?
>
> I want to, so I believe, repent of it.
> Repent! Alas, what would I be doing?   (4124–5)

Although Jean is careful to maintain the continuity of the two parts of the poem, the tone of the continuation is very different from that of the earlier poem: the cadence is extremely quick, often due to a large number of lines beginning with simple connectives such as 'et,' 'ne,' and 'si.' The rapidity of the pace is accentuated by rapid exchanges between characters within a single line (e.g., 4223–4). Jean allows the lover to vacillate regarding his planned course of action for only a short time before he reintroduces Reson. She appears when she hears the lover's complaint, just as Natura appeared to the sorrowful narrator at the beginning of the *De planctu Naturae*, and embarks upon a long disquisition on love, the first of many such lengthy monologues in Jean's *Rose*. She offers to take the place of the rose, suggesting that she is a more suitable companion for Amant. In this connection, it is necessary to recall how Guillaume de Lorris used the term *deduit* to represent both the joyful diversion of amorous desire and the optical phenomenon of refraction. In the commentary on the *Aeneid* attributed to Bernardus Silvestris, the rainbow is interpreted as the senses, while the sun, opposed to it, represents reason (see chapter 3). Here, Jean once again places the lover between sensuous pleasure and reason, and, once again, the lover fails to choose Reson.

Although Jean begins his continuation at line 4029, not until more than 6000 lines later does he reveal that there is a second author. He places this admission at the midpoint of the poem as a whole, thereby reinforcing the reader's impression that the work is complete only with the continuation attached. This strengthens Jean's claim that the first *Rose* is merely a fragment and that Guillaume discontinued the poem involuntarily when his work was cut short by illness or death. In fact, the narrator of Guillaume's poem simply stops at the end of his narrative, as he comes to recognize that the object of his love is a reflection of himself. He 'dies,' fictionally, just as Narcissus dies after he realizes that his love is futile and it is himself he adores. In his continuation, Jean de Meun quotes the last three lines of Guillaume's poem, and states that 'Ci se reposera Guillaumes' ['Here Guillaume shall rest' (10531)]. With this phrase, Jean describes not the death of Guillaume the historical figure, but rather that of Guillaume in the persona of narrator/lover within the poem. This poetic death makes it possible for Jean to take the place of his predecessor, and incorporate the earlier allegory within his new *Rose*.

In his study of Jean's adaptation of Guillaume de Saint-Amour's *De Periculis* in the midpoint speech of Faux Semblant (10931–12014), Kevin Brownlee argues that 'In a very literal sense, Guillaume de Saint-Amour has replaced Guillaume de Lorris.' By inserting the Guillaume de Saint-Amour material, 'Guillaume de Lorris ceases to function as privileged model and point of departure'; his work is 'displaced.'[8] At least some medieval readers of the *Rose* were aware of the nature of Jean's continuation. Hult writes that the early scribe and redactor Gui de Mori considered it to be a perversion of Guillaume's completed work, and adds that Gui 'interprets Jean's assertions of the incompleteness of the first *Roman de la rose* as a mere pretext, an excuse for his own continuation.'[9] Huot notes that, in one manuscript rubric, Gui de Mori recorded his belief that Jean de Meun had actually removed Guillaume de Lorris's original ending to the *Rose* in order to add his own.[10]

Perhaps Jean's rewriting of Guillaume's poem is nowhere more evident than in the climactic storming of the fortress. Douglas Kelly points out the importance of an alteration made by Gui de Mori, who had noticed a striking discrepancy between the ending projected by Guillaume and the ending actually brought about by Jean. Gui rewrote his own adaptation of the *Rose* to rectify the inconsistency: he changed Guillaume's reference to the castle 'Qu'Amors prist puis par ses esforz' to read 'Que Venus prist par ses effors' (3486).[11] Kelly does not explain exactly why, for Jean, it was necessary to have Venus rather than Amors lead the assault on the castle: in fact, Venus must assume a dominant role in order to make Jean's identification of the lover with Pygmalion, instead of Narcissus, complete.

Narcissus and Pygmalion

Guillaume's Amant is caught in a self-reflective narcissistic trap, and therefore the rose he sees is a phallic image. He can gaze at the object of his desire, but he can never possess it, reenacting the experience of Narcissus who endlessly admires his own image. The allegorical meaning of the Narcissus myth in Guillaume's poem is that courtly love is necessarily narcissistic, based on love for the self and embodying a desire that can never be fulfilled. This meaning is reinforced by the structural allegory of the poem, where images of enclosure are mirrored by corresponding images on each side of the central episode of the poem, the lover's gaze into the fountain of Narcissus. In Jean de Meun's continuation of the *Roman de la rose*, the myth of Pygmalion is recounted and compared with that of Narcissus in order to emphasize the fruitfulness of Pygmalion's fertile, outwardly directed love, in contrast with the sterile, self-directed love of Narcissus. Jean's lover takes on the role of Pygmalion, loving an object outside himself. Therefore, he can not only look at but touch and even sexually possess the rose, which is no longer seen as phallic, but is instead a receptive, exuberantly female sexual image.

Before he can replace Narcissus with Pygmalion, however, Jean must first defuse the import of the Narcissus story in the first *Rose* by redefining the significance of Echo. In Guillaume's poem, the narrator attempts to draw a parallel between ladies who ignore their faithful lovers and the relationship of Narcissus and Echo. Ironically, the lover himself is the one who reenacts the experience of Narcissus, while Amors and other personifications resemble Echo. Just as William of Conches associates visual reflections and aural echoes, so Guillaume de Lorris relates personifications such as Amors to Echo in order to emphasize how these personifications reflect the lover's own identity: how *deduit* as refraction produces a multiplicity of images that resemble each other and their original exemplar. Jean is aware of Guillaume's strategy and must redirect the reference to Echo in order to incorporate Guillaume's poem within his own. Therefore, he draws an alternative parallel between Reson and Echo: Reson concludes her attempt to persuade the lover to love her instead of the rose by comparing herself to Echo (5804–8). Additionally, before he counters the Narcissus myth with the story of Pygmalion, Jean offers another counterexample, that of Adonis. Though, as Brownlee notes, Adonis, like Narcissus, metamorphosed into a flower, 'Adonis' transformation ... is effected through the direct intercession of Venus – as will be the case with Amant's final attainment of the rose in the romance.'[12] Yet the Adonis story is not itself sufficient to counterbalance the role of Narcissus in the first *Rose*, for Adonis cannot serve as a model for the lover in his effort to take the rose. So instead the myth of Pygmalion is introduced: if the lover models himself after Pygmalion, his love will be successful. He, like Pygmalion, will be able to embrace the object of his desire, sharing a love that proves to be fruitful.

In Guillaume's poem, the god of love attempts to counter the sterility necessarily associated with the fountain of Narcissus:

> sema d'Amors ici la graine
> qui toute acuevre la fontaine.
> ...
> Por la graine qui fu semee
> fu ceste fontaine apelee
> la Fontaine d'Amors par droit.
>
> here Amors sowed the seed
> that entirely covers the fountain.
> ...
> Because of the seed that was sown,
> this fountain has been called
> the Fountain of Amors by right.   (1587–8, 1593–5)

The seed that represents the sexual power of Amors is meant to transform the self-absorbed love of the fountain of Narcissus into the outwardly directed love of the fountain of Amors.[13] The fountain's fertility contrasts strongly with the sterile self-love of the fountain of Narcissus; it is Amors's 'par droit,' its rectitude starkly opposed to the physical manifestation of self-love, which Amors says is a practice that is 'sanz droiture' (2162).

The sterile quality of the narcissistic love of Guillaume's narrator is reinforced by a short parable that appears near the close of the poem. The narrator says

> Je resemble le païsant
> qui giete a terre sa semance
> et a joie grant ou comance
> a estre bele et drue en herbe;
> mes avant qu'il en cueille gerbe,
> l'empire, tel eure est, et grieve
> une male nue qui lieve
> quant li espi doivent florir,
> si fet le grain dedenz morir
> et l'esperance au vilain tost,
> qu'il avoit eüe trop tost.

I am like the peasant who casts his seed on the earth and rejoices when it begins to be fair and thick when it is in the blade; but before he collects a sheaf of it, the weather worsens and an evil cloud arises at the time when the ears should sprout, and damages them by making the seed die within and robs the wretch of the hope that he had had too soon. (3932–42)

The narrator likens himself to the peasant whose seed is scattered but bears no fruit in order to emphasize the sterility of his narcissistic love. His love, like the peasant's grain, is not fruitful or in any way productive. The passage recalls the description of the peasant in Mark 4:26–9 who scatters seed and, 'when the grain is ripe, at once he goes in with his sickle, because the harvest has come.'[14] In the gospel, the ripe grain signifies the kingdom of God; in Guillaume's parable, the lost grain signifies lost hope, the impossibility of satisfying his desire.

In Jean de Meun's continuation, images of fertility replace those of sterility. Jean's character of Genius describes those who 'vont ... / arer en la terre deserte / ou leur semance vet a perte' ['go off to plow in desert land where their seeding goes in waste' (19614–16)]. While the sexual metaphor of the plow is adapted from the *De planctu Naturae* of Alanus de Insulis, it also recalls the unsuccessful efforts of Guillaume's peasant. Those who sow unsuccessfully are condemned, while Genius goes on to urge his audience toward fruitful reproduction: 'Arez, por Dieu, baron, arez' ['Plow, for God's sake, barons, plow!' (19671)]. This passage is immediately followed by a synopsis of the story of Cadmus, whose 'bone semance' ['good seeding' (19719)] generates enough persons to populate a city. Cadmus serves here as a model of fruitful sowing that Genius urges his audience, including the lover, to follow. In Jean's continuation, by means of reproductive metaphors and sexual symbolism, Amant is steadily directed toward the fulfilment of his desire, that is, sexual satisfaction and, ultimately, insemination of the rose.

The model for his successful love lies in the myth of Pygmalion which is recounted near the end of Genius's sermon, just prior to the successful storming of the castle and the lover's taking of the rose. The myth is introduced almost incidentally, as the figure that appears in the tower is compared somewhat disparagingly to that of Galatea: it is 'comme de souriz a lion' ['as a mouse is to a lion' (20786)].[15] The figure in the tower becomes the target for the lover's desire, replacing the undefined, shifting loci of desire found in the first *Rose*. In Guillaume's portion of the poem, the character of Bel Acueil gradually becomes the focus of the lover's desire and the one to whom he addresses his complaint. But in his continuation Jean stresses the feminine aspects of Bel Acueil, so that the homoeroticism evident in the first part of the poem vanishes in the second.[16]

Pygmalion takes great care of his beloved statue, delicately stitching her clothing with an 'aguille ... / d'or fin' ['golden needle' (20969–70)]. This recalls one of the first actions of the narcissistic narrator of Guillaume's poem, who stitches his own sleeves with a needle of silver (91). By replacing silver with gold, Jean implies that Pygmalion's adoration is of a higher grade than that of Narcissus. Pygmalion himself states this explicitly, comparing his love to that of Narcissus:

> [M]aint ont plus folement amé.
> N'ama jadis ou bois ramé,
> a la fonteine clere et pure,
> Narcisus sa propre figure
> ...
> puis an fu morz. ...
> Don sui je mains fos toutevois,
> car, quant je veill, a ceste vois

> et la praign et l'acole et bese,
> s'an puis mieuz souffrir ma mesese;
> mes cil ne poait avoir cele
> qu'il veait en la fontenele.

Many have loved more foolishly. Didn't Narcissus, long ago in the branched forest, fall in love with his own face in the clear, pure fountain? ... he afterward died of his love. ... Thus I am in any case less of a fool, for, when I wish, I go to this image and take it, embrace it, and kiss it; I can thus better endure my torment. But that one could not possess what he saw in the fountain. (20845–58)

While the object of Pygmalion's love is a statue, causing his love to seem unnatural, it is not as futile as the love of Narcissus for his own reflection. At least Pygmalion's object of adoration is outside himself, a form that can be touched, while Narcissus loves only a shadow of himself. It has been stated that Pygmalion's love is better than that of Narcissus because it is 'a full and mutual passion.'[17] Pygmalion's love is superior not because it is mutual, but because it is directed toward an external object; as Camille puts it, 'Pygmalion can touch and feel the object of his desire, whereas all that stares back from Narcissus's watery mirror is the empty surface of the gaze.'[18] While Narcissus loves the insubstantial reflection he sees in the fountain, Pygmalion loves an 'ymage' made of ivory that can, with the help of Venus, become flesh. Pygmalion's love is fully realized when Venus answers his prayer and brings the statue to life. Similarly, Jean's lover gains access to his beloved through the efforts of Venus, who leads the assault on the tower.[19] Venus's efforts meet with success after Amors's attempts have proved to be futile. She is described as a 'bone archiere' ['good archer' (20761)], for she uses arrows, the usual weapons of Amors, to reach her target. Hill notes the allegorical significance of Venus's assumption of Amors's role: 'It is only when Amors is assisted by Venus, when the refined courtly aspect of sexuality is aided by natural sexuality, that the rose is won.'[20]

Just as Guillaume's narrator reenacts the experience of Narcissus at the fountain, thus becoming capable only of a sterile, self-directed love, so Jean's narrator follows the model of Pygmalion. Like Pygmalion's, his love is capable of being physically expressed, as Jean embodies the rose by means of the feminine figure in the tower. The love-object of Jean's narrator is described in the most explicit physical terms, in contrast with the ambiguous, ethereal description of the beloved in Guillaume's portion of the poem. The love of Pygmalion and Galatea is clearly a fertile one: Jean says 'Tant ont joue qu'ele est anceinte' ['They played so well that she became pregnant' (21154)]. Similarly, there is some implication at the poem's end that Jean's rose, like Galatea, will bear a child.[21] Camille stresses the positive, procreative aspects of the Pygmalion myth as it appears in the *Rose,* where the birth of Paphus 'conclud[es] the story not with death but with the hope of new generation,'[22] paralleled by the 'generacion' urged on the lover by Genius. But there is also a sinister side to this fecundity: Amant mentions that Pygmalion's lineage includes both Adonis and Myrrha, persons whom Alanus de Insulis names along with Narcissus in the *De planctu Naturae* as figures of a love that is contrary to nature.[23] The lover states that his mention of Myrrha and Adonis is a digression, but indicates that it is not an insignificant one: 'Bien orroiz que ce senefie / ainz que ceste euvre soit fenie' ['Before this work is finished you will hear what it means' (21183–4)].

Le Miroër aus Amoreus

In his analysis of the Pygmalion story, Camille illuminates the difference between Pygmalion's desire and that of Narcissus: 'Statues are, by contrast [with mirrors], experienced "face-to-face" as part of our space and not the phantasmata of the enigmatic surface, which made them explicit sexual as well as sacred substitutes.'[24] Camille's allusion to Corinthians is apt, for just as the image seen dimly in a mirror is replaced by the beatific vision of the kingdom of God, so the elusive image of Narcissus's reflection is replaced by the ivory image embraced by Pygmalion, and so the allegorical mirror embodied in the structural allegory of the first *Rose* is replaced by Nature's catalogue of substantial, functional mirrors and lenses. In this respect, the two parts of the *Rose* illustrate a shifting literary perspective: from the merely visible to the tangible, from the formal and spiritual to the material and carnal. Correspondingly, the language of allegory changes as well, moving from the reliance on structural allegory and consistently polysemous terms found in vertical allegory to the more digressive structure and the regenerative, shifting puns found in horizontal allegory.

Gunn states that 'Jean de Meun's return in Nature's discourse to the subject of mirrors is ... a structural link with Guillaume's composition.'[25] But this link serves not to follow from and complement the preceding part, but rather to overshadow and enclose it. Jean's intention to focus on and expand Guillaume's central theme of the mirror is apparent in the title he proposes to replace that of the *Roman de la rose*: 'le Miroër aus Amoreus' ['the Lovers' Glass' (10621)].[26] He counters Guillaume's focus on different types of vision as seen in dreams and mirrors within the long digression by Nature that immediately precedes Genius's sermon. The scientific sources he uses are far more sophisticated than those which would have been available to Guillaume, permitting him, in the person of Nature, to display a sophistication and erudition that completely eclipse the more subtle references made by Guillaume to images appearing in dreams and mirrors.

Jean places Nature in the position of a powerful authority, based on her lofty position in Latin philosophical allegory, especially in the *De planctu Naturae*, and enhanced by the scientific learning with which he adorns her discourse. Jean's Nature uses terminology derived from the Greek language – 'ydoles' from *eidola*, 'fantasie' from *phantasia* (18231–7) – in order to highlight her scientific and philosophical knowledge. Her authority permits her to discount Guillaume's meaningful *somnium* as an insignificant *insomnium* and to emphasize the false and deceptive aspect of mirrors. Further, Nature is uniquely qualified to denounce dreams and mirrors as producers of false images, for she is a personification whose own role is explicitly reproductive. She is the handmaiden of God, who is Himself a generative mirror: Genius calls him 'li biaus mirouers ma dame' ['the fair mirror of my lady' (19870)], and Nature states that all those living came from this eternal mirror (17441–2). Nature's authority, like her reproductive power, comes from God, and therefore is absolute, making her denial that the narrator's dream is a *somnium* and her affirmation of the deceptive qualities of mirrors completely irresistible.

Knoespel, who rightly stresses the importance of optical metaphors in both Guillaume's and Jean's *Rose*, states that 'By calling attention to Alhazen and his newer analytical optics, Jean criticized Guillaume's less sophisticated account of vision.' He compares the optical source he suggests influenced Guillaume with those cited by Jean: 'In contrast to earlier optical discussion such as Hunain's *Liber de oculis*, Alhazen's *Perspectiva* contained not only more sophisticated integration of geometry and physiological investigation of the eye, but more detailed interest in the process by which visual images were assimilated by the internal senses.'[27] In his edition of the *Rose*, however, Lecoy suggests that it is at best misleading to take Jean's citation of his sources at face value.[28] Wetherbee rightly points out that Jean de Meun's relation to the twelfth-century allegorists is not one of simple imitation, but rather 'a matter of highly complex poetic allusion, rather than adherence to their philosophic ideas.' Specifically, concerning Reson's discussion with Amant following her exposition of the myth of the castration of Saturn, Wetherbee states that 'at this point the *Roman* significantly parts company with the *De planctu*.'[29] The point Wetherbee makes concerning Jean's ambivalent relation to the twelfth-century allegorical tradition also holds true for his citation of scientific texts. Jean's sources on optics are not those he cites explicitly, such as Euclid, Ptolomy, and Alhazen, but others whom he does not name: the perspectivists who drew upon those writers on geometrical optics, specifically, Robert Grosseteste and Roger Bacon.

Patricia Eberle rightly points out the importance of optics as a unifying theme in Jean's continuation, stating that, while some readers have assumed that Jean's 'design ... appears to make digression an end to itself,' Nature's discourse is in fact 'not a digression, but the very centre of her self-revealing monologue.' Eberle states that, in Jean's citation of 'Alhacem' (18036), he 'may well be making reference to one of the Latin translations of Alhazen that were beginning to be available in the thirteenth century as *De aspectibus* or *Perspectiva*, but his reference is of the most general sort and does not involve any knowledge of the precise details of Alhazen's theories.'[30] Eberle is right to dispute Lecoy's suggestion that Jean heard of Alhazen through Pecham or Witelo, who 'come rather late to have had major influence on Jean de Meun.'[31] I would, however, modify Eberle's suggestion that 'Grosseteste's work, supplemented by contributions of [*sic*] Albertus Magnus, influenced the poetic imagination of Jean de Meun.'[32] Eberle's emphasis on the importance of Grosseteste is the result of her reliance on Crombie's assessment of Grosseteste's role as 'the herald of a new era in Western optics,' and his characterization of Grosseteste's philosophy as a 'metaphysics of light.'[33] As Lindberg has shown, however, Grosseteste's position may not have been quite as central as Crombie suggests (though he certainly laid the foundations for Bacon's development of the science of *perspectiva* as formulated by Alhazen).[34] In addition, Lindberg has argued that Crombie's conception of Grosseteste's 'metaphysics of light' is misleading, and suggests that one might more accurately refer to Grosseteste's 'philosophy of light.' Finally, Eberle's reading of the optical context is impaired by her assumption (in keeping with the Robertsonian view prevalent in the 1970s) that Jean's poem is supplementary, an aid to the interpretation of Guillaume's poem.

Eberle usefully articulates the plural, multiple nature of Jean's continuation as compared to Guillaume's work: Jean's poem is 'a complex optical instrument ... designed to supplement the single perspective offered in Guillaume's dream-vision of love with a multiplicity of perspectives on the subject of love and on the dream-vision itself.'[35] As was the case with Knoespel's reading of Guillaume de Lorris's optical allegory, however, Eberle is right to seek out a scientific context, but wrong in her choice of sources. To contextualize Jean's account of optical phenomena, it is necessary to look to the writings of Roger Bacon, who much more than Albertus Magnus was influenced by Grosseteste's work and who, unlike Grosseteste, discusses Alhazen's theories in depth.[36]

Nature's digression begins as an explanation of the reflective properties of the moon, which she likens to a mirror. This causes her to give a detailed description of the process of making a mirror (16825ff.), a physical object whose tangibility contrasts strongly with the ethereal mirror that appears both physically, on the water in the pool of Narcissus, and metaphorically, in the symmetrical structural allegory of Guillaume's poem. She goes on to mention different kinds of mirrors and the various distorting effects they can have, including inversion and multiplication of images (18143–66; 18183–90), and emphasizes their deceptive character (18166; 18201; 18206; 18208). In Nature's discourse, Jean refers to 'Alhacem, li nieps Huchaÿn' (18004), whose 'livre des *Regarz*' is indispensable to anyone who would understand optics:

> Lors porra les causes trover,
> et les forces des mirouers,
> qui tant ont merveilleus pouers
> que toutes choses tres petites,
> letres grelles tres loing escrites
> et poudres de sablon menues,
> si granz, si grosses sunt veües
> et si pres mises aus miranz,
> car chascuns les peut choisir anz,
> que l'an les peut lire et conter
> de si loign que, qui raconter
> le vorroit et l'avroit veü,
> ce ne porroit estre creü
> d'ome qui veü ne l'avroit
> ou qui les causes n'an savroit.

There he will discover the causes and the abilities of mirrors that have such marvellous powers that all things that are very small – thin letters written very long, tiny grains of sand – are perceived to be so big, and placed so near the observers, that each person can distinguish an individual one, so that one can read them and count them from so far away that, if someone who had seen it wanted to recount the event, he could not be believed by anyone who had not seen it for himself, or by anyone who did not know the causes behind it. (18014–28)

Alhazen does indeed note how handwriting can be enlarged or distorted by means of magnifying lenses; however, the example of the magnified grains of sand does not appear in his *Perspectiva*.[37] Jean also associates the 'livre des *Regarz*' with the rainbow (18008), a topic which Alhazen treated only in another treatise that was never translated into Latin.[38] Jean goes on to include a discussion of burning mirrors (18137–42), which Alhazen discussed not in the *Perspectiva* but in his *De speculis comburentibus*, a work that became available in Latin translation only late in the thirteenth century.[39]

These topics, however, are central to the writings of Roger Bacon: burning mirrors are the subject of his *De speculis comburentibus*,[40] and Bacon discusses the rainbow extensively in his *Opus maius*.[41] Each of the phenomena described by Nature that appear in the source Jean cites, Alhazen's *Perspectiva*, can also be found in Bacon's writings. These include mirrors which produce 'fantosmes' ['phantasms'] in the air[42] and those which make one thing into two, three into six, or four into eight.[43] Most telling, however, is Bacon's conclusion to book five of his *Opus maius*, a passage echoed in Nature's discourse in the *Rose*:

The wonders producible by refraction are even greater, for it is easily evident from the foregoing rules that the very large can be made to appear very small and vice versa, and that distant objects can be made to appear very close and close things distant. ... Thus from an incredible distance we would be able to read the smallest letters and count particles of dust and sand, on account of the size of the angle under which we see them. And we would scarcely be able to see enormous bodies from nearby because of the smallness of the angles under which we see them. ... Thus a boy could appear to be a giant and a man to be a mountain of whatever size we choose, since we can see the man under as large an angle as the mountain and as near as we wish.[44]

Jean's 'letres grelles tres loing escrites / et poudres de sablon menues' (18018–19) are a translation of Bacon's 'litteras minutissimas, et pulveres ac harenas,' while the mirrors in which a dwarf appears 'plus granz que .X. geanz' (18194) are drawn from Bacon's statement that 'sic posset puer apparere gigas.'

From the passages cited above, it is evident that Jean de Meun drew upon Bacon's account of *perspectiva* for the details of Nature's discourse on optics. The dating of these works, though necessarily imprecise, supports the argument: Jean's continuation of the *Rose*, according to the text's most recent editor, was begun around 1265 and not completed until the end of 1278.[45] Although Bacon's *Opus maius* was not completed until 1267, it was, as Lindberg notes, 'patched together from writings on various subjects that Bacon had prepared in the past, with new sections where they were required for continuity.'[46] More specifically, the fifth book of the *Opus maius*, on *perspectiva*, is dated around 1263 by Easton and Lindberg.[47] Additionally, it is known that Bacon lectured in the faculty of arts at Paris during the 1240s, and it is possible that he remained there until as late as 1251 or after.[48] There is therefore a possibility of an oral as well as a written influence. Such oral dissemination of Bacon's work was apparently prevalent, for Lindberg cites a passage in the *Opus minus* in which Bacon mentions that he was kept under guard to prevent his writings from being 'divulged to others.' Additionally, in another treatise surviving only in part, Bacon mentions that fragments of his work also circulated freely in written form: 'I had sometimes compiled in cursory fashion certain chapters now on one science and now on another at the request of friends ... And those things that I wrote I no longer have [in my possession], for because of their imperfection I did not take care of them.'[49]

Bacon was above all a synthesizer, unifying disparate optical theories into a single unified extramission theory of vision, based primarily on the work of Alhazen. His greatest optical innovation, however, was the notion of the multiplication of species. Building on Grosseteste's explanation of the propagation of light by means of the reproduction of the individual forms of light, Bacon argued that all visual perception takes place similarly as each visible form reproduces itself many times over as it passes through a transparent medium.[50] But the multiplication of species is the medium, not only of vision, but of all sense perception; and it takes place not only outside the body but also within, mediating between the different faculties of the mind. In addition, the multiplication of species also manifests itself in the reproduction of kind: as Bacon puts it, the 'species produces every action in the world, for it acts on sense, on the intellect, and on all matter of the world for the generation of things.'[51] In other words, the multiplication of visible species has its parallel in the multiplication of the human species, that is, in sexual reproduction. This is why Jean de Meun places the discourse on optics in the mouth of Nature, the personification who endlessly stamps form on matter to propagate the species. Jean's optical allegory is intimately connected to the climactic event of the poem, the insemination of the rose.

Between his two adaptations of passages from Bacon, Jean places an account of the plight of Mars and Venus, who while lying in bed together were trapped in the nets placed there by Vulcan. If they had only had a magnifying mirror, says Nature, they would not have been caught in those 'laz soutilz et delïez' ['fine, thin nets' (18038)]; they would have escaped even if Vulcan 'les eüst fez d'ouvraigne / plus soutille que fil d'iraigne' ['had made the nets of finer work than spider web' (18041–2)]. Nowhere does Nature seem more prone to digression than in this passage, where the relevance of magnifying mirrors seems peripheral at best.[52] I suggest that here Jean offers an optical pun meant to parallel the optical pun on *deduit* that Guillaume offered in the first *Rose*. The nets 'finer ... than spider web' refer to an important structure of the eye, the retina or *aranea* (Lat. spider web): 'The eye has three tunics or coats, three humours, and a web resembling a spider's web. And its first tunic ... branches out like a concave net. And this first part is therefore called the "rete" or "retina."'[53] Bacon's account of the *aranea*, like others of his time, does not describe the function we now know the retina to have. Medieval physiologists considered the *aranea* to be the posterior portion of a spherical membrane which contains the glacial humours, the front of which is the iris. The watery mass of the vitreous humour and the crystalline humour at its centre (what we now call the lens) are contained (*continetur*), as Bacon specifies later in his text, by the 'spider's web.'[54] The *aranea* performs the function of enclosure, like Jean's continuation itself: just as the *aranea* encloses the watery fluid of the eye and the crystalline lens, so Jean's redirection of Guillaume's allegory encloses and nullifies the significance of the fountain of Narcissus and the crystals at its base. In keeping with his decentering of the structure of Guillaume's *Rose*, Jean's optical emblem is not the crystalline humour at the centre of the eye, but the spider web at its periphery.

As Nature apparently draws her lecture to a close, she states that she will discuss neither mirrors nor apparitions any further. She will not tell

> ou tex ydoles ont leur estre,
> ou es mirouers ou defores,
> ne ne raconterai pas ores
> d'autres visions merveilleuses,
> ... s'eles sunt foraines
> ou, san plus, en la fantasie.

> where such images have their being,
> either inside the mirror or outside.
> I will not now tell about
> other marvellous visions,
> ... whether they come from outside
> or only from fantasy. (18231–7)

Nature's claim that she will abbreviate her discourse is, of course, untrue. She continues to discuss these false images, describing apparitions, and stating that their source is from within the dreamer:

> mainte diverse figure
> se font parair en eus meismes
> autrement que nous ne deismes
> quant des mirouers parlions,
> don si briefmant nous passions,
> et de tout ce leur samble lores
> qu'il sait ainsinc por voir defores.

> They make many diverse images appear inside themselves, in ways other than those we told about when we were speaking of mirrors just a short time ago. And it seems to them then that all these images are in reality outside them. (18320–6)

These apparitions exist only inside the individual, just as the images which appear in a mirror exist only within it. They are said to be 'manconge, / ainsinc con de l'ome qui songe ... con fist Scipion jadis' ['lies, just as with a man who dreams ... as did Scipio long ago' (18333–7)]. Thus Jean alters Guillaume's association of mirrors and dreams:

while for Guillaume, following the model of Alanus de Insulis in the *De planctu Naturae*, the dream is a mirror, for Jean the dreamer's mind is the mirror. Just as false images appear in mirrors, so false images appear in the fantasy of the dreamer. Such a dream is necessarily (to return to Macrobian definitions) an *insomnium*, a dream having no hidden significance, generated by physical stimuli such as fatigue or overeating.[55] By having Nature describe only this type of dream, Jean implies that the narrator's dream is similarly an *insomnium*. Various details of the *insomnia* described by Nature correlate with Guillaume's description of his narrator's situation, making the identification more explicit. For example, she states that some, deceived by their dreams, get dressed, put on shoes, and depart on journeys (18274–96), recalling the actions of Guillaume's narrator at the poem's opening (89–93). Nature also groups Scipio's dream, which Macrobius clearly states to have been a *somnium*, with other *insomnia* (18337), thus implicitly questioning Macrobius's status as an authority on the significance to be found in dreams. Despite Nature's insinuations, it is clear that Guillaume defines his dream as a *somnium*, one with an externally verifiable truth. But by having Nature describe only one kind of dream, Jean manages to imply that this dream too is a false one, generated by an external cause and containing no truth.

Contreres Choses

At long last, Nature ends her confession so that Genius may begin his sermon. Nature first must deny the substantive reality of the images generated by mirrors and the apparitions that can be seen in dreams so that Genius can enjoin mankind to carry out the physical task of procreative insemination; or, to put it another way, Narcissus's intangible beloved must be denied to make way for the tangible, penetrable woman beloved by Pygmalion. Nature bases her account of optics on the theory of the multiplication of species in order to prepare the way for Genius's sermon on the need to multiply the human species. Genius urges his listeners toward physical reproduction, using images of sexual fertility borrowed from Alanus de Insulis. In the midst of his sermon, he offers alternative images to replace the garden, fountain, and crystals central to Guillaume's poem. Economou has described how 'Genius associates the Fountain of Life ... with Natura's truthful mirrors,' but connects 'the perilous fountain of Narcissus ... with Natura's false, distorting mirrors.'[56] Having redefined Guillaume's *somnium* as an *insomnium*, Jean goes on to have Genius make an unfavourable 'comparaison' (20255) between the two gardens and their contents, a contrast which parallels the comparison between Narcissus and Pygmalion appearing shortly afterward. Jean's garden is 'voir' ['true'], while Guillaume's is 'fable' (20258), 'pardurable' ['eternal'] while Guillaume's is 'corrumpable' ['corruptable' (20321)]. In every way, this garden is said to be superior to the garden of Deduit. Even harsher criticism is directed toward the fountain of Narcissus, whose negative effect on the lover is emphasized in Jean's poem: 'quant il s'i mira, / maintes foiz puis an soupira' ['when he looked at himself in it, he sighed many times thereafter' (20387–8)]. In Jean's continuation, the fountain is described sarcastically:

> Dex, con bone fonteine et sade,
> ou li sain devienent malade!
> Et con il s'i fet bon virer
> por soi dedanz l'eve mirer!

God! What a good and pleasing fountain, where the well become sick! And how good it is to bend over and look at oneself in the water!

(20391–4)

Knoespel says that 'Jean subverted the physical allegory [of the fountain] by approaching it as if it were a simple literal description. Rather than an eye, Jean had his audience think of the fountain as a cloudy well.'[57] Jean paraphrases Guillaume's description of elements of the poem in order to rebut them point by point and finally replace them with images of his own. The barren pine growing by Guillaume's fountain is replaced by a fruitful olive, and the stone engraved with a reference to Narcissus is replaced by a scroll promising salvation. The water sources and crystals are faulted because there are two of each, and so they are replaced by trinitarian imagery.

The greatest fault of the fountain and crystals, however, is that they are not self-sufficient: their source, according to Jean, is external, for the water comes from outside (20395–400) and the crystals require the light of the sun in order to reflect the garden (20409–30). In place of Guillaume's transparent 'cristaus merveilleus' (1547), Jean offers a red stone, 'uns carboncles merveillables / seur toutes merveilleuses pierres' (20498–9). This stone needs no sunlight to make it shine; on the contrary, it illuminates the whole park in place of the sun (20524–8). In his

De mineralibus, Albertus Magnus describes the luminescent properties of the carbuncle: 'When it is really good it shines in the dark like a live coal, and I myself have seen such a one.'[58] Jean de Meun selects the carbuncle to replace Guillaume's crystals for several reasons: first, in order to nullify the symmetrical allegory of reflection in the first *Rose,* he must replace the twin crystals with a single stone bearing three facets recalling the three persons of the Trinity; second, in order to obscure the pun on *deduit* as *detuitio,* he must replace the crystals that become multicoloured by the light of the sun with a carbuncle that generates its own light; third, he exchanges Guillaume's clear stones for one that is deep red, the colour of amorous love and of the rose itself. In his description of the carbuncle, Albertus Magnus goes on to classify three types of the stone as '*balagius, granatus,* and *rubinus,*' and notes that the *granatus,* which derives its name from the ruby-red seeds of the pomegranite, 'is the most excellent of these; but jewellers consider it less valuable.'[59] While the fountain of Narcissus in Guillaume's *Rose* is said to be sown with the 'graine' of Amors (1587), the fountain praised by Genius is illumined by the *granatus.* The reference to seed in its name recalls the figurative grain that Genius urges his audience to plant when he shouts 'Arez, por Dieu, baron, arez!' (19671). Again, in the case of the carbuncle as in the story of Pygmalion offered as a model for Amant, fertility replaces sterility, the act of carnal love replaces unfulfilled longing.

Jean was surely aware of the centrality of the theme of *deduit* as both pleasurable diversion and optical refraction to Guillaume's poem, and of its relationship to narcissism, futile desire for the self. Consequently, in Jean's continuation, digression and diversion are instead related to the quest for an external object of love, where the lover's desire can be fulfilled. The multiple objects of desire in Guillaume's poem – the reflected rosebushes, the rosebud, Amors, Bel Acueil – are replaced by Jean with a single object: the physically attainable rose.[60] Eberle revealingly states how the frequent digressions in the poem actually 'divert the lover from the direct pursuit of his chosen path,'[61] demonstrating that the theme of diversion is maintained, but it is incorporated within the rhetorical form of the writing rather than in the allegorical content of the poem. Wetherbee specifically remarks of Nature's wildly digressive monologue that it is meant 'to mirror the real and illusory attractions, the variety of impulses and the many misleading paths by which the quest of the Rose has proceeded.'[62] The many impulses and paths that have characterized the lover's quest for the rose are ultimately drawn together by Genius's sermon, which sets the stage for the conquest of the fortress and the penetration of the rose.

As we have seen, in his continuation of the *Roman de la rose,* Jean de Meun replaces each of the significant themes of Guillaume's poem, replacing each of the emblems found in the first *Rose* with alternative ones embodying the principles of fertility and tangibility. He prepares the way for this replacement by first making Guillaume's allusions to other texts (most importantly Alanus de Insulis's *De planctu Naturae* and Andreas Capellanus's *De arte honeste amandi*) seem unobtrusive by Jean's own ostentatious quotation of them and of a panoply of other authorities. In his long treatise on mirrors contained within Nature's discourse, Jean makes use of the most contemporary theories of optics, overshadowing Guillaume's subtle allusion to the three different kinds of vision as formulated by the twelfth-century master William of Conches. Finally, the myth of Narcissus, which serves as the pattern for the experience of the narrator of the first part of the poem, is replaced by that of Pygmalion, which is the model for Jean's protagonist. This mythic replacement and consequent redirection of the narrator from unfulfilled desire to consummated love permits Jean to completely recreate the story written by Guillaume de Lorris.

If one reads only the first four thousand lines of the poem, which comprise Guillaume's contribution, the great distance between the allegorical significance of the two parts of the *Roman de la rose* is evident. This is why one early reader of the poem, Gui de Mori, who only discovered Jean's continuation after he had read and commented on Guillaume's portion of the poem, was certain that Jean's assertion of the sudden death of his predecessor was merely an excuse to expand the earlier work.[63] Interestingly, Gui de Mori included a curious riddle in one of his continuations of the *Rose*:

> E se de mon nom veult avoir
> Aucuns aucune cognoissance,
> Ne l'en ferai or demonstrance
> Autrement fors que par mos teus,
> C'on entre par moi es osteus.

And if anyone wishes to have some information concerning my name, I will not now reveal it to him in any other way than by the following words: for it is through me that one enters one's lodging.[64]

Hult proposes that the riddle's solution can be found 'in a word used in the very text of the *Rose* and which describes the portal through which the Lover gains entry into the garden: "*gui*chet" ... through this clever strategem our scribe/author's name becomes for the Lover, and at the same time for the reader, the gateway to the allegorical world of Deduit's garden.'[65] If Hult's solution is correct, Gui's riddle shows that he understood just how central the theme of enclosure and the penetration of barriers is to Guillaume de Lorris's poem.

Without comprehending each part of the *Roman de la rose* separately from the other, it is not possible to understand the work as a whole. Jean signals this in a frequently cited passage that is interposed within his account of the taking of the rose:

> Ainsinc va des contreres choses,
> les unes sunt des autres gloses;
> et qui l'une an veust defenir,
> de l'autre li doit souvenir

> Thus things go by contraries,
> one is the gloss of the other;
> and he who wants to define one of the pair
> must remember the other.   (21543–6)

On one level, this passage signifies that it is first necessary to understand the allegorical meaning of Guillaume's poem in order to grasp the significance of Jean's redirection of the narrator and ultimate rewriting of the narrative. On another, as several readers have noted, it illuminates Jean's own reformulation of the genre of allegory. Regalado uses this passage as a key to the significance of the mythological *exempla* that appear throughout Jean de Meun's *Rose*. She writes that the scholastic terminology surrounding this passage introduces Aristotelian categories of definition, where 'il est nécessaire d'avoir la connaissance de son *contraire*,' and adds that Jean employs this rhetorical strategy in his definition of love at the beginning of his continuation.[66] Regalado asserts that the yoking together of 'contraires choses' is central to Jean's project, part of a 'système compréhensif' which manifests 'la relation nécessaire entre *coilles* et *reliques*, entre le monde créé et la "volenté benigne" de Dieu.'[67] Regalado's allusion to '*coilles* et *reliques*' concerns Reson's discussion with Amant, in which she refers to 'Saturnus ... / cui Jupiter coupa les coilles, / ses filz, con se fussent andoilles / ... / puis les gita dedanz la mer, / donc Venus la deesse issi' ['Saturn, whose son Jupiter cut off his testicles as though they were sausages and threw them into the sea, causing the goddess Venus to be born' (5506–11)].

The story of Saturn's castration reappears twice more in the *Rose*, once briefly in Genius's reference to his mother Venus, who was engendered by Saturn (10797–800), and again when Genius mentions the golden age when Saturn reigned, adding that 'Jupiter fist tant d'outrages, / ... et tant le tourmenta / que les coillons li souplanta' ['Jupiter committed a great outrage and tormented him a great deal when he cut off his testicles' (20004–6)]. The repeated references to Saturn's *coilles* accomplish two things: first, as Regalado rightly states, they 'fournissent à Jean de Meun le vocabulaire nécessaire pour parler littéralement de la partie de l'expérience humaine qui échappe au langage, l'expérience érotique. ... L'*exemplum* de Saturne et de Vénus ... parle ouvertement de la castration et nomme les organes ... '[68] Second, they allow Jean once again to stress the preeminent importance of masculine fertility, not only in the practical sense of reproduction, but as it constitutes the male identity.[69]

The term *coilles* appears in Jean's poem not only in connection with Saturn, but also in Nature's reference to Origen (17022–4) and Amis's brief discussion of Abelard (8729–802). In each case, the term is used only in reference to castration. Hult has pointed out that each of these episodes of mutilation, paradoxically, bears fruit: Saturn's castration generates Venus, while for Origen and Abelard, 'castration allows for a philosophical or pedagogical development.'[70] But it is also noteworthy that, in each case, emasculation is linked to association with women: Origen castrates himself so that he may teach women without scandal, Abelard is castrated as a consequence of his marriage to Heloïse, and Saturn's castration gives rise to Venus, goddess of love. Yet the relationship of cause and effect remains unclear: must a man be castrated in order to associate with women (Origen), or is he castrated as a result of his association with women (Abelard)? Is his castration the source of women's sexual power (Saturn)? And what does this imply with regard to the ultimate fate of the lover who inseminates the rose?

Jean's allusions to castration are pertinent not only within the context of medieval misogyny but also with regard to theories of language. Clearly, this aspect is particularly important in relation to Jean's use of allegory, as several readers have noted. For example, Poirion suggests that Jean's reference to Abelard's castration is related to 'la crise nominaliste dont Pierre Abélard, à tort ou à raison, était devenu le symbole.' Although this crisis manifested itself in theological and philosophical thought of the twelfth century, it did not appear in literary texts until over a century later. According to Poirion, nominalism's dissolution of the necessary link of words and things caused medieval readers and writers to experience 'une certaine angoisse de voir le langage ainsi coupé de la réalité; obsession que pourrait figurer le châtiment infligé à Abélard.'[71] Abelard's castration is literally a punishment for his involvement with Heloïse; but figuratively, in the myth that develops concerning his relationship with Heloïse, it is a punishment for his involvement in the dissolution of the essential link between word and thing.

In the course of his lecture on the ills of marriage, Amis gives a detailed account of Heloïse's character and learning,[72] but recounts only a few facts regarding Abelard: his marriage to Heloïse, his castration, his entry into the monastery of Saint Denis and his founding of the Paraclete. He does, however, specifically mention both Abelard's 'Vie' (that is, the *Historia calamitatum*) and the couple's exchange of 'espitres' (8772, 8784), both of which Jean went on to translate.[73] In his translation, Jean again uses the vulgar term *coilles* in place of the more formal Latin 'genitalibus privati.'[74] The extreme change in register from the Latin phrase to the vernacular epithet is striking. Jean's use of the term here strengthens Poirion's assertion, based on its appearance exclusively in the context of castration in the *Rose*, that there is a link between the punishment of Abelard and nominalism's dissolution of the link between word and thing. Just as *nomen* and *res* are divided from one another, so 'fu la coille a Pierre tenue' (*Rose* 8766). Heloïse's fate, however, is just the opposite: 'puis qu'el fu ... d'Argentuell nonain revestue, / fu la coille a Pierre tenue' ['as soon as she was veiled as a nun at Argenteuil, the testicles were taken from Pierre' (*Rose* 8763–6)]. Abelard's fate embodies the split between word and the thing it signifies; Heloïse's embodies the wrapping of meaning in the word, the veiling of allegorical significance in the *integumentum*. If Abelard represents the word according to nominalism, Heloïse represents the word according to realism.

Hult has recently built upon Poirion's examinations of Jean's perspective on the link between word and thing, suggesting that Jean's poem is 'very much a polemical response ... to a particular way of thinking that involved a univocal approach to the relationship between word and thought.'[75] Nowhere is this response more evident than in Amant and Reson's argument over the use of the word 'coilles.' Curiously, this passage received little attention from most medieval readers of the *Rose*: Huot states that 'Raison's defense of "plain speech" is nearly always passed over in silence by both annotators and rubricators ... It may be that many medieval readers found the language and tenor of Raison's argument disconcerting.'[76] The readers' lack of comment suggests that they were, by and large, not particularly interested in Jean's theory of language. As Huot notes, 'the point was not to grasp the poem in its entirety, but rather to mine it for items of particular interest.'[77] Amant objects to Reson's use of the term in connection with the castration of Saturn on the basis that it is not courteous for her to use such language: 'ne sai con nomer les osastes, / au mains quant le mot ne glosastes / par quelque cortaise parole' ['I do not know how you dared name them, at least when you did not gloss the word with some courteous turn of phrase' (6903–5)]. Reson responds that testicles are God's creation and therefore good: 'Je puis bien nomer ... apertement par propre non / chose qui n'est se bone non' ['I can well name openly, by its own name, a thing which which is nothing if not good' (6915–18)]. The lover replies that, even if God made the things, man made the words.

Reson, in her role as personification of man's rational faculty, declares that 'Je fis les moz, et sui certaine / qu'onques ne fis chose vilaine; / et Dix ... tient a bien fet quan que je fis' ['I made the words, and I am certain that I never made any vulgar thing; and God considers everything that I have made to be well done' (7089–92)]. Here, Reson seems to affirm that the link between words and things is accidental, not essential. Some readers take this to mean that the argument between Reson and the lover can be expressed in terms of the debate between nominalist and realist theories of language. But, as Hult points out, Reson's position is not nearly so clear-cut: despite her apparent 'admission of arbitrariness in the relation between word and thing,' her status as 'the mythical founder of human language' gives her act 'the same sort of originary status that would have characterized God's own putative language.'[78] In fact, Reson implies that the link between words and things is not arbitrary, for she states explicitly that words can be used to signify a higher truth. The problem, for Reson, is euphemism.

Reson justifies her use of plain speech in two ways: first, she declares that it is better to speak 'apertement' or openly (7121), and second, she insists that she meant what she said, not literally, but figuratively (7128). Quilligan argues that the 'conflict between the Lover and Raison is the major turning point in the poem,' after which 'the language of the poem degenerates into increasingly lewd euphemisms while a jumble of superficially "polite" metaphors carry the action of seduction.' Quilligan states that 'euphemism subverts the allegorical use of metaphor,'[79] making the climax of Jean's *Rose* not allegorical at all, because there is no hidden significance but rather only euphemism. It is in this sense that, as Hult puts it, 'Jean de Meun manages, *in practice*, to prove ... [the] nominalist point' Reson had apparently made earlier in her argument with the lover.[80]

In the end, however, I would hesitate to call Jean de Meun's strategy in the *Rose* 'nominalist.' Thomas Maloney has shown that, in his *Compendium studii theologiae* and his *De signis*, Roger Bacon stresses 'the freedom involved in the act of naming' and acknowledges that names are arbitrarily assigned to the things they signify; but, at the same time, 'he grants that there is a necessary relationship between the name of something and the concept of that something.'[81] For Bacon, the distinction hinges on the difference between word as name and word as vocal sound. It is possible that Jean de Meun's theory of language is more closely related to that of Roger Bacon, whose optical writings we know Jean relied upon, rather than to philosophical trends of the thirteenth century that can be loosely described as 'nominalist.' In her statement that she speaks figuratively (7128) and her suggestion that Amant read the 'integumanz aus poetes' (7138), Reson affiliates herself with an earlier form of allegory, one which steadfastly subordinates the *integumentum* to its meaning, the personification to the abstraction it embodies, the literal to the allegorical. When Amant rejects her offer to be his 'amie' instead of the rose, Jean simultaneously rejects the type of allegory she advocates. Instead, he proposes a new version of the genre, one which revels in the tangible pleasures of the literal level and promises the reader pleasure rather than knowledge. Through his repeated references to castration, Jean suggests that language has a potent power of fertility that can be used, not only when the genitals are attached to the body, but even after they have been separated from it. Hult aptly names this mode of writing a 'poetics of dismemberment,' for it both permits 'the free play of words in their material sense' and provides 'the very basis of metaphor.'[82]

Yet when the genitals are, metaphorically speaking, detached from the body, their reproductive power is not what it was before. When Saturn couples with Rhea, according to the commentary on Martianus Capella attributed to Bernardus Silvestris, the four elements are generated;[83] when Saturn's testicles are cast upon the water, Venus rises from the foam. The former is constructive, for the elements are the basic units of all material things; the latter is destructive, for Venus represents the dangerous power of carnal love. Correspondingly, when the ability of the word to convey numerous meanings functions within the structure of a vertical allegory (that is, a unified allegory that, as a whole, refers to another level of meaning), the word is polysemous and serves to convey an otherwise inexpressible truth; when the ability of the word to convey numerous meanings functions within a horizontal allegory (that is, a discursive allegory without a continuous figurative level of meaning), the word is euphemistic and generates only the slippery play of signification. In brief, when the word is attached to the allegorical body, it produces meaning; when the word is removed from the allegorical body, it can only reproduce reality.

Jean's view of the nature of the link between word and thing is perhaps most obvious in his use of personifications. Jon Whitman points out that there is always a tension between the figurative meaning of the personification and its status as a character within a story. Because their 'rapport' is always 'imperfect,' it 'produces pressures to reconcile the fictional personification with the human personality.'[84] This tension is particularly evident in Jean de Meun's personifications due to the way he reformulates the genre of allegory. Zumthor notes that allegory 'promotes an abstract *nomen* to the (grammatical) status of *res*.'[85] As I have shown above, Jean's interest in the dissolution of the bond between words and things is central to his continuation of the *Rose*. Therefore, he cannot use personifications in the same way that they appear in twelfth-century philosophical allegory or even in Guillaume's *Rose*, for he has finally undone the link between name and thing, between personification and the abstraction it is supposed to embody. That is not to say that the link between personification and abstraction is as indivisible in Guillaume's poem as it is, for example, in Alanus's *De planctu Naturae*. In the earlier poem, Nature's characteristics consistently conform to the abstraction she represents. Her accidental qualities mirror her essential being. But in the first *Rose*, the absolute correspondence has already begun to break down. This is in part because, unlike the *De planctu Naturae*, Guillaume's *Roman de la rose* is a psychological allegory: the personifications are generated from qualities within the narrator's mind, not from ideas existing in the divine mind, as is the case in Alanus's poem.

In the *Roman de la rose*, whether the personified qualities are those of the lady (as some critics argue) or those of the narcissistic lover (as I have suggested), they are nonetheless generated from within the mind. For that reason, some readers see the personifications' imperfect conformity to the qualities they are supposed to signify as indicative of what Muscatine has described as 'the emergence of psychological allegory.'[86] Yet a personified quality can resemble or behave like another only when the personification has come to be, as it were, a persona: no longer abstraction, but rather character. This certainly does happen in later medieval allegories, as I will show below in my discussion of Chaucer's *Tale of Melibee*. But something quite different transpires in Guillaume's poem: personifications fail to conform to the abstractions they embody, not because they have developed into independent personae or characters, but rather because they are each a reflection of the narcissistic self. Bel Acueil is described as 'dangereus' (3425), a quality that by rights should apply only to Dangiers. Because Bel Acueil is a reflection of Amant, they begin by sharing many characteristics and become increasingly similar. Bel Acueil becomes 'dangereus' only because 'Dangiers' is another reflection of Amant, as noted in the previous chapter: both Dangiers and Amant are 'vilains.' The process can be seen in the passage where Bel Acueil, Amant, and Dangiers virtually collapse into a single identity. Bel Acueil rebuffs Amant's advances toward the rose, thus doing the work of Dangiers. At the same time, by calling Amant 'vilains' (2899), Bel Acueil likens the lover himself to Dangiers, who has just been described as 'uns vilains' (2809). At that moment, when both Bel Acueil and Amant stand in the place of Dangiers, the normal boundaries of personification have been blurred: the personifications have no independent existence, but rather are reflected images of the self. But the ambiguity is only momentary, for 'Atant saut Dangiers li vilains / de la ou il s'estoit muciez' ['immediately the villain Dangiers leapt out from where he had been hidden' (2904–5)]. Dangiers was hidden both literally, concealed in the hedge, and figuratively, concealed within Amant the 'vilains' and the 'dangereus' Bel Acueil.

The transitional nature of Guillaume's use of allegory can also be seen in the blurred distinction between symbol and personification.[87] The most obvious example of this is the rose itself, which Kenneth Knoespel asserts 'never becomes a personification but remains a symbol throughout Guillaume's poem.'[88] The rose is a symbol or emblem in the same way that the mirror is in the first *Rose*; yet at the same time it is, at least temporarily, the object of Amant's desire, leading many readers to interpret the rose as a personification of the beloved lady. The same ambiguity is evident in Guillaume's representation of Biauté as at the same time one of Amors's arrows (935–8) and a lady taking part in the dance (992–3).[89] She is simultaneously both an emblem and a personification. While Guillaume's treatment of Bel Acueil, the rose, and Biauté demonstrates his evasion of the traditional norms of personification, his representation of Deduit shows his conformity to his twelfth-century models, where the personification is simultaneously person and abstraction, and the literal level that characterizes the person is subordinated to the figurative level that conveys the abstraction. Oiseuse names Deduit as 'cil qui est cist jardins' ['he who is this garden' (589)] well before he appears personified as a sanguine man bedecked with flowers (802–28). The verb Oiseuse uses is significant, for by stating that Deduit is he who *is* this garden, she indicates that Deduit is (at least) two things at the same time. Deduit is the man who personifies the quality, and Deduit is the garden enclosing the carolers. Additionally, deduit is the abstract quality of pleasurable diversion embodied by the man, and deduit (as *detuitio*) is the optical property manifested by the crystals in the fountain of Narcissus.

Guillaume's personification of Deduit also reveals something about the relationship of allegory and art. He states that Deduit the man 'resembloit une pointure' ['resembled a painting' (810)], while deduit as refraction appears in the crystals in which the whole garden appears 'con s'ele ert ou cristal portrete' ['just as though it had been painted upon the crystal' (1568)]. For Guillaume, as in the twelfth-century tradition, allegory is like painting. Of course, this connection goes back to the very roots of personification, which 'arises from the ancient practice of elevating abstract concepts to the condition of personalized gods' that could be represented as icons.[90] Yet Guillaume's relation of allegory and painting is close to that evident in the *De planctu Naturae* when Alanus describes the rose as 'picta fideliter, / A vera facie devia paululum' ['faithfully depicted, differing but little from its actual appearance'].[91]

While for Guillaume the allegorical significance of the personification is primary, for Jean the personification's status as a fictional character is privileged. The personification ceases to be primarily an abstract quality and secondarily a person, instead becoming a richly fictional character which only secondarily signifies the qualities traditionally associated with that personification. This change is evident in Jean's depiction of Genius and Nature. While Fleming characterizes Genius as the concupiscence of a 'vitiated nature,'[92] Wetherbee connects Jean's Genius more closely to the twelfth-century Neoplatonic conception of Genius as 'the power which relates the natural order to the divine.' He states that Genius's priestly offices are taken from the *De planctu Naturae*, while his role in relation to procreation is based not only on the cosmic Oyarses but also the sexual *genii* found in Bernardus Silvestris's *Cosmographia*.[93] Although Wetherbee's analysis of Nature and Genius within the context of twelfth-century Neoplatonism is exhaustive, he does not consider the extent to which Jean deviates from twelfth-century models of personification allegory. Like Fleming, Wetherbee underestimates the significance of Jean's relegation of the abstract significance of the personification to a second-ary level. This is evident in Jean's depiction of Nature, whom Wetherbee calls 'more woman than divine power, garrulous, subjective and easily distracted.' Most tellingly, Wetherbee refers to the 'personality' of Jean's Nature, and notes that Reson 'takes over much of the moral instruction performed by Nature in the *De planctu*, enabling Jean de Meun to elaborate the passionate feminine character of the goddess herself.'[94] In Alanus's poem, Natura's femininity is a consequence both of the femi-nine gender of the noun and of her fundamental association with birth. Although her status as a woman limits the roles she can fill, requiring her priest Genius to read the sentence of excommunication against those who practise unnatural love, Alanus certainly does not highlight her 'passionate feminine character.'

Conversely, Jean's Nature is not just a personification but a 'personal-ity.' Because he stresses Nature's fictional character at the expense of the allegorical significance she historically embodies, he is able to use her authoritative voice to proclaim his own scientific erudition. In Jean's *Rose*, the voice of the personification and that of the author are not entirely distinct, leading some medieval readers to dispute heatedly the question of whether one can blame an author for a view espoused by one of his personifications. To Pierre Col's statement that 'maistre Jehan de Meung ... fait chascun personnaige parler selonc qui luy appartient' ['master Jean de Meun makes each personification speak according to what is appropriate to it'], Christine de Pizan replies indignantly that 'en tous personnaiges ne se peut taire de vituperer les fames' ['in all his personifications he could not stop himself from harshly condemning women'].[95]

Jean alters not only personifications adapted from the preceding liter-ary tradition, but also those appearing in Guillaume's part of the *Rose*. Lewis notes Jean's emphasis on Bel Acueil's feminine nature: he 'forgets for thousands of lines together than Bialacoil is a "young bachelor."' Only the name, with its masculine gender, 'survives to remind us of the absurdity. The allegory has broken down.'[96] But Jean does not 'forget' that Bel Acueil is male; on the contrary, his emphasis of Bel Acueil's feminine qualities is necessary in order to suppress the threat of homo-eroticism present in Guillaume's poem, and is one component of his larger redirection of Amant from sterile narcissism to fertile love of another. Jean also denudes Oiseuse of the role she played in the first *Rose*. Richards notes that Oiseuse is mentioned only thirteen times in the poem, nine of those times in the first part: 'Oiseuse's disappearance in the last ten thousand lines must underscore her relatively insignificant role for Amant in Jean de Meun's part of the romance.'[97] More exactly, Oiseuse is insignificant in Jean's continuation precisely because she is so important in the first *Rose*, bearing a mirror that emblematically foreshadows the central mirror in the fountain of Narcissus. Jean negates Guillaume's symbolism of the mirror by replacing, as Richards points out, Oiseuse's mirror with a banner:[98] 'Dame Oiseuse ... i vint o la plus grant baniere' ['Lady Oiseuse came there with the largest banner' (10419–20)].

Allegory and Translation

Jean's transformation of allegory into a genre that elevates the literal at the expense of the allegorical is manifested, not only in his use of personification, but in his repeated refusal to provide the promised gloss to an apparent *integumentum*. The absence of an explicit statement of the figurative meaning of the poem should not in itself disqualify Jean's *Rose* from being considered just as allegorical as, for example, Bernardus Silvestris's *Cosmographia* or Alanus's *Anticlaudianus*. In both of these twelfth-century works, the allegorical significance is not overtly stated, but is instead left to the able reader to discern for himself. Similarly, in Guillaume de Lorris's *Rose*, the meaning of the structural allegory of the mirror, the optical allegory of the three types of vision, and the mythic allegory of Narcissus is not explicitly stated, but merely left open for the reader to discover.

But Jean de Meun's omission of the promised gloss is quite different. Reson indicates that there is a figurative meaning to the myth of the castration of Saturn, but declines to reveal it, telling Amant to gloss it for himself (7128–50). Amant promises his readers that 'Bien savrez lors d'amors respondre ... quant le texte m'orrez gloser' ['you will know well how to reply about love when you have heard me gloss the text' (15118–20)]; but, of course, Amant never offers an interpretation of his dream. Similarly, he promises at the end of the story of Pygmalion that 'Bien orroiz que ce senefie / ainz que ceste euvre soit fenie' ['By the time you have finished this work, you will have heard what it means' (21183–4)]. In these passages, Jean never delivers the gloss because there is no hidden allegorical significance, at least none that rivals the importance of the literal narrative which culminates in the lover's intercourse with the rose.

Wetherbee cites Jauss's statement that 'Jean de Meun "no longer took seriously" the allegorical mode which is his ostensible vehicle,' adding that 'the unflinching literalism of la Vieille and Genius dominates, and reveals how thoroughly Jean has "de-allegorized" his materials.' Wetherbee concludes that there is a great distance between the philosophical allegory of the twelfth century and Jean's allegory: 'the allegorical world of the *De planctu* is finally the structure Jean destroys in order to construct his own world out of the actual facts of experience.'[99] Similarly, it is necessary for Jean to first destroy the structure – that is, the structural allegory, the optical allegory, and the mythic allegory of Narcissus – of Guillaume's *Rose* in order to put forward a new model of the genre, one which is based on translation. In the *Rose,* Jean had mentioned that there was a need for a translation of the *De consolatione Philosophiae* (5007–10), a need which he himself filled by producing *Li Livres de Confort de Philosophie,* along with many other translations. In the preface to his translation of Boethius, Jean states that, after completing the *Roman de la rose,* he translated 'le livre Vegece de Chevalerie[100] et le livre des Merveiles de Hyrlande[101] et la Vie et les Epistres Pierres Abaelart et Heloys sa fame et le livre Aered de Esperituelle Amitié.'[102] He goes on to state the reason for producing such translations, telling the king 'Ja soit ce que tu entendes bien le latin, mais toutevois est de moult plus legiers a entendre le françois que le latin' ['although you know Latin very well, nonetheless it is much easier to understand French than Latin'].[103] Such a justification seems to support Dembowski's characterization of Jean as a 'service-translator,' that is, a writer of relatively exact, literal translations, intended to facilitate the reading of Latin texts by those whose understanding of the language was inadequate.[104] But there is another way to read Jean's statement of purpose: he writes, not that it is easier to read (*lire*) French than Latin, but that it is easier to understand (*entendre*). It may be that here Jean is implying that the vernacular is innately superior, a case argued later by Dante in his *De vulgari eloquentia.*

Eric Hicks comments that 'La carrière de Chrétien de Troyes débute par la traduction et s'achève dans le roman: celle de Jean de Meun débute par le roman et s'achève dans la traduction.'[105] Yet this statement draws a false dichotomy between the enterprises of translation, or writing derivatively, and fiction, or writing creatively.[106] In 1402, Jean Gerson accused Jean de Meun of taking quotations out of context in the *Roman de la rose*: they are 'translatés, assemblés et tirés come a violance et sans propos autres livres plusseurs, tant d'Ovide come des autres.'[107] As is evident from Gerson's use of the term 'translaté,' it is deceptive to divide Jean's work up into two categories, the translations and the fictional works (the *Roman de la rose, Testament,* and *Codicille*), for translation is present in all his works. Within the *Rose,* he translates (in the simplest sense of the term) Ovid, Roger Bacon, and a host of other authorities. As Kelly puts it, Jean's *Rose* is 'a patchwork ... of bits and pieces, snippets of learning ... Jean de Meun was both a romancer and a translator.'[108] Yet Jean's translation in the *Rose* is not limited to his stitching together of various authorities. Kelly divides *translatio* into three varieties: 'translation as such, including scribal transmission; adaptation; and allegorical or extended metaphorical discourse.'[109] Jean practises the first of these in his translations of Boethius, Abelard, and so on, and the second in his incorporation of passages from various authorities that frequently appear as the famous 'digressions' of the continuation. The third sense of *translatio* was prominent among Latin writers of the twelfth century, when to understand something 'translate' meant to interpret it figuratively; for this reason, 'translate' is often itself rendered by modern translators as 'allegorically.'[110] In Jean's hands, *translatio* has lost the meaning current in the twelfth century, and has become something else altogether.

To Jean, translation was nothing less than the reproduction of an original: not reproduction in the sense of producing an identical copy (something much easier for modern than for medieval copyists to achieve), but in the procreative sense. Genius's injunction 'Arez, por Dieu, baron, arez' is meant to urge, not only physical insemination, but literary dissemination. Genius's audience wields both the figurative and the literal stylus: 'greffes avez, pansez d'escrire ... Pansez de vos molteplier' ['you have styluses; think about writing ... Think about multiplying yourself' (19763–71)]. Jean figuratively multiplies himself by means of his writings; this is the allegorical sense in which he understands *translatio.* At the same time, Jean questions whether reproduction can ever generate truth. Because all generation takes place by means of the multiplication of species, the original is always degraded by being transmitted. The multiplication of the visible species entails the degradation of the image as it is reproduced during its passage through the medium; the multiplication of the human species entails the degradation of mankind as they degenerate further from the perfection they enjoyed in Eden, from the Golden Age when Saturn ruled; and the multiplication of the species in language entails the degradation of meaning. In Bacon's theory of language, the species of language is the concept;[111] Jean de Meun builds on this model of language, and so in his last work, the *Testament,* he emphasizes the importance of reproducing literary progeny.

Many echoes of the *Roman de la rose* appear in the *Testament*, a work of Jean's old age.[112] For example, Genius's command to reproduce is reiterated in Jean's declaration that one should 'Toute riens veut et [n']aimme son pareil par nature; / Pour ce di je que fame et homs se desnature / Qui [n']aimme a ceste fin humainne creature, / Car raison s'i acorde e Dieu et escripture' ['There is nothing which does not naturally desire and love its peer; therefore I say that a woman or man denatures himself who does not love another human creature before the end, for Reason agrees with this, and both God and Holy Scripture' (61–4)].[113] Other textual echoes from the *Rose* also appear, including the reappearance of Dangiers's warning to the lover to flee ('Fuiez, vassaus, fuiez, fuiez, / fuiez, dist il' [*Rose* 14797–8]) as Jean's warning to avoid the temptations of this world: 'Fuions mauvés amour, fuions mauvés avoir, / Fuions toutes les choses que nous poons savoir / Qui desplaisent a Dieu' ['Let us flee evil love, let us flee evil possessions, let us flee all things that we know are displeasing to God' (*Testament* 230–1)].

Another verse of the *Testament* alludes to the *Roman de la rose* more obliquely:

> Les fames sont diverses et li homme felon,
> Pour ce s'entraimment il de l'amour Ganelon;
> Agnés n'aimme Hubert, non fait Perrot Belon:
> Il ont nom 'Fol s'i fie,' s'a droit les appelon.

Women are variable, and men villainous; for this reason they love each other with the love of Ganelon [that is, with a traitorous love]. Agnes does not love Robert, nor does Perot Belon; they have the name 'fool-who-believes-it,' if we name them rightly. (*Testament* 489–92)

These men and women are epitomized in Jean's *Rose* by the lady and the lover. The rose – who emphatically does represent a real woman in Jean's continuation, or at least a real woman's body – is 'diverse' in her changing receptivity to Amant, while Jean does not try to represent the lover as anything but a churl. In the *Testament*, Jean calls these people by an epithet which indicates that they are fools to give credence to the traitorous love they share, but at the same time conceals a pun: the name 'Fol s'i fie' also means 'philosophie.' In his rejection of earthly concerns in the *Testament*, Jean apparently dismisses carnal love and learned philosophy in the same breath, instead advocating the abandonment of temporal concerns in favour of renewed devotion to God. Correspondingly, the tactile imagery of the *Rose* is supplanted in the later work by a strong emphasis on vision. Jean evokes the figure of the mirror found in Corinthians, stating that he longs for the 'très douce vision, / Où l'en voit face à face Diex sans division' ['very sweet vision where one sees God face to face, without division'].[114] He goes on to explicitly associate this vision *facie ad faciem* with the unwrapping of the *integumentum* of allegory:

> Sacrement et articles seront la descouvert,
> Qu'a nostre congnoissance n'i aura riens couvert;
> Quancque ci nous est clos, nous sera la ouvert.

> Sacraments and articles will be revealed there,
> for there, nothing will be covered from our understanding.
> Whatever is closed to us here will be open to us there.   (1845–7)

Guillaume de Lorris began the *Roman de la rose* with just such a promise, stating that men dream of 'maintes choses covertement / que l'en voit puis apertement' (19–20). In his continuation, however, Jean replaces Guillaume's promise that the reader will finally see the meaning with an explicit account of the physical penetration of the body of the rose. Yet by the time he writes the *Testament,* Jean has come full circle with regard to the relationship of vision and knowledge. He concludes the work by invoking the Virgin Mary, 'Dame en la qui biauté tout paradis se mire, / Dame la qui bonté homs ne puet descrire' ['Lady in whose beauty all paradise sees itself reflected, Lady whose goodness man cannot describe' (2073–4)]. Unlike the woman whose body Jean verbally anatomizes as the rose, Mary cannot be represented by man's art. Unlike the rose that has to be taken by force and deception, Mary knows all that is in her lover's heart, and sees face to face:

> Tu sces ma volenté, tu sces m'entention,
> Pour ce te suppli je par grant affection
> Que tu si nous empetres vraie remission
> Et lassus avec toy parfaitte vision.

> You know my will, you know my intention;
> for which reason I beg you, with great affection,
> that you obtain for us true redemption,
> and, on high with you, perfect vision.   (*Testament* 2089–92)

Chapter Five

DANTE'S *VITA NUOVA* AND *CONVIVIO*

Dante's *Commedia* is many things to many people. It has been read as a paradigmatic allegory, a true Christian vision, a political treatise, and a *summa* of late medieval philosophy, to name just a few.[1] The primary framework for the interpretation of Dante's work has been theology, whether couched in terms of Augustinianism or Thomism, on the one hand, or in terms of the literature of contemplation, on the other. Regardless of which theological framework is drawn upon, however, vision remains at the centre of the inquiry. As Pertile puts it, for Dante, the spiritual progression is simultaneously visual, 'affect[ing] the pilgrim's eyes as much as his mind and heart.'[2] A number of readers have noted the development of the act of vision during the course of the *Commedia*, and some have even sought to classify it in terms of the three types of vision differentiated by Augustine.[3] As Barolini observes, however, such a schema can at best function only in terms of metaphor, for the types of vision in the three parts of the *Commedia* are not ontologically distinct.[4] Nonetheless, the mechanism of vision in each of the *cantichi* is different, and it varies for two important reasons: first, it illustrates the development of the narrator's acuity, both visual and intellectual; second, it establishes with increasing precision the nature of the relationship between man and God, a relationship which must be defined in terms of subject and object – yet which resists every effort to construct it in such terms. As discussed in chapter 1, pseudo-Dionysius had early on wrestled with this difficulty, and later, during the thirteenth century, Aquinas too sought to express the apparent paradox of divine transcendence, which nonetheless maintains a connection to that which it creates, in terms of the radiation of light.[5] In the scientific writings of Aquinas's master, Albertus Magnus, Dante found an exposition of alternative optical systems that (unsurprisingly) dovetailed neatly with Thomistic theology, and which emphasized on a physical level the role of the intermediary in the communion of subject and object, just as Aquinas did on the theological level.[6]

Many sophisticated interpretations of the allegorical significance of vision in the *Commedia* have already been offered, and some readers have even sought scientific texts that may have been sources of Dante's optical imagery.[7] Yet these studies are limited in that they fail to consider both the possibility that Dante's knowledge of optics may have become more specific during the course of his career, and the likelihood that Dante chose among different optical models depending upon the nature of the fiction he wished to construct. In other words, it seems improbable that Dante believed himself to be limited to using only the most scientifically verifiable theory of vision in composing a figurative description of love, whether human or divine. That is not to say that Dante never uses optical models that were considered scientifically valid in the early fourteenth century. On the contrary, in the last cantos of the *Purgatorio* and in the *Paradiso*, Dante relies upon perspectivist theory as formulated by Witelo in the late thirteenth century. He also frequently employs the theories of Albertus Magnus, whose role as one of Dante's most important sources for scientific information was long ago demonstrated by Toynbee and Nardi.[8] Interestingly, in the *Inferno*, as in certain of his lyrics, Dante uses an antiquated, Platonic extramission theory of vision, one which Albertus Magnus himself had most emphatically and vigorously refuted. Superficially, it might seem that, as Dante's knowledge of optics became more detailed, he began to use increasingly sophisticated models of vision in the *Commedia*.

This cannot be the case, however, for long before the *Commedia*, in the *Vita nuova* and the *Convivio*, Dante showed himself to be aware of both the terminology of perspectivist optics and the problems perspectivist theory posed to any effort to link metaphorically seeing and knowing. In the *Vita nuova*, Dante dramatizes the problem of mediated vision: how subject and object can never be truly linked as long as some 'mezzo' or 'simulacrum' comes between them. Jean de Meun had formulated this problem in terms of the 'multiplication of species,' using the name 'species' (visible form) popularized by Roger Bacon. Bacon, we recall, had suggested that the form of the object seen is transmitted by a series of 'species,' forms that reproduce in great number along an imaginary visual ray between object and subject. The final visible species then creates an imprint upon the eye of the subject, becoming a sensible species that is transmitted to the common sense, and then is reflected on the mirror of the imagination. When the sensible species is apprehended by the rational faculty, it becomes the intelligible species. Because Dante uses a different optical source (that is, not Bacon but Witelo and Albertus Magnus), he uses different terminology for the visible form, and stresses not the degradation of the image that is an inevitable consequence of the multiplication of species, but rather the danger that an imperfect medium can degrade the form as it is transmitted from object to subject.

In this chapter and the one that follows, I will analyse Dante's use of figurative language based on the act of vision in the *Vita nuova*, the *Convivio*, and the *Commedia*. Although the first two of these three works cannot, strictly speaking, be defined as allegories, both show the author fundamentally preoccupied with the question of how to use figurative language to express higher truths, meanings that cannot be conveyed through ordinary discourse. More to the point, in both works Dante not only uses personifications (Amore in the *Vita nuova*, Filosofia in the *Convivio*) but also reflects upon their use, asking how we can speak of an 'accident in a substance' as though it were itself a 'substance'; that is, how we can talk about a quality or characteristic as though it were a thing existing independently. Throughout his works, Dante moves between different models of vision: he uses an extramission model based on Plato's theories when he wishes to stress the power of the seeing subject, and intromission models such as those of Albertus Magnus and Witelo when he wishes to stress the power of the object seen.

Giving up Simulacra: The *Vita nuova*

In the *Vita nuova*, Dante begins the story of his life by telling how his love was kindled within by vision: 'Nove fiate già appresso lo mio nascimento era tornato lo cielo de la luce quasi a uno medesimo punto, quanto a la sua propria girazione, quando a li miei occhi apparve prima la gloriosa donna de la mia mente' ['Nine times already since my birth the heaven of light had circled back to almost the same point, when there appeared before my eyes the now glorious lady of my mind' (2.1; Musa 3)].[9] The phenomenon of love being caused by sight is not uncommon in medieval literature; for example, Andreas Capellanus states that love ordinarily arises 'ex visione et cogitatione.'[10] But in the *Vita nuova*, the lover's vision is not a fleeting glance of the real woman, but rather a permanent gaze at her image stored in memory and flashed upon the mirror of imagination: 'la sua imagine, la quale continuatamente meco stava, fosse baldanza d'Amore a segnoreggiare me' ['her image, which remained constantly within me, was Love's assurance of holding me' (2.9; Musa 4)]. The lover desires to see his beloved not only within his own mind but also directly, so that he cannot keep from staring at her when they are both in church. Others notice him staring; but instead of discovering Dante's love for Beatrice, they suppose that he is attracted to another woman seated between Dante and Beatrice.

Dante is pleased by their error, because it enables him to keep the true nature of his love a secret. He determines to perpetuate their false belief, making this 'gentile donna' into a 'schermo de la veritade' ['a screen to hide the truth' (5.3; Musa 8)]. Dante uses the language of optics to describe the encounter that others see and misinterpret:

> nel mezzo di lei [Beatrice] e di me per la retta linea sedea una gentile donna di molto piacevole aspetto, la quale mi mirava spesse volte, maravigliandosi del mio sguardare, che parea che sopra lei terminasse. Onde molti s'accorsero de lo suo mirare. ... io intesi che dicea di colei che mezzo era stata ne la linea retta che movea de la gentilissima Beatrice e terminava ne li occhi miei.

> Halfway between her [Beatrice] and me, in a direct line of vision, sat a gentlewoman of a very pleasing appearance, who glanced at me frequently as if bewildered by my gaze, which seemed to be directed at her. And many began to notice her glances in my direction. ... I realized that the lady referred to was the one whose place had been half-way along the direct line which extended from the most gracious Beatrice, ending in my eyes. (5.1–2; Musa 8)

The phrase 'retta linea,' which Dante repeats twice, is a technical phrase that is used in geometrical descriptions of vision as a model for the motion of the visible ray. For example, Albertus Magnus uses the Latin version of the phrase, 'linea recte,' several times in his explanation of how rays are reflected in a mirror.[11] The geometrical precision of the line is stressed by the fact that it extends from the 'mezzo' of Beatrice and of Dante, that is, from the middle or central point of each of their bodies.

The Italian phrase 'retta linea' is only the most obvious of the optical terms in Dante's description of his experience in the church. The lady sitting between Beatrice and Dante has a pleasing 'aspetto,' a term that, in general, means 'appearance' but in optical texts refers to the surface of a three-dimensional object. Dante uses the term 'terminare' ('terminasse,' 'terminava') twice, just as he does 'rette linea'; and, like that phrase, 'terminare' is used in treatises on optics to describe the motion of the visual ray. Albertus Magnus uses the term repeatedly to explain how colour is made visible by the termination of the light ray at some depth within the surface of the object. Albertus states that this phenomenon is most readily observed in bodies of water: 'when it is far away, the sight of it is terminated [*terminatur*] at the surface of the water upon which, when it is smooth, a great deal of light is diffused; thus it appears exceedingly whitish. When, however, it is near, the sight of it does not appear superficially, where light is greatly diffused, but penetrates into the depths, where there is less light; and thus it appears darker.'[12] Albertus refers to an opaque body as 'corpus terminatum,' that is, a body that terminates the progress of the light ray and, by reflecting it back from a deeper or more superficial depth, generates various colours.[13] By using terminology from geometrical optics, Dante draws attention to the importance of vision in this encounter: not just his own gaze directed toward Beatrice, but that of the 'donna gentile' who repeatedly turns to look at Dante ('mi mirava spesse volte'), and that of the onlookers who, by noting the direction of her gaze, presume to know the direction of his.[14] At the same time, Dante draws attention to the deceptive nature of vision, the extent to which what is apparent differs from what is really true. He does this most obviously by showing how the lady acts as a 'schermo' or screen, deceiving the onlookers; the 'rette linea' only seems ('parea') to terminate above her.

Yet Dante also begins to suggest the limitations of his own vision, limitations that will be brought out explicitly later in the *Vita nuova*. Although he names the 'retta linea' twice, it appears in a different context each time: in the first instance, the 'donna gentile' is said to be seated in the middle of a line defined by Dante's 'sguardare'; in the second, she is in the middle of a line that comes from Beatrice and terminates in Dante's own eyes ('ne li occhi miei'). The change in direction is reflected in the slight variation of the phrase: the first time, it is a 'retta linea'; the second, a 'linea retta.' Dante thus generates ambiguity regarding where the power of vision resides: is the line of sight generated by the viewer or by the object seen? In addition, Dante alludes to that which prevents there being any perfect union of subject and object: the intermediary or 'mezzo.' He twice states that the 'donna gentile' is 'nel mezzo,' in the middle; although this is a common phrase, it takes on additional significance in connection with the lady's role as 'schermo' or screen. In optical writings, the Latin cognate of 'mezzo,' *medio*, appears as a synonym for the diaphanous, the medium that conducts light and the visible species.[15] As 'mezzo,' the 'donna gentile' does not facilitate the passage of the visual ray; instead, she is a 'schermo,' hiding truth rather than revealing it.

In this passage, Dante implies that vision is limited, and that its limitations are derived from the role of intermediaries in vision; several chapters later, Dante makes the problem more explicit. In a dream, he sees Amore who solemnly addresses him in Latin, saying 'Fili mi, tempus est ut pretermictantur simulacra nostra.' Musa translates this to mean, 'My son, it is time to do away with our false ideals'; but Amore's words can be translated more literally, indicating that it is time to omit simulacra, that is, to omit those intermediaries of vision that figuratively represent intermediaries of love.[16] Yet, paradoxically, Amore concludes the exchange with Dante by suggesting that he write poems to convey his love: 'Queste parole fa che siano quasi un mezzo, sì che tu non parli a lei immediatamente, che non è degno' ['Let your words themselves be, as it were, an intermediary, whereby you will not be speaking directly to her, for this would not be fitting' (12.8; Musa 18)]. In other words, Amore commands Dante to abandon one 'mezzo' for another: he should leave behind the intermediaries of vision, and instead rely upon the 'mezzo' of poetic language. In spite of the command to abandon *simulacra*, it remains impossible to do away with mediation entirely, for Dante may not approach Beatrice 'immediatamente': his poetry will represent him, acting as a *simulacrum* of the lover.[17]

Amore offers to accompany the poem and explain its meaning to Beatrice, thus acting as yet another 'mezzo' in the chain between lover and beloved. A comparable relationship is carried out on a textual level in the *Vita nuova*, this time between the poet and the reader. While Amore interprets poems for Beatrice, the *ragioni* interpret the poems for the reader: on Dante's behalf, both explain 'come tu fosti suo tostamente de la tua puerizia' ['that ever since you were a boy you have belonged to her' (12.7; Musa 18)]. It is for this reason that Dante's digression concerning the personification of Amore is so important to an understanding of the structure of the work: as explicator of poetry, Amore represents the text in which the poems are embedded, the body of the *Vita nuova* itself. As a result, the justification for the personification of Amore will correspond to the justification for composing the *Vita nuova*.

Dante offers the digression out of concern for the reader who may wonder at his speaking of love 'come se fosse una cosa per sé, e non solamente sustanzia intelligente, ma sì come fosse sustanzia corporale' ['as if it were a thing in itself, as if it were not only an intellectual substance, but also a bodily substance' (25.1; Musa 54)]. He acknowledges that love 'non è per sé sì come sustanzia, ma è uno accidente in sustanzia' ['does not exist in itself as a substance, but is an accident in a substance']; that is, love is a state of being, rather than something which exists in itself independently. In acknowledging this, Dante follows his 'first friend' (3.14) Guido Cavalcanti, whose famous *canzone* 'Donna me prega' begins with the avowal that the poet will 'dire / d'un accidente che sovente è fero / et è sì altero ch'è chiamato amore' ['speak / Of an accident that is often unruly / And so haughty that it is called love'].[18] Guido's *canzone* is an important subtext for the *Vita nuova*, not only because it is among the first poetic efforts to use the language of philosophy and science to describe how love affects both body and mind, but also because it defines love in terms of vision.[19] Yet Guido does not associate the experience of love with the experience of sight; on the contrary, love is equated with blindness rather than sight, and darkness rather than light.

Many readers have recognized the sharp contrast between Guido's theory of love and that presented by Dante in the *Vita nuova*, particularly in his *canzone* 'Donne ch'avete intelletto d'amore.'[20] Guido begins his analysis of love with a metaphor of light:

> In quella parte dove sta memora
> prende suo stato, sì formato, come
> diaffan da lume, d'una scuritate
> la qual de Marte.

> In that part where memory resides,
> [Love] takes its place – given its form, just as
> Transparency is by light, by a darkness
> That comes from Mars.   (15–18)

Love is kindled in the same way that an object is illuminated; but while form is made visible by light's penetration of the diaphanous medium, love is generated by 'scuritate' or darkness. This opposition is reinforced by their different sources: light comes first of all from the sun, while the 'scuritate' that gives rise to love comes from Mars, the planet that represents the god of war. By characterizing love in terms of darkness, Guido reinforces his assertion that love is a purely sensate phenomenon, divorced from the radiant light of the intellect. He states that love 'è creato ed ha sensato nome, / d'alma costume e di cor volontate' ['is created and has a sensate name, / Is a habit of the soul and an intention of the heart' (19–20)]. Whereas love is created, the intellect is eternal;[21] whereas the will that governs love comes from the heart, the will that governs the intellect resides in the mind; whereas love is given its form by the darkness exuded from the red planet of war, the intellect is given its form by the intelligible sun. Guido goes so far as to conclude that love 'luce rade' (68), that it eliminates light, swallowing up the intellect in the obscurity of the senses.

Dante collapses this false dichotomy of love and intellect most concisely in the first line of the first *canzone* of the *Vita nuova*: 'Donne ch'avete intelletto d'amore.' 'Intelletto' does not simply mean understanding in general, but specifically refers to the intellectual faculty of the mind; Dante makes this evident by referring at the beginning of the first stanza of the *canzone* to human 'intelletto d'amore,' and at the beginning of the second to the divine ['divino intelletto']. The ladies' understanding of love mirrors God's own understanding; thus, just as He perceives Beatrice as a 'un'anima che 'nfin qua su risplende' ['a radiant soul whose light reaches us here' (19.7; Musa 32)], so the ladies perceive Beatrice in terms of light. They know, as does Dante himself and all those who have 'intelletto d'amore,' that

> De li occhi suoi, come ch'ella li mova,
> escono spirti d'amore inflammati,
> che feron li occhi a qual che allor la guati,
> e passan sì che 'l cor ciascun retrova.

> her eyes, wherever she may choose to look,
> send forth their spirits radiant with love
> to strike the eyes of anyone they meet,
> and penetrate until they find the heart.   (19.12; Musa 33)

In Guido Cavalcanti's 'Donna me prega,' light is associated with the intellect, love with darkness; in Dante's 'Donne ch'avete,' light is associated not only with the intellect but with love. Consequently, intellect and love are not opposing forces, but rather two complementary parts of perfect vision, as Dante finally demonstrates in the *Paradiso*.

Dante's most important source for describing love in terms of vision and light is, ironically enough, Guido Cavalcanti himself. Although in 'Donna me prega' Guido explicitly dissociates light from love, in several of his other poems he describes the effect of the luminous beam projected by the eyes of the beloved lady.[22] For example, in his 'Era in penser d'amor,' a sympathetic girl says to the lover, "L tuo colpo, che nel cor si vede, / fu tratto d'occhi di troppo valore, / che dentro vi lasciaro uno splendore / ch'i' nol posso mirare' ['Your wound, that can be seen [in] your heart, / Was drawn by eyes of overwhelming strength / Which left within a radiance / Such that I cannot look at it']. After the lover names the lady whose eyes inflicted this injury, the girl's companion adds that 'La donna ... / dentro per li occhi ti mirò sì fiso, / ch'Amor fece apparire' ['The lady ... / Looked inside you through the eyes so intently that she made Love appear'].[23] Thus love is generated not just by the gaze of the lover upon some beloved object; instead, it is generated by the power of the mutual gaze. This theory of love, distinct from what Guido Cavalcanti proposes in 'Donna me prega,' is the basis of Dante's theory of love in the *Vita nuova*. Both are based on a Platonic extramission theory of vision, where the eye emits a fiery beam that, as it were, reaches out, touches the object, and returns to the eye, carrying the form of the object back to the seeing subject. In Dante's time, this theory was not accepted as an adequate scientific explanation of the process of vision, although certain writers on optics (including Robert Grosseteste, writing in the early thirteenth century) put forth theories of vision that combined extramission and intromission. But the Platonic extramission of theory could be quite useful as a metaphor for the operation of love, which is how Guido Cavalcanti used it: the fiery beam causes damage to the heart of the lover and creates the burning heat of passion.

Dante follows Guido in using a Platonic extramission theory of vision to explain how love is kindled; but he differs from Guido in asserting that the fiery beam generates not only heat but also light. In addition, Dante suggests that the light of love is not different from the light of the intellect, radically departing from Guido's declaration in 'Donna me prega' that love is a purely sensory phenomenon. Having adopted light as a metaphor for both love and intellect, and vision as a metaphor for their full experience, Dante goes on in the *Vita nuova* and his subsequent writings to explore the implications of the metaphor, including the variations imposed upon it by different optical theories. Those theories explored by Dante include antiquated Neoplatonic extramission theories; primitive intromission theories such as that of Democritus, as described by Albertus Magnus; and the more sophisticated intromission theories of Albertus and the perspectivist Witelo.

Even though vision is the vehicle of the lover's experience of Beatrice in the *Vita nuova*, it is also the vehicle of his experience of another woman. Sorrowing after the death of Beatrice, Dante 'levai li occhi per vedere se altri mi vedesse' ['looked around to see if anyone were watching me']. He meets the glance of 'una gentile donna ... la quale ... mi riguardava sì pietosamente, quanto a la vista, che tutta la pietà parea in lei accolta' ['a gracious lady ... who was looking down at me so compassionately, to judge from her appearance, that all pity seemed to be concentrated in her' (35.2; Musa 74)]. Dante believes that his attraction to the 'gentile donna' is a betrayal of his love for Beatrice, and blames his eyes for the transgression: 'Io venni a tanto per la vista di questa donna, che li miei occhi si cominciaro a dilettare troppo di vederla ... Onde più volte bestemmiava la vanitade de li occhi miei' ['The sight of this lady had now brought me to the point that my eyes began to enjoy the sight of her too much ... So, many times I would curse the wantonness of my eyes' (37.1–2; Musa 76)]. Although the desire for both women is born of love, their images as perceived by the lover differ: one comes from the actual body of the 'gentile donna,' while the other comes from the image of Beatrice stored in the lover's memory. Dante uses this difference to express the conflict generated by his desire for two different objects: 'maggiore desiderio era lo mio ancora di ricordarmi de la gentilissima donna mia, che di vedere costei, avvegna che alcuno appetito n'avessi già, ma leggiero parea' ['my greatest desire was still that of remembering my most gracious lady rather than of gazing at this one – even though I did have some desire for her then; but it seemed slight' (38.6; Musa 79)]. In other words, the image of Beatrice comes from *ricordare*, while the image of the 'gentile donna' comes from *vedere*.

This distinction may seem slight; but if Dante's experience is examined in light of medieval faculty psychology, the difference is significant, for one image is perceived directly, along a straight line of vision, while the other is perceived only by reflection, along a line of vision bent obliquely. The image of the 'gentile donna' comes from her form, imparted to the visible species which travels through the diaphanous, reaches the lover's eye, is transmitted to his common sense, and strikes the mirror of his imagination. Like that of the 'gentile donna,' the image of Beatrice came originally from a form transmitted in a direct line to the lover's imagination; but her image was then stored in memory from which it must be retrieved to reappear upon the lover's imagination. When Dante contemplates the image of the 'gentile donna,' he looks at her directly; when he contemplates the image of Beatrice, he contemplates a second-hand image generated by a bent ray.[24] The problem is resolved, if only temporarily, by 'una forte imaginazione,' a powerful vision through which Beatrice is made immediately present to the inner eye (39.1; Musa 80). She appears just as she did the first time Dante saw her, making the intervening years seem to disappear: the linear passage of time is collapsed into a single point, and the frailty of memory is no longer a problem. This visual experience obeys Amore's earlier command to do away with 'simulacra' (12.3; Musa 17), for Beatrice is seen directly in the imagination, without being transmitted by species or passing through any medium. Yet just as, in the earlier passage, abandoning visual simulacra meant simply moving to the 'mezzo' of poetic language, so here the direct vision of Beatrice leaves the lover with the dilemma of how that vision can be expressed through language and still be perfect.

Therefore Dante concludes the *Vita nuova* by not concluding it; that is, he leaves the end deliberately unfinished, a testimony to the impotence of language in the face of the supernatural power of love.[25] Yet he carefully chooses the last poem of the work, the last three words of which appeal to 'donne mie care' ['dear ladies'], recalling the 'donne' addressed in the first *canzone*. Those ladies' 'intelletto d'amore' reappears in the final poem, but here in the form of a 'intelligenza nova,' a new intelligence that informs the lover's 'sospiro,' sending it on a celestial pilgrimage 'Oltre la spera che più larga gira' ['Beyond the sphere that makes the widest round' (41.10; Musa 85)]. This voyage anticipates that of the pilgrim of the *Commedia*, for the sigh 'vede una donna, che riceve onore, / e luce sì, che per lo suo splendore / lo peregrino spirito la mira' ['sees a lady held in reverence, / splendid in light; and through her radiance / the pilgrim spirit looks upon her' (41.11; Musa 85)].[26]

'Oltre la spera' also reworks another of Guido Cavalcanti's poems, 'Veggio negli occhi de la donna mia.'[27] As in Dante's poem, the lady's ability to inspire love is figuratively described as light: her eyes give off 'un lume pien di spiriti d'amore' ['a radiance full of spirits of love' (line 2)]. Her 'spiriti d'amore,' named in the poem's first lines, generate comparable spirits in the lover, 'sospiri / che dicon: "Guarda; se tu coste' miri, / vedra' la sua vertù nel ciel salita"' ['sighs / That say: "Look; if you gaze on that one / You will see her essence gone up to heaven"' (lines 18–20)]. But while Guido's vision of his lady in heaven is only possible and conditional ['se tu coste' miri'], Dante's is truly experienced, albeit through the medium of the sigh: the lover in Guido's poem knows that *if* he gazes upon his beloved, he will see her in heaven; but the lover in Dante's poem actually sees the lady ['vede una donna,' line 6]. Guido's poem concludes with the lover's sigh, symptom of the awful power of love and of the lady; Dante's begins with the lover's sigh ['sospiro,' line 2], continuing the account of love from the point of the nadir of the lover's fortunes.

While Guido emphasizes the extent to which the sensation of love is separate from the intellect, Dante allies love and the intellect in this, the final poem of the *Vita nuova,* just as he had done in 'Donne ch'avete,' the first *canzone* of the *Vita nuova.* Guido states that his experience of love is something that he cannot 'a lo 'ntelletto dire' ['express to the intellect' (line 6)], and that his lady is 'sì bella ... che la mente / comprender no la può' ['so beautiful ... that the mind / Cannot grasp her' (lines 8–9)]. Conversely, Dante indicates that his own experience is motivated by an 'intelligenza nova' ['new intelligence' (line 3)]. Like Guido, Dante suggests that the lady's power is embodied within her very name; but, unlike Guido, Dante names her explicitly. The 'peregrino spirito' ['pilgrim spirit'] of the sigh 'spesso ricorda Beatrice, / sì ch'io lo 'ntendo ben, donne mie care' ['often says "Beatrice"; that much I understand well, dear ladies' (41.13; Musa 85)].[28] By speaking her name, the poet comes as close as he can to conveying her essence for, as Dante notes earlier in the *Vita nuova,* 'Nomina sunt consequentia rerum' ['Names are the consequences of the things they name' (13.4; Musa 22)]. Dante follows 'Oltre la spera' by stating that, after he wrote the poem, he experienced yet another vision; but this one made him determined not to write any more about his love for Beatrice 'infino a tanto che io potesse più degnamente trattare di lei' ['until I would be capable of writing about her in a nobler way' (42.1; Musa 86)]. 'Oltre la spera' is thus the eternally penultimate poem: it precedes some imaginary poem that could 'più degnamente' describe Beatrice. It is significant, then, that the poem that 'Oltre la spera' builds upon, 'Veggio negli occhi,' is the last poem Guido Cavalcanti wrote before composing 'Donna me prega.'[29] In this sense, Guido's famous *canzone* has its counterpart not in Dante's 'Donne ch'avete,' but rather in the imaginary work that the *Vita nuova* can only point toward.

The Consolation of Philosophy: Vision in the *Convivio*

Although the *Convivio* was written after the *Vita nuova,* it does not fulfil the criteria set out in the closing chapter of the earlier work. Yet the two works are connected: Dante explicitly links them when he recalls the conflict he experienced as a result of his desire for the 'gentile donna' and his desire for Beatrice. In the later work, as in the earlier, he stresses the difference between the women's images, one received directly and one indirectly: 'l'uno era soccorso de la parte dinanzi continuamente, e l'altro de la parte de la memoria di dietro' ['one drew help continually from what lay before ... while the other drew its help from what lay behind, that is, from my memory' (2.2.4)].[30] But in the *Convivio* there is no supernatural apparition to reassert Beatrice's place as the focus of Dante's love. Instead, Dante expresses his love for Beatrice's rival, the 'gentile donna' who 'vene a consolar la nostra mente' ['comes with consolation for our mind' (*Vita nuova* 38.9; Musa 79)], now explicitly named as Filosofia. Some readers suggest that Filosofia can be identified with Beatrice, for she is characterized in language very similar to that used of Beatrice in the *Vita nuova*.[31] But at the same time, she is explicitly identified with the 'gentile donna' of the *Vita nuova* who had distracted the lover from his desire for Beatrice. Dante addresses her in the first line of the first *canzone* of the *Convivio* as 'Voi, che 'ntendendo il terzo ciel movete' ['You who move the third heaven by understanding'], thus linking love and intellect as he had done in the first line of the first *canzone* of the *Vita nuova*, 'Donne ch'avete.' But in the *Vita nuova*, Dante places love above intellect, Beatrice above the 'gentile donna'; in the *Convivio*, he places intellect above love, Filosofia above Beatrice. In abstract terms, the dilemma concerns whether love or intellect occupies a higher place in the human soul.

Dante describes the lover's desire for and experience of Filosofia in terms of vision, just as he had done with regard to Beatrice in the *Vita nuova*:

> Cose appariscon ne lo suo aspetto,
> che mostran de' piacer di Paradiso;
> dico ne li occhi e nel suo dolce riso,
> ....
> Elle soverchian lo nostro intelletto,
> come raggio di sole un frale viso;
> e perch'io non le posso mirar fiso,
> mi conven contentar di dirne poco.

Things appear in her countenance which show some of the pleasures of Paradise; these things appear in her eyes and in her sweet smile. ... They overwhelm our intellect, as a ray of sunlight overwhelms weak sight; and since I cannot steadily gaze on them, I must be content with saying little of them. (3.canzone 2, lines 55–62)

Dante expresses Filosofia's effect on vision explicitly, testifying to the power of her visible form and the radiant light she emits; but he also expresses it implicitly, by placing the passages in which he refers to vision at numerically significant points in the *Convivio*. Dante thus makes Filosofia's power as visible to the reader as it is to himself. The passage in which Dante most specifically describes the meaning of vision occupies the entire ninth chapter of the third book, where Dante elucidates the literal meaning of the last lines of the *canzone* 'Amor, che ne la mente mi ragiona,' found at the beginning of the third book of the *Convivio*. That *canzone* has a numerical structure based on multiples of nine,[32] a number to which Dante had assigned great symbolic significance in the *Vita nuova*. There, the number denoted Beatrice: 'ella era uno nove, cioè uno miracolo, la cui radice, cioè del miracolo, è solamente la mirabile Trinitade' ['she was a nine, or a miracle, whose root, namely that of the miracle, is the miraculous Trinity itself' (29.3; Musa 62)].

Dante expresses Beatrice's essence through number because, as he says in the *Convivio*, number is itself an enigma: 'l'occhio de lo 'ntelletto nol può mirare; però che 'l numero, quant'è in sé considerato, è infinito, e questo non potemo noi intendere' ['the eye of the intellect cannot gaze on this, because number, considered in its intrinsic nature, is infinite, and this we human beings cannot directly apprehend' (2.13.19)]. Dante associates arithmetic with the sphere of the sun, for just as corporeal sun, due to its overwhelming brilliance, blinds the corporeal eye, so number, in its essence, blinds the intellectual eye. Yet number is only one manifestation of this quality: Dante states that certain things 'nostro intelletto abbagliano, in quanto certe cose affermano essere che lo intelletto nostro guardare non può (cioè Dio e la etternitate e la prima materia)' ['dazzle our intellect, in that they affirm the existence of certain things on which our intellect cannot look, namely, God, eternity and prime matter' (3.15.6)]. The vision of Beatrice in heaven shares these qualities, and so remains inaccessible to the human mind; the same is true of the vision of Filosofia.

The purpose of allegory is to make such enigmas visible to the human mind, clothing the brilliant light of these 'certe cose' in a veil of figurative language that makes them perceptible. The *canzoni* of the *Convivio* attempt to do exactly this, which is why Dante follows each of them with explanatory commentary that is 'litterale e allegorica' (2.1.2). In explaining the purpose of his commentary, Dante stresses the importance of the literal level: 'lo litterale dee andare innanzi, sì come quello ne la cui sentenza li altri sono inchiusi, e sanza lo quale sarebbe impossibile ed inrazionale intendere a li altri, e massimamente a lo allegorico' ['the literal sense must always be accorded primacy, as the one in whose meaning all others are contained, and without which it would be impossible and irrational to attend to the other senses, especially the allegorical' (2.1.8)]. Dante obeys this interpretive principle in the subsequent chapters, stating, for example, that it is unnecessary to offer a full allegorical interpretation of the final stanza of the second *canzone*, since the passage 'per la litterale esposizione assai leggermente qua si può ridurre' ['can readily be interpreted in the present context from what was said in the literal commentary' (3.15.19)].[33]

Dante concludes the literal part of his exposition of the first *canzone* of the *Convivio* with what he calls a 'disgressione' concerning vision. The digression is meant to help the reader to interpret what Dante refers to as the third part of the *canzone*, the fifth stanza or *tornada*, where he writes:

> Canzone, e' par che tu parli contraro
> al dir d'una sorella che tu hai;
> che questa donna, che tanto umil fai,
> ella la chiama fera e disdegnosa.
> Tu sai che 'l ciel sempr'è lucente e chiaro,
> e quanto in sé, non si turba già mai;
> ma li nostri occhi, per cagioni assai,
> chiaman la stella talor tenebrosa.
> Così, quand'ella la chiama orgogliosa,
> non considera lei secondo il vero,
> ma pur secondo quel ch'a lei parea:
> ché l'anima temea,
> e teme ancora, sì che mi par fero
> quantunqu'io veggio là 'v'ella mi senta.

Canzone, it seems that you contradict what is said by one of your sisters, in that this lady, whom you describe as so humble, is called by her harsh and disdainful. You know that the sky is always bright and clear, and in itself is never stormy; and yet for many reasons our eyes from time to time say that the stars are darkened. So when the soul says that the lady is proud, it is not speaking of her as she really is, but only as she appears to it, for the soul was fearful, and indeed still is, so that whatever I see when this lady perceives me seems harsh. (3.canzone 2, lines 73–86)

In other words, the lover's perception of the lady is altered by his own fear, just as our perception of the stars is altered by certain conditions.

But the analogy actually runs deeper, for while the lover's situation and that of the person who sees the stars seem at first to be merely similar, they turn out in fact to be identical. The identity becomes apparent if the reader understands, according to the optical science of the early four-teenth century, how the form of a visible object enters the mind of the subject.

In elucidating the literal meaning of the *tornada*, Dante refers to what he says are Aristotle's writings on optics in the *De anima* and *De sensu et sensato*; but in fact he follows Albertus Magnus's works of the same name, as well as the *De homine*. Since these works are, according to Albertus himself, merely interpretations of and glosses on Aristotle's writings, Dante's attribution is probably not meant to be misleading. It is, on the contrary, reminiscent of the way Jean de Meun refers to Alhazen as his source when he in fact uses Roger Bacon. Because Albertus Magnus and Roger Bacon defer to the authorities who provide the foundation of their own work, Dante and Jean de Meun follow them in alluding to Aristotle and Alhazen. Despite Albertus's modest claim, his *De anima* and *De sensu et sensato* actually go far beyond Aristotle's original texts, as does the *De homine*, which contains a summary of various optical theories concluding with Albertus's own version of intromission theory based largely (though not entirely) on Aristotle's writings. Albertus offers a much more detailed rebuttal of the extramission theory of vision origi-nally formulated by Plato, taking into account medieval discussions of vision by Augustine, Gregory of Nyssa, Alfarabi, al-Kindi, Averroës, and Avicenna, among others. He even refutes what seems to be the modified extramission theory offered by Robert Grosseteste in the early thirteenth century, though he does not refer to Grosseteste by name.[34]

Dante begins by specifying that, strictly speaking, colour and light alone are visible.[35] The form of the object is transmitted to the eye 'non realmente ma intenzionalmente' (3.9.7), reiterating Albertus's statement that species 'non proprie qualitates sunt, sed intentiones.'[36] I must (like Dante) pause here to digress on the mechanism of vision, for Albertus's treatment of species differs from that of Roger Bacon, discussed in the previous chapter. Bacon is the great perspectivist synthesizer: he adopts the model of the visual cone developed by the extramissionists (without accepting the notion that a fiery beam emitted from the eye is the organ of vision) and combines it with Alhazen's intromission of theory, itself based in part on the Greek atomist theory of vision. Bacon uses the term species to denote visible form, and asserts that vision is generated by the multiplication of species in a diaphanous medium between subject and object.

Albertus, on the other hand, is not at all comfortable with the notion of species, which he does not treat as a synonym for form, as Bacon does. In the earliest of the three works in which he treats vision, *De homine,* Albertus explains that the species is, as it were, a form copied from another form: it is 'not itself an image or form, but rather the species of an image or form.' The species exists only 'in the soul as a habit, in a mirror as a disposition, and similarly in air.'[37] Albertus goes on to emphasize that species do not exist materially: 'visible species are accidental: therefore they are not in the eye or in the medium [i.e., the diaphanous].' Yet Albertus concludes his discussion of the visible species by admitting that it is sometimes material, for there are three kinds of species. The first consists of a form impressed upon the watery substance of the eye, and this kind of species does not exist in air. The second exists when the eye first knows the object of vision, apprehending the visible species as a habit or disposition; this kind, too, does not exist in air. The third, however, occurs both in the eye and in air, being 'terminated in the eye, but *en route* in the medium; and thus it is in one mode in the eye, that is, actual, but in another in the medium, that is, potential.'[38] In *De sensu et sensato,* Albertus seems to continue to distinguish between the material and immaterial visible species as he had done in *De homine.* In his later *De anima,* however, Albertus begins to use the term species to refer only to what he had defined in *De homine* as the third kind, that is, to the visible form that is in some sense material. Albertus uses *forma* to refer to the visible form transmitted through the diaphanous, but species to refer to the visible form received by the eye.[39] At the same time that he refines his terminology regarding the form that acts as intermediary in vision, Albertus focuses more attention on the role of the diaphanous medium, asking whether air retards or facilitates the passage of the visible form. He concludes not only that air facilitates such passage, but that vision cannot occur in a vacuum.[40]

In the *Convivio,* Dante gives an account of vision according to the theory of intromission, stating that visible forms travel to the eye 'per lo mezzo diafano ... sì quasi come in vetro transparente' ['along the diaphanous medium ... as if they were entering transparent glass' (3.9.7)]. The movement is terminated when it strikes the watery surface of the eye, just as a ray is terminated when it strikes the surface of a mirror: 'E ne l'acqua, ch'è ne la pupilla de l'occhio, questo discorso che fa la forma visibile per lo mezzo, sì si compie, perché quell'acqua è terminata – quasi come specchio, che è vetro terminato con piombo' ['The movement of the visible form along the medium is completed in the water found in the pupil of the eye, because the water is backed by a surface, rather like a mirror, which is made of glass backed with lead' (3.9.8)]. Albertus refers several times to the watery surface of the eye, stating that 'water is the dominant [substance] in the eye, for it is in this part that impressions are made.'[41] Interestingly, however, Dante's statement that the watery surface acts like a mirror most closely resembles not the primarily Aristotelian notion of how the eye receives form given by Albertus in *De homine*, but rather the account Albertus gives in *De sensu et sensato* of Democritus's theory of vision. Democritus is the best known proponent of the earliest and most extreme theory of vision 'intra suscipientes,' that is, by intromission. Only when he summarizes Democritus's theory does Albertus discuss the role of the pupil which, like the interior parts of the eye, is composed primarily of water.[42] But the pupil also acts as *pupilla*, that is, 'little doll,' showing within itself an image of what the eye sees. Albertus explains that the 'little doll' is generated by a process separate from the act of perception: the eye perceives the object insofar as it is the instrument of vision, but it reflects an image (the 'little doll') insofar as it is smooth and terminates rays, like a mirror.[43] Thus in his interpretation of the *tornada* of the second *canzone* of the *Convivio*, Dante employs the strictest intromission theory of vision formulated, following the account given of it by Albertus Magnus.

Dante chooses a curious metaphor to describe the impact of the visible form upon the mirroring surface of the eye: he says that 'a modo d'una palla percossa, si ferma' ['like a ball that strikes something, it [the species] comes to a halt' (3.9.8)]. This is exactly what perspectivist theories of vision claim does not happen, as Roger Bacon states explicitly: 'If one speaks of the rebound of species from a body, it is evident from what has been said that this does not occur by violence; but when it is forbidden to pass through [the body] by the density opposing it, the species generates itself in a direction open to it. For if it were to be driven back like a ball from a wall, it would be necessary for it to be body.'[44] For the metaphor to be accurate, the visible form that strikes the eye as a ball bounces against a wall must be substantial or corpuscular. Thus in this detail as well as in the preceding description, Dante adheres to the strict, rather primitive intromission theory of Democritus: according to Democritus and the other atomists, the visible form is said to be substantial,[45] while later proponents of intromission such as Albertus and the perspectivists clearly state that the visible form or species is a *habitus*.

Dante explains that the 'spirito visivo' or visual spirit then transmits the form from the pupil to 'la parte del cerebro dinanzi' ['the front part of the brain' (3.9.9)]. The concept of a visual spirit comes originally from Galen, and was adopted in later Neoplatonic theories of vision by extramission. Later proponents of intromission included the visual spirit as it operates within the eye in their explanations of the act of vision.[46] Of all medieval proponents of intromission, Albertus Magnus most strongly emphasizes the importance of the 'spiritu[s] visibili[s],' which is sent 'ad oculos per nervos opticos et concavos ab interiori parte cerebri' ['along concave optic nerves from the interior part of the brain to the eyes'].[47] The visual spirit then returns, impressed with form, to the common sense, located in the front compartment of the brain. In a way, the visual spirit is another 'mezzo' for the transmission of form, like the airy diaphanous.[48] This is why, according to both Albertus and Dante, it is so important that both media be perfectly transparent. As Dante puts it,

> acciò che la visione sia verace, cioè cotale qual è la cosa visibile in sé, conviene che lo mezzo, per lo quale a l'occhio viene la forma, sia sanza ogni colore, e l'acqua de la pupilla similemente; altrimenti si macolerebbe la forma visibile del color del mezzo e di quello de la pupilla.

> for seeing to be truthful, that is, to represent faithfully the visible thing as it is in itself, the medium through which the form comes to the eye must be quite colorless, as must the water in the pupil; otherwise the visible form would be tainted by the color of the medium and of the pupil. (3.9.9)

This conclusion reveals the problem at the centre of the metaphor of vision, the problem of the 'mezzo.' Whether the medium is the visible form, the diaphanous, or the water of the pupil, it must be, as Albertus Magnus puts it, 'clar[us] et pur[us]'[49] in order to mediate perfectly between subject and object. Similarly, the poet's figurative language also must be 'clear and pure' in order to transmit meaning, so that the reader 'sees' it as it truly is.

Having concluded his explanation of the process of vision, based on an extreme version of the intromission theory, Dante at last relates optical theory to the metaphor of stars in the *tornada* of 'Amor, che ne la mente mi ragiona.' Because the diaphanous medium and the internal visual spirit must be perfectly transparent in order to transmit form accurately, it is easy to see how a star may appear more or less bright depending upon the state of these media. In the case of the external *mezzo*, an abundance of light in the diaphanous generated by the sun overwhelms the light of the stars, making them invisible even though they are still luminescent. Alternatively, the sky may be so cloudy that 'lo quale mezzo ... transmuta la imagine de la stella, che viene per esso' ['the medium changes the image of the star which passes through it' (3.9.12)]. In the case of the internal *mezzo*, infection or fatigue may impair the transmission of the image, causing things to appear

> non ... unite ma disgregate, quasi a guisa che fa la nostra lettera in su la carta umida: e questo è quello per che molti, quando vogliono leggere, si dilungano le scritture da li occhi, perché la imagine loro vegna dentro più lievemente e più sottile, e in ciò più rimane la lettera discreta ne la vista. E però puote anche la stella parere turbata.

> no longer firm but blurred, in much the same way as does our writing on a damp page; this is why many people when they wish to read hold the writing away from their eyes to allow the image from it to enter more smoothly and finely, so that the writing remains more distinct to their sight. In this way, too, the experience of the star can be distorted. (3.9.14)

By alluding to the almost illegible 'carta umida,' Dante reinforces the analogy of the act of vision and the acquisition of knowledge through texts. Just as the person seeing an object will see it more or less accurately depending upon the integrity of the *mezzo* without and the *mezzo* within, so Dante's reader will understand the truth allegorically veiled within the *canzone* depending upon the integrity of the *mezzo* of the poem's language and the *mezzo* of the reader's own capacity for understanding.

Dante concludes this section of the *Convivio* by emphasizing, not the metaphorical meaning of vision, but the actual physical experience of it. He remarks that he himself formerly experienced impairment of the visual spirit as a result of excessive reading ['studio di leggere'], recalling Albertus Magnus's warning that the visual spirit can be injured by excessive 'vigiliis et studio.'[50] During that time, Dante writes, 'le stelle mi pareano tutte d'alcuno albore ombrate' ['the stars appeared to me to be covered by a kind of white haze']. But Dante hints that the metaphorical meaning of vision is still to be sought when he notes that he experienced this impairment 'in the same year as this *canzone* was conceived' (3.9.15). When Dante states that too much reading impairs the visual spirit, causing the stars to seem dim, he metaphorically indicates that gazing at Filosofia impairs the eye of the mind, causing her eyes to appear other than they truly are. In this metaphor, however, which is the image and which the reality conveyed through the image? To put it another way, which is the vehicle and which the tenor?[51] Conventionally, metaphor is used to signify abstract truth through figurative language: a vivid image is invented to take the place of some reality that cannot be expressed literally. But in his use of the metaphor of vision, Dante takes the qualities of the vehicle and assigns them to the tenor: the vehicle is real experience, and the tenor is a vivid, poetic image. This is, I think, the first example of a technique Dante employs throughout the *Commedia* in which inexpressible truth is conveyed, not through a beautiful fiction, but through real experience, what is demonstrably true. Dante does this most evidently in the *Commedia* on a large scale, where he uses real people and known history to describe the fate of souls after death. But he also does this on a smaller scale, employing extended metaphors and similes in which both vehicle and tenor are true. In the later parts of the *Commedia*, Dante goes even farther, at times making it difficult to tell which is the tenor and which the vehicle, at times making tenor and vehicle apparently identical.

Dante concludes his 'disgressione' by restating the fundamental problem with vision: its inability to mediate perfectly between subject and object: 'E così appaiono molte cagioni, per le ragioni notate, per che la stella puote parere non com'ella è' ['It is evident, then, from the explanations given that there are many reasons why a star can appear other than it actually is' (3.9.16)]. Having fully elaborated on the vehicle, Dante goes on in the next chapter to explain the tenor, that is, how sight of the lady injured his mind's eye. He states that she seemed 'disdegnoso e fero' (3.10.3), not because she was truly 'disdainful and harsh,' but rather 'per infertade de l'anima, che di troppo disio era passionata' ['on account of a sickness of the soul which was impassioned by excessive desire' (3.10.1)]. Dante explains that such a condition is a failure of reason, causing one's judgment to be that of an animal rather than a human being (3.10.2). Seemingly, Dante is describing a purely emotional disturbance, his passion causing him to misinterpret the lady's behaviour as haughty.[52] In fact, Dante is carrying on the metaphor of vision all the time, in great detail and with great precision. Albertus Magnus describes exactly what happens to the sense of vision when a person experiences great passion: 'vision is obscured in angry persons, for anger causes blood to gather about the heart, and causes vapors to rise up to the head.'[53] In other words, the visual spirit is physically altered by emotions such as anger, fear, and even love. Thus Dante is at once describing a disturbance of the figurative visual spirit (that is, the spirit governing the eye of the mind) and of the actual visual spirit. Both occurrences really took place, as Dante assures us by describing the eyestrain he experienced 'in the same year as this canzone was conceived' (3.9.15); therefore, the metaphor has no fictional component, no beautiful lie clothing the truth.

Dante carries the optical metaphor even further, based on the image of the *pupilla*. He emphasizes the importance of that term by repeatedly referring to the pupil when he might just as accurately have referred to the eye in general (3.9.8–10). He does this in the ninth chapter so that, in the tenth, the reader will recall the term and consider the role of the pupil in determining the lady's 'apparenza' (3.10.1). Albertus Magnus states that several internal causes can disturb vision: for example, if the brain is unusually warm due to a disturbance of the emotions, the visual spirit generated in the brain becomes turbulent. When that spirit is sent to the eye by way of the optic nerve, its turbulence impairs the transmission of the visible form.[54] But vision can also be disturbed by the size of the hole in the eye, that is, the pupil: 'Indeed, when it [the *uvea* or iris] is pushed inward and goes toward the interior of the eye, so that opening is narrowed together, then things are seen smaller than they are in reality. At other times it goes toward the exterior, so that the opening is dilated, and things are seen larger than they are.'[55] It was well known, then as now, that fear causes the eyes to be dilated; thus the fear that Dante says afflicted him ['l'anima temea'] affects his vision not only by disturbing and clouding the visual spirit, but by changing the size of the *pupilla*. When the hole becomes larger, so does the little doll reflected on its surface as in a mirror. The lady seems grand and overwhelming only because the image entering the lover's eye is larger than it would be if it were transmitted accurately.

Dante places his digression on vision in the ninth chapter of the third book not only because the position is numerically significant, as I have indicated above, but also because it serves to link Dante's *Convivio* to Boethius's *De consolatione Philosophiae*. The ninth chapter of the third book of the *Convivio* is the digression on vision; the ninth chapter of the third book of the *De consolatione Philosophiae* is the famous summary of Neoplatonic cosmology, 'O qui perpetua mundum ratione gubernas.'[56] In many respects, Dante's use of vision as a metaphor for understanding in the *Convivio* as in his other works is drawn from the Neoplatonic tradition most widely disseminated through Boethius's work.[57] It is nothing new to note that Dante specifically models his *Convivio* after the *De consolatione Philosophiae*, both in adopting the *prosimetrum* style and in making the figure of Filosofia the focus of the writer's attention. It is new, however, to observe that Dante's explanation of why the lady's appearance differs from her actual being is based on Boethius's representation of Philosophia. Like Dante, Boethius indicates that the sight of Philosophia can be affected by the state of the viewer's eyes: 'my sight [lit. 'pupil'] was so dimmed with tears that I could not clearly see who this woman was.'[58] Yet the state of the viewer's eyes can not only make the visible form appear cloudy, but also distort its size: 'It was difficult to say how tall she might be, for at one time she seemed to confine herself to the ordinary measure of man, and at another the crown of her head touched the heavens; and when she lifted her head higher yet, she penetrated the heavens themselves, and was lost to the sight of men.'[59] Boethius's description of Philosophia's appearance is often interpreted as meaning that philosophy is in some respects easily accessible to the human intellect, in others quite remote from mortal understanding. But her variable appearance can also be indicative of the state of the person gazing upon her. When Boethius sits weeping in his cell, she appears to be 'mulier tam imperiosae auctoritatis,' a 'woman of ... commanding authority' (1.pr.1.45–6); when Dante is afflicted by fear, she seems 'fiero et disdegnosa,' 'harsh and disdainful.' Boethius provides Dante with a model not only for the weakness of the narrator's eyes, but for the strength of Philosophia's gaze: 'oculis ardentibus et ultra communem hominum valentiam' ['her burning eyes penetrated more deeply than those of ordinary men' (1.pr.1.4–5)]. Dante similarly stresses the power of Filosofia's gaze; but by doing so he seems to violate the theory of vision he so carefully expounded in the digression. In the *canzone*, Dante declares that his soul is still fearful, 'sì che mi par fero / quantunqu'io veggio là 'v'ella mi senta' ['so that whatever I see when this lady perceives me seems harsh' (3.canz.2.85–6)]. The lady's gaze is the source of the lover's fear, which in turn distorts the image received by his eye.

According to the theory of vision Dante offers in 3.9, however, there is no extramission: no beam is ever projected from the eye, whether from the eye of the lover or from the eye of the lady. Thus in the last lines of Dante's exposition of the literal level of the *canzone*, he seems to violate the theory of vision he has just elaborately presented: he reverses the direction of the act of vision so that the flow of forms from object to subject (intromission) becomes the emanation of a beam from subject to object (extramission). But there is no contradiction if the lady's eyes are not body but light. Dante says that the rays emitted by the lady 'come se fusse stato diafano, così per ogni lato mi passava' ['passed through me on all sides, as if I were something diaphanous' (3.10.4)]. Thus the lady's eyes become the sun, and Dante himself becomes the diaphanous medium; her light passes through him and illuminates him completely, just as the sun's light illuminates air. Dante is no longer located at the end point of the 'linea recta' of the visual ray, as he was in the *Vita nuova*; instead, located in the centre, he is the medium of light. The poet's new role as medium is reflected in the structure of the chapter in which he describes that role. Dante signals that 3.10 is a transitional chapter, mediating between the digression on vision and the allegorical exposition to follow:

> Partendomi da questa disgressione, che mestiere è stata a vedere la veritade ...

> Having finished with this digression, which was necessary to bring out the truth [lit. 'to make the truth visible'] ... (3.10.1)

> ... l'ordine de l'opera domanda a l'allegorica esposizione omai, seguendo la veritade, procedere.

> ... the plan of this work requires that I now proceed to the allegorical exposition, in which the truth will be made plain. (3.10.10)

The quotations above are the first and last words of this chapter. Note that the first word looks back to what came before ['Partendomi'], while the last word looks forward to what will come next ['procedere']. In one sense, what came before and what will come next are different, for one is part of the literal exposition and one is part of the allegorical; but in another sense, they are similar, for both have 'veritade' as their object. But even in this they differ: the literal exposition makes truth visible, while the allegorical only follows after the truth ['seguendo'], perpetually seeking understanding. In keeping with this deferral of truth, Dante does not openly explain the meaning of his declaration that the lady's eyes passed through him as through he were diaphanous. Instead, he promises that the phenomenon could be explained, 'ma basti qui tanto avere detto; altrove ragionerò più convenevolemente' ['but here let it suffice to have said this much – I shall discuss the subject later at a more appropriate point' (3.10.4)]. Needless to say, he does not return to the subject. This deferral is a restatement of Dante's vow at the end of the *Vita nuova* to write no more of Beatrice until he can 'più degnamente trattare di lei'; in the face of a mystery, it is better to keep silent.

Chapter Six

DANTE'S *COMMEDIA*

In his continuation of the *Roman de la rose*, Jean de Meun includes a long digression in which Nature describes the various images generated in mirrors. The reproduction of the visible species in the mirror proves to be a model for the reproduction of the human species, that is, the lover's insemination of the rose, which is itself a metaphor for the poet's act of creation. In the *Commedia*, Dante too includes a digression on the reflective properties of mirrors, and like Jean he places the lecture on optics in the mouth of a woman modelled after Boethius's Philosophy. But while Jean uses the mirror's capacity to reflect as a metaphor for physical reproduction, Dante uses it as a metaphor for the reflection of divine light by human souls.[1] In the first sphere of paradise, Dante passes through the diaphanous moon. His description of the planet suggests that it is neither material nor immaterial:

> Parev' a me che nube ne coprisse
> lucida, spessa, solida, e pulita,
> quasi adamante che lo sol ferisse.
> Per entro sé l'etterna margarita
> ne ricevette, com' acqua recepe
> raggio di luce permanendo unita.

It seemed to me that a cloud had enveloped us, shining, dense, solid and polished, like a diamond smitten by the sun. Within itself the eternal pearl received us, as water receives a ray of light, itself remaining uncleft. (Pa.2.31–6)[2]

Insofar as it is like a cloud, the moon is insubstantial; insofar as it is like a diamond or pearl, it is substantial. It is unclear, too, precisely how light acts upon the moon: does it refract light, like a diamond, or reflect it diffusely, like a pearl? Or does it do both at the same time? The ambiguity indicates the narrator's uncertainty about the phenomena he experiences, and prepares the reader to welcome the instruction Beatrice will offer shortly.

In response to Dante's query, Beatrice explains why the moon's surface is unevenly illuminated by the light of the sun. Dante supposes that what appears 'diverso' is caused by 'corpi rari e densi' [rare and dense matter' (Par.2.59–60)], that is, the sun's light is reflected differently depending on whether a given point on the moon's surface is glassy or rough. But Beatrice refutes this explanation by noting that rare matter is diaphanous; therefore, if Dante's hypothesis were true, when a solar eclipse occurs, the sun's light would pass through those areas of the moon. Beatrice further demonstrates that the mottled appearance of the moon's surface could not be caused by areas of rare matter extending only to a limited depth, thus reflecting light from a greater depth than does the dense matter on the surface. She proves this using a model drawn from experimental optics:

> Da questa instanza può deliberarti
> esperïenza, se già mai la provi,
> ch'esser suol fonte ai rivi di vostr' arti.
> Tre specchi prenderai; e i due rimovi
> da te d'un modo, e l'altro, più rimosso,
> tr'ambo li primi li occhi tuoi ritrovi.
> Rivolto ad essi, fa che dopo il dosso
> ti stea un lume che i tre specchi accenda
> e torni a te da tutti ripercosso.
> Ben che nel quanto tanto non si stenda
> la vista più lontana, lì vedrai
> come convien ch'igualmente risplenda.

From this objection experiment, which is wont to be the fountain to the streams of your arts, may deliver you, if ever you try it. You shall take three mirrors, and set two of them equally remote from you, and let the other, even more remote, meet your eyes between the first two. Turning toward them, cause a light to be placed behind your back which may shine in the three mirrors and return to you reflected from all three. Although the more distant image may not reach you in so great a quantity, you will there see it must needs be of equal brightness with the others. (Pa.2.94–105)

If the light source is, for example, the flame of a candle, the more distant mirror will reflect a smaller image of that flame than those nearer; but the brightness of the light reflected will not be diminished. This is not intuitively obvious because the total illumination of the more distant mirror will indeed be less than that of the nearer mirrors, because the illuminated area (the image of the flame) is smaller.[3] If, however, a large light source and small mirrors are used, so that the entire surface of each mirror is covered by the reflected image of the light source, the phenomenon is more readily apparent. Since the diminished size of the image in the more distant mirror is not evident, the total illumination offered by each mirror is identical. The experiment shows that, though the visible form is altered by being transmitted through the medium, the intensity of the light is not similarly altered. The implications are enormous: while the visible species degenerates as a result of its multiplication through the diaphanous, light is invulnerable.

Beatrice's explanation is not only an accurate account of physical science; it is also a metaphor for the subject of the poem, which is, as Dante states in the *Epistle to Can Grande*, 'the state of souls after death.'[4] The mirrors are a model for souls in a state of grace: just as mirrors reflect the physical sun, they reflect the divine Sun. But here the metaphor becomes inadequate, for souls do not reflect light the way a simple mirror does, but rather the way the moon does: they reflect light more or less brightly, depending upon how God's 'bontate' is alloyed with the potential goodness of the person. Then, Beatrice says, 'la virtù mista per lo corpo luce / come letizia per pupilla viva' ['the mingled virtue shines through the body, as gladness does through a living pupil' (Pa.2.143–4)]. The souls reflect divine light, but they do so not like a planar mirror, but rather like a diaphanous body.

Here Beatrice distinguishes between the two different states of matter that cause reflection. The first occurs when a transparent medium is backed with something opaque, such as when, as Beatrice puts it, 'color torna per vetro / lo qual di retro a sé piombo nasconde' ['color returns from the glass that hides lead behind itself' (Pa.2.89–90)]. This is how the mirrors in Beatrice's experiment reflect light, and how Dante formerly thought the moon reflected light. The second occurs when a transparent medium is entirely filled with light, causing it to reflect light from its surface. This is how the moon, according to Beatrice, actually reflects light; it is also how reflection occurs in transparent stones, in clouds, and in water. It is in the context of these two different types of reflection that Beatrice's simile of the pupil must be understood: God's 'bontate,' mingled with the potential for good in the human soul, shines 'come letizia per pupilla viva' ['as gladness does through a living pupil' (Pa.2.144)]. Dante carefully chooses the word 'pupilla,' just as had done in the *Convivio* (3.9), in order to play on the double meaning of the term: it is both the aperture of the eye and the 'little doll' that appears on the surface of the viewer's eye. But when the soul gazes upon God, there is no 'little doll'; the pupil reflects only divine light.

Beatrice uses the same term to describe the generation of God's 'bontate' that Grosseteste and the perspectivists Bacon, Pecham, and Witelo use to describe the generation of species: it is '*multiplicata* per le stelle' ['multiplied through the stars' (Pa.2.137)]. More specifically, the species (as it were) of God's 'bontate' is multiplied like the species of light, for as Beatrice just demonstrated experimentally, unlike the visible species, the species of light is not diminished by its passage through the medium. Some readers have suggested that Dante derives this analogy from what is often called the 'metaphysics of light' of Robert Grosseteste, or from the related theory of light offered by ibn Gabirol (Avicebron).[5] As Parronchi recognized, however, it was by no means necessary for Dante to go to these sources for the basis of the analogy, for perspectivist writings also emphasize the special role of the species of light as the paradigm for the multiplication of sensible species. In addition, unlike Grosseteste, the perspectivists offer detailed examinations of the role of mirrors and lenses in the reflection and refraction of light. Dante seems to have some knowledge of perspectivist optics; but the fact that he never uses the term species suggests that he may not have been acquainted with the writings of Bacon or Pecham (or, for that matter, Grosseteste), who use the term constantly. The only one of the medieval perspectivists who does not use the term species is Witelo, who wrote in Italy near the end of the thirteenth century. In avoiding the term, Witelo resembles Albertus Magnus, who was an important source for Dante's knowledge of the various competing optical theories circulating prior to the perspectivist synthesis, and who used the term species only infrequently in his writings on optics.

In a prefatory letter to his *Perspectiva* addressed to his friend William of Moerbeke, Witelo describes a 'metaphysics of light' almost as mystically as did Grosseteste himself:

Light is the diffusion of the supreme corporeal forms, applying itself through the nature of corporeal form to the matter of inferior bodies and impressing the descended forms of the divine and indivisible artificers along with itself on perishable bodies in a divisible manner, and ever producing by its incorporation in them new specific or individual forms, in which there results through the actuality of light the divine formation of both the moved orbs and the moving powers. Therefore because light has the actuality of corporeal form it makes itself equal to the corporeal dimensions of the bodies into which it flows and extends itself to the limits of capacious bodies, and nonetheless since it always contemplates the source from which it flows according to the origin of its power, it assumes *per accidens* the dimension of distance, which is a straight line, and thus it acquires the name 'ray.'[6]

Witelo's emphasis on the corporeal nature of light and his description of how light fills bodies is consistent with Dante's use of light as a metaphor for God's 'bontate' in the *Paradiso*. Further, Witelo (like Dante) associates the emanation of light with the act of creation; and Witelo's choice of terminology, particularly his avoidance of the term species and his frequent use of the term ray (*radius*), corresponds with Dante's use of optical terms in the *Commedia*. Finally, Witelo bases his account of reflection in the brief summary which prefaces his *Perspectiva* on the reflective properties of the moon. Like Beatrice, he characterizes the moon as a body which reflects the rays of the sun more or less brightly depending upon the rarity of the moon itself ('raritatem lunaris corporis').[7]

Immediately after she refutes Dante's hypothesis regarding the cause of the mottled surface of the moon, Beatrice uses a metaphor based on light to prepare Dante for her description of how divine goodness is disseminated throughout the universe:

> Or, come ai colpi de li caldi rai
> de la neve riman nudo il suggetto
> e dal colore e dal freddo primai,
> così rimaso te ne l'intelletto
> voglio informar di luce sì vivace,
> che ti tremolerà nel suo aspetto.

Now – as beneath the blows of the warm rays the substrate of the snow is left stripped both of the color and the coldness which it had – you, left thus stripped in your intellect, will I inform with light so living that it shall quiver as you look on it. (Pa.2.106–11)

Beatrice informs Dante in two ways: she imparts information to him, and she recreates him, imposing form on his notion of the nature of the universe. In disabusing Dante of his wrong supposition regarding the moon, Beatrice dis-informed him, that is, she removed an idea (literally, a form) from his intellect; similarly, the sun removes form from snow, making it into water. Then, by informing him correctly, Beatrice illuminates Dante: like the water that was previously opaque snow, he is now a diaphanous medium. In being filled with this living light, Dante fulfills the potential hinted at in the *Convivio*, where Filosofia's eyes beamed rays so powerful that they passed through the lover as though he were 'diafano' (3.10.4). But Dante is not the only soul filled with living light; on the contrary, all souls in a state of grace are filled similarly, as Beatrice indicates in her experiment of the three mirrors. Filled with light, they reflect that light to a greater or lesser extent depending on the alloy of their potential for goodness with the active Goodness they receive, just as the moon reflects the light of the sun. Beatrice had anticipated the revelation of this enigma early in her explanation of the moon's properties: if the moon did have rare and dense matter, she says, 'come comparte / lo grasso e 'l magro un corpo, così questo / nel suo volume cangerebbe carte' ['just as fat and lean are distributed in a body, it would alternate the pages in its volume' (Pa.2.76–8)]. By likening the moon to a living body, Beatrice anticipates the enigma she will unveil, making Dante himself a vessel of living light at the very moment when she explains how all souls in a state of grace are such vessels. Simultaneously, by likening the moon to a book, Beatrice makes the *Commedia* itself a vessel of living light at the very moment the reader understands the text: Beatrice, Dante, text, and reader become a series of mirrors all reflecting the same divine light.

Even before she appears to Dante in purgatory, Beatrice is the medium that links him to God. In the second canto of the *Commedia*, Virgil tells Dante how he was sent to lead him on his journey: a 'gracious lady,' whom the reader will recognize as Mary, became aware of Dante's plight and called Lucy; Lucy, in turn, informed Beatrice, who then went to Limbo to ask Virgil to be Dante's guide. While Lucy goes to Beatrice ['si mosse' (Inf.2.101)], and Beatrice goes to Virgil (2.71), and Virgil goes to Dante (2.50), Mary alone does not move; instead, she calls Lucy to her ['chiese Lucia in suo dimando' (2.97)], indicating that Mary is the origin of the impulse that is transmitted from her to Lucy to Beatrice to Virgil to Dante. In this context, Mary is identified with love, for Beatrice tells Virgil that 'amor mi mosse' ['Love moved me' (2.72)]. This love is transmitted through their several souls just as light is transmitted through the multiplication of its form, the phenomenon described by Beatrice in the *Paradiso* (Pa.2.137).[8]

Beatrice is more than a single link in a causal chain linking Dante to God: she is *the* link. Christ is the universal mediator between man and God; but for the individual, Dante suggests, there can also be a kind of personal mediation, one suited particularly to the individual who approaches God. For Dante, the figure in the *Commedia* who by synecdoche represents all of humanity, the mediator is Beatrice. It is in this context that Virgil's superficially perplexing comment about Beatrice's role in salvation can be understood. He calls her 'donna di virtù, sola per cui / l'umana spezie eccede ogne contento / di quel ciel c'ha minor li cerchi sui' ['Lady of virtue, through whom alone mankind rises beyond all that is contained by the heaven that circles least' (Inf.2.76–8)]. Virgil seems to be implying that Beatrice is the universal mediator, which cannot be the case since that is uniquely the role of Christ. But the remark can be understood differently if we consider the double meaning of 'spezie': it is the 'species,' in the modern sense, only secondarily; primarily it is the form of humanity, the prototype or exemplar that stands for the whole group. Insofar as Dante is Everyman, he is 'l'umana spezie,' the part that stands for the whole.

The other person who stands for the human race is, of course, Adam. He is the prototype for all humanity, and so most of the half-dozen passages in the *Commedia* in which Dante uses the term 'spezie' refer to 'quell' uom che non nacque' ['that man who never was born' (Pa.7.26)]. In the *Inferno*, Dante uses the term only twice, once in reference to Beatrice's role in salvation (2.77), and once to refer to the damned souls who curse 'Dio e lor parenti, / l'umana spezie e 'l loco e 'l tempo e 'l seme / di lor semenza e di lor nascimenti' ['God, their parents, the human {species}, the place, the time, the seed of their begetting and of their birth' (3.103–5)]. The souls curse both their own individual origins and their collective origin: both the time and place of each person's creation, and the creation of 'l'umana spezie,' Adam, in Eden. Dante's use of the term in the *Paradiso* is similarly both general and particular: when Beatrice says that heaven is the 'loco / fatto per proprio de l'umana spece' ['place made for humankind as its proper abode' (Pa.1.56–7)], she means that heaven is the home of all humanity, and of the first human. Dante will discover at first hand that this is true when he meets ''l patre per lo cui ardito gusto / l'umana specie tanto amaro gusta' ['that Father because of whose audacious tasting the human race tastes such bitterness' (Pa.32.122–3)]. Because Adam is the primary expression of the species of humanity, every subsequent expression of the species is affected by his experience: therefore, what Adam tastes, we all taste. Even here, the only place in the *Commedia* that Dante uses the word 'spezie' solely in the collective sense, referring to all human beings, he emphasizes the relationship of part to whole.

At another point in the *Paradiso*, Dante uses the word 'spezie' polysemously. It refers to a category, all human beings; to an idea, the impression of form that makes one human; and to the first expression of that form, Adam:

> [Q]uell' uom che non nacque,
> dannando sé, dannò tutta sua prole;
> onde l'umana specie inferma giacque
> giù per secoli molti in grande errore,
> fin ch'al Verbo di Dio discender piacque
> u' la natura, che dal suo fattore
> s'era allungata, unì a sé in persona
> con l'atto sol del suo etterno amore.

[T]hat man who never was born, in damning himself damned all his progeny; wherefore the human [specie] lay sick down there for many centuries in great error, until it pleased the Word of God to descend where He, by the sole act of His eternal love, united with Himself in person the nature which had estranged itself from its Maker. (Pa.7.26–33)

As a result of Adam's transgression, both the individual human species (Adam himself) and the collective human species (his descendants) spent centuries in hell. Adam's creation is mirrored in the Incarnation, when Christ united the human form with the divine 'in persona,' that is, in His own person. Yet the redemption of humanity is not complete in the Incarnation; at the Crucifixion, Christ descended in order to remove souls from hell and carry them to paradise. In the Incarnation, Christ embraces human form figuratively, becoming a second Adam; in the harrowing of hell, He embraces it literally when he touches Adam.[9] Both the Incarnation and the Crucifixion occurred due to 'l'atto sol del suo etterno amore' ['the sole act of His eternal love']. By referring to this 'atto *sol*,' Dante recalls Virgil's initial greeting of Beatrice '[donna di virtù, *sola* per cui ...' (Inf.2.76)]. In this context 'virtù' is love, and it moves both Christ and Beatrice. The human species is redeemed collectively by Christ; but it can also be redeemed individually by blessed souls who mirror Christ. Beatrice is one such.[10]

Because Beatrice is such a powerful mediator, many readers are tempted to identify her with one or another quality, whether it be Love, Theology, or Wisdom; in other words, they see her as a personification analogous to Boethius's Philosophia. It is clear that Dante intends us to make such an identification, for the Beatrice of the *Commedia* (as distinct from the Beatrice of the *Vita nuova*) in many respects resembles the lady of the *Convivio*, Filosofia. At no time does Beatrice resemble a traditional personification more than when she lectures on the reflective properties of the moon, recalling not only Philosophia's instruction of Boethius but also Nature's lecture on optics in the *Roman de la rose*. Yet Beatrice is more than a personification: she does not always represent the same thing (Love, Theology, Wisdom), but rather reflects whatever quality is rayed upon her by God. She is a mirror to reflect divine light, a vessel to be filled. In this sense, Beatrice most resembles not Raison or Nature but rather the rose in the *Roman de la rose*. In that poem, the rose is a reflection of the lover's desire: in Guillaume's portion of the work, that desire is the narcissistic love of self, projected upon an only apparently external object; in Jean's continuation, that desire is the carnal love of another, physically expressed through penetration and insemination. But in the *Commedia*, Beatrice is a reflection of God's desire for man. Dante believes that he loves Beatrice; but he later discovers that what he has really loved all along is the divine as it appears in her. Beatrice is a fluid symbol like the rose, rather than a fixed personification like Nature or Philosophia. But at the same time she is undoubtedly very different from the rose, for she is in some sense a real woman, Beatrice Portinari whom Dante has loved (as he tells us in the *Vita nuova*) since they were children.[11]

Beatrice takes the place of traditional personifications such as Wisdom, Contemplation, Philosophy, and Nature; but she is not alone in doing so. At times Beatrice represents Contemplation; but so does Rachel. At times she represents Love; but so does Lucy. Like every other soul, Beatrice always reflects some aspect of the divine, but she does not always reflect the same aspect. It is misleading to say that Dante simply transforms traditional personifications into 'real' female characters;[12] because none of these characters represents a fixed quality, they are more than personifications. In addition, by replacing traditional personifications with historical figures, Dante furthers the effort begun in the

Convivio to create a kind of figurative language in which both vehicle and tenor are true, in which there is no beautiful lie.

Vision in the *Commedia*: Inferno

In the previous chapter, we saw that, in the *Convivio*, Dante illustrates how the viewer's emotional state can cloud his ability to see, both physically and intellectually. He draws upon the intromission theory of vision as described by Albertus Magnus in his *De sensu et sensato* to show how the viewer's turbulent emotional state taints the visual spirit, and hence interferes with the transmission of the image from the object seen to the mind of the viewer. In Dante's *Commedia*, the act of sight is characterized with the same careful detail we observed in the *Convivio*, yet the mechanism of vision varies throughout, ranging from the Platonic extramission model in the *Inferno*, to the atomist intromission model seen in the *Purgatorio*, to the perspectivist model seen in the closing cantos of the *Purgatorio* and in the *Paradiso*.

In the encyclopedic survey of theories of vision found in his *De homine*, Albertus Magnus classifies them according to whether they are based on extramission or intromission. (He again distinguishes them in terms of whether the medium is primarily fire [extramission] or primarily water [intromission] in his *De sensu et sensato*.)[13] Albertus associates extramission with fire, because of the fiery visual beam projected from the eye, and intromission with water, because of the role of the diaphanous.[14] According to most intromission theories, the diaphanous medium can transmit form only because it is humid: thus the diaphanous medium is always a substance with some degree of moisture in it, whether water, air, cloud, or stone. In addition, in his own version of the intromission theory, Albertus emphasizes the role of the watery humour within the eye; though Galen had been the first to stress this point, his observations were incorporated into the intromission theories of later proponents of the Aristotelian model, including Avicenna and, of course, Albertus himself.

Dante follows Albertus in classifying the two fundamental scientific models of vision in terms of fire and water. Thus hell is illuminated only by dim fires, compelling Dante and Virgil to find their way by means of the fiery beam projected from their eyes. In order to traverse purgatory, Dante's eyes must be bathed in the rivers that flow there both at the beginning, before he enters the gate of purgatory proper, and at the end, before he leaves the earthly paradise. In the *Purgatorio*, Dante plays on the intromission model by, as it were, magnifying the narrator's experience: Dante himself becomes the species journeying through the diaphanous medium, a medium made active by the presence of Virgil's (and, later, Beatrice's) light. Finally, in the *Paradiso,* the terminology of perspectivist optics comes to the fore: Dante emphasizes the role of the visual ray and how it is changed by being reflected or refracted. Dante synthesizes the fire of extramission and the water of intromission in the 'miro gurge,' the stream that casts off droplets of water that are simultanously fiery sparks, and in the final vision of God as a double rainbow bound by a fiery belt.

At the beginning of the first canto of the *Inferno,* Dante alludes to the power of the sun to guide those who are lost: he sees the 'raggi del pianeta / che mena dritto altrui per ogne calle' ['rays of the planet that leads men aright by every path' (Inf.1.17–18)]. On a literal level, this means that the sun helps travellers to find the east or west, and thus know what direction they should travel in; on an allegorical level, it means that God helps the soul find its way to him. In the *Convivio,* Dante refers to the 'sole spirituale e intelligibile, che è Iddio' ['spiritual and intelligible sun, that is, God' (Conv.3.12.6)] and, similarly, in the *Purgatorio* refers to God as 'l'alto Sol' (Pg.7.26). Hell, however, is a place where the light of the sun is extinguished. Dante emphasizes its darkness, noting the 'fioco lume' ['dim light' (Inf.3.75)]. The flickering fires in the lighthouse are only signals to call Phlegyas; they offer no illumination (8.4–6). Even the inscription at the entrance of hell is of 'colore oscuro' (3.10); that is, since there is little light, the colour is difficult to make out. The abyss below is so 'oscura e profonda' that, Dante says, 'io non vi discernea alcuna cosa' ['I could make out nothing there' (4.10, 12)]. The residents of hell see no better for, deprived of God's light, they have only the light of reason. Thus Limbo, where Aristotle and Plato, Homer and Euclid, are gathered, has 'un foco / ch'emisperio di tenebre vincia' ['a fire, which overcame a hemisphere of darkness' (4.68–9)].[15]

The majority of those in hell do not have even that dim light, having 'perduto il ben de l'intelletto' ['lost the good of intellect' (3.18)]. Hell is, as Virgil calls it, the 'cieco mondo' (4.13), the blind world. The circle of lust is 'd'ogne luce muto' ['mute of all light' (5.28)], for its inhabitants 'la ragion sommettono al talento' ['subject reason to desire' (5.39)]. In the circle of gluttony, Ciacco's eyes are 'diritti' or straight as long as he speaks to Dante; as soon as Ciacco stops talking, his eyes become 'biechi' or crossed; he then falls down among the other 'ciechi' or blind ones (6.91, 93). In the case of Ciacco, the defective eye of the intellect is mirrored in the defective eye of the aerial body.[16] But even for those inhabitants of hell whose eyes are not distorted, vision is poor. Those in the circle of sodomy must squint to make out the features of the travellers: 'ciascuna / ci riguardava come suol da sera / guardare uno altro sotto nuova luna; e sì ver' noi aguzzavan le ciglia / come 'l vecchio sartor fa ne la cruna' ['each looked at us as men look at one another under a new moon at dusk; and they knit their brows at us as the old tailor does at the eye of his needle' (15.17.21)].

In the circle of the heretics, Farinata describes how it is that the inhabitants of hell are able to see events before they come to pass: 'Noi veggiam, come quei c'ha mala luce, / le cose ... che ne son lontano; / cotanto ancor ne splende il sommo duce' ['Like one who has bad light, we see the things ... which are remote from us: so much does the Supreme Ruler still shine on us' (10.100–2)]. Singleton's translation is misleading here, for the light in hell does not shine upon the inhabitants, but rather illuminates things far away from them. This is why, as Farinata says, 'Quando s'appressano o son [le cose], tutto è vano / nostro intelletto' ['when [the things] draw near, or are, our intelligence is wholly vain' (10.103–4)]. As long as the event is far away, those exiled in hell can see it, for it is illuminated by a distant light; once the event is nearer, their own lack of light makes it impossible to perceive. They have neither the light of the intelligible sun, nor the internally generated light of the intellect.

Dante and Virgil, however, do have some light to guide them on their journey. It is not an external source, but rather a visual beam that comes from within. In keeping with the Platonic notion of vision as an act of physical contact between subject and object, analogous to touch, Dante reaches things with his eye ['con l'occhio attinghe' (18.129)]. Like the English word 'look,' the Italian 'viso' can be taken in either a passive or an active sense: it can mean either how an individual appears to others, or how an individual gazes. Dante sometimes uses 'viso' in the former sense (e.g., 23.83); but he frequently uses it in the latter sense, to describe the ray emanating from the eye (e.g., 4.11; 20.10).[17] Although Virgil's line of vision is not emphasized as much as Dante's (probably because the reader is meant to see from Dante's point of view), he too sees by means of a visual beam. Yet the power of that beam is limited in the face of the darkness of hell: Virgil, halted at the gates of Dis, discovers that 'l'occhio nol potea menare a lunga / per l'aere nero e per la nebbia folta' ['his eye could not lead him far through the dark air and the dense fog' (9.5–6)]. As Dante descends lower into hell, vision becomes almost tangible. The beam projected from the eye is implicitly likened to a sword: Dante's exchange of glances with Venedico Caccianemico is a kind of repartee waged with visual rays,[18] while another soul urges Dante to sharpen his eye ['aguzza ver' me l'occhio' (29.134)]. The sense of sight is said to be satisfied ['sazie' (18.136)], as though it experiences hunger; one can 'fast' by avoiding the sight of something ['vorrebbe di vedere esser digiuno' (28.87)]. Dante's eyes become 'inebrïate' (29.2), drunk with unshed tears. Particularly in the lower parts of hell, the eye is an organ that perceives sensibly rather than intelligibly; the opposite will be true in paradise.[19]

In describing the visual beam or 'viso,' Dante often uses some version of the phrase 'ficcar lo viso' (4.11; also 12.46, 15.26); usually translated 'to peer intently,' the phrase literally means 'to thrust one's gaze.' The phrase implies that the viewer is in a position of control, his gaze penetrating through space to fix the object like a butterfly on a wall. The aggressive, penetrating aspect of the 'viso' is brought out in the canto in which Dante meets his old friend and teacher, Brunetto Latini in the circle of the sodomites. There, flakes of fire fall like snow, burning the flesh of the aerial bodies so severely that Brunetto is recognized only after, as Dante says, 'io ... ficcaï li occhi per lo cotto aspetto' ['I ... fixed my eyes on his scorched face' (15.26)]. John Ahern has argued that Brunetto's mention of the 'dolce fico' ['sweet fig' (15.66)] in his prophecy regarding the future of Florence reflects on his own sin of sodomy.[20] The 'fico' is certainly a phallic image, as Dante reminds us later in the *Inferno* when Vanni Fucci makes an obscene gesture: 'il ladro / le mani alzò con amendue le fiche, / gridando: "Togli, Dio, ch'a te le squadro!" ['the thief raised up his hands with both the figs, crying, "Take them, God, for I aim them at you!"' (25.1–3)].[21] By using the phrase 'ficcare lo viso,' proximate to Brunetto's reference to the 'dolce fico,' Dante underlines the piercing, even phallic aspect of the gaze.

Brunetto's sin is one of self-imposed sterility: he chooses not to 'be fruitful' in a physical sense, and not to be fruitful in a literary sense, writing in French instead of his native language.[22] In his *Testament,* Jean de Meun showed that one could produce literary progeny, followers who would write in the tradition he had established; by writing outside his own vernacular, Brunetto chooses not to establish a Italian literary lineage.[23] Because Brunetto chose to write in French, and incorporated parts of the *Roman de la rose* in his own *Tresor,* it is striking that, when they meet, Dante includes metaphors of sterility found in the *Rose.*[24] Dante declares, 'giri Fortuna la sua rota / come le piace, e 'l villan la sua marra' ['let Fortune whirl her wheel as pleases her, and the yokel his mattock' (15.95–6)]. In his note to the line, Singleton comments that the phrase 'seems to echo some unknown proverb'; in fact, it echoes the closing lines of Guillaume de Lorris's portion of the *Roman de la rose*. In that passage, the unhappy lover likens himself to a 'paisant' who loses his crop to bad weather; just as the peasant loses his 'esperance,' so the lover laments, 'Je criens ausi avoir perdue ... m'esperance' (*Rose* 3932–44).[25] Both losses, he continues, are caused by Fortune and her 'roe qui torne,' her turning wheel (*Rose* 3958).

The connection between the *Roman de la rose* and the canto concerning Brunetto Latini goes beyond a common concern with the role of Fortune in human expectation. Guillaume's peasant is a figure of misdirected fertility, for he 'giete a terre sa semance' (3933): in one sense, he performs a hopeful act, sowing seeds of wheat; in another sense, he sins, casting his own seed upon the ground.[26] The sodomite commits the same sin; as Alanus de Insulis puts it, 'He imprints on no matter the stamp of a parent-stem: rather his ploughshare scores a barren strand.'[27] Dante further underscores the sodomites' sterility by placing them on a sandy plain (15.117; 16.40), a surface on which nothing can grow made up of grains that can generate no fruit.[28] Sterility is metaphorically expressed in the *Fiore*, tentatively attributed to Dante, in just the same way: in the passage which corresponds to Guillaume's description of the peasant's fruitless sowing, the *Fiore*'s narrator inhabits not a barren field but a desolate, sandy beach.[29] The barrenness of the sodomites is highlighted by the excessive fertility of the usurers, who are consigned to the same circle: Dante emphasizes that each usurer carries a 'borsa' (17.59). The purse filled with coins is a common euphemism for the testicles, for each repository contains material stamped with form that has the potential to reproduce itself. In fact, Jean de Meun has Reson use 'borse' as a euphemism for the testicles (*Rose* 7113), while in the *Testament* he uses it to refer to a purse.[30] Thus in this circle of hell, Dante recalls both extremes of the *Roman de la rose*: the sterility of the first part of the *Rose* appears in his account of the sodomites, while the fertility of the second part appears in his account of the usurers.

Deeper in hell, in the circle of the schismatics, Dante encounters another poet: Bertran de Born, the Provençal poet who sowed discord between father and son. In this canto, Dante stresses the role of vision, both his own and that of the souls he meets there. When he sees Mohammed, Dante reflects the violence wrought by the schismatics in his gaze, 'attacking' him with his visual beam: 'tutto in lui veder m'attacco'

(28.28). But this attack is at the same time a means of communication through the mutual gaze, for Mohammed returns his look ['guardommi' (28.29)] and invites Dante to look at him in return: 'Or vedi com' io mi dilacco! / vedi come storpiato è Mãometto!' ['Now see how I rend myself, see how mangled is Mohammed!' (28.30–1)].

Dante's power of vision continues to be stressed throughout the canto, especially as he approaches his meeting with Bertran de Born. He introduces the terrible sight of Bertran's body by declaring, 'Io vidi certo, e ancor par ch'io 'l veggia' ['Truly I saw, and seem to see it still' (28.118; cf. 'riguardar,' 28.112; 'vidi,' 28.113)]. Dante saw it with the eye of the body; he continues to see it with the eye of the mind:

> Io vidi ...
> un busto sanza capo ...
>
> ...
>
> e'l capo tronco tenea per le chiome,
> pesol con mano a guisa di lanterna:
> e quel mirava noi e dicea: 'Oh me!'
> Di sé facea a sé stesso lucerna,
> ed eran due in uno e uno in due.

I saw ... a trunk without the head ... and it was holding the severed head by the hair, swinging it in hand like a lantern, and it was gazing at us and saying: 'O me!' Of itself it was making a lamp of itself, and they were two in one and one in two. (Inf.28.118–25)

By likening the head to a lantern, Dante implies that light shines from it, presumably through the holes in the lantern, that is, the eye sockets. Dante shows that vision functions by extramission in hell in several other passages as well, not only those naming the visual beam or 'viso,' described above, but also those in which he refers to the eyes as 'luci' ['lights' (29.2)]. But by likening the head to a lantern, Dante emphasizes the theory of vision operating here even more, for extramission theories commonly refer to the eyes as 'lucerna' or lamps. Albertus Magnus states that 'in Plato's *Timaeus*, it is written that vision occurs when fire exits from the eye through the permeable tunics of the eye, just as light issuing from a lamp passes through glass or through permeable skin.'[31] In fact, Dante uses this precise analogy elsewhere in the *Inferno*, describing the eyes of two sinners as 'lucerne empie' ['baleful lamps' (25.122)].

In that passage, Dante likens one of those souls as it is transformed to a snail retracting its horns. But the word he uses, 'lumaccia' (25.132), literally means 'little light,' so named because some varieties of snail are luminescent. The snail is significant because it is an example of matter that seems to give off its own light. Proponents of extramission theory say matter can give off light, as in the case of the eyes of animals that apparently can see in the dark, such as cats and wolves; it is because their eyes give off a beam of visual fire, say the Platonists, that these animals can see despite the absence of an external source of light. In his refutation of the Platonic extramission theory, however, Albertus Magnus states that these animals actually receive light from outside that is reflected and multiplied within their own eyes. The glow that seems to come from their eyes is merely reflected light, as is also the case with other bodies that seem to glow, such as 'capita piscium et corpora putrefacta' ['the heads of fish and putrid corpses'].[32] Thus the tiny 'lumaccia' is proof, for the proponents of extramission, that bodies can glow; but it is also part of Albertus's proof that such luminescence is simply the reflection of rays received from an external source. Even while the mechanism of extramission is clearly in motion, its failings as an adequate account of what really happens are just barely apparent.

Vision in the *Commedia*: Purgatorio

In the *Purgatorio*, Dante continues to use seeing as a metaphor for knowing; but the mechanism of vision is increasingly characterized in terms of intromission rather than extramission, as Dante's narrator becomes increasingly aware of the subordination of the individual human subject to the all-encompassing sight of God. In his *De homine*, Albertus Magnus divides optical theories into two kinds: extramission, in which 'the eye is dominated by fire or light,' and intromission, in which 'the eye is dominated by water.'[33] Albertus's distinction between fiery extramission and watery intromission underlies Dante's depiction of vision in the *Purgatorio*. Dante's purgatory is a place of rivers and streams, denoting both the salvific washing away of sin carried out there and the receptive eye described by Albertus. This can be seen on the several occasions when the narrator has occasion to have his sight made clean: in antepurgatory, Cato tells Virgil that he must immediately bathe Dante's 'viso,' 'ché non si converria, l'occhio sorpriso / d'alcuna nebbia, andar dinanzi al primo / ministro, ch'è di quei di paradiso' ['for with eye dimmed by any mist it would not be fitting to go before the first minister of those of Paradise' (Pg.1.95, 97–8)]. Virgil bathes Dante's eyes; but he also bathes his 'viso,' his gaze or visual power. Correspondingly, near the end of purgatory, Dante's visual power is revived ['virtù ravviva' (33.129)] by Matelda. She offers a 'dolce ber' ['sweet draught' (33.138)] from the rivers Lethe and Eunoe, satisfying Dante's thirst for vision – at least for the moment.

In the *Purgatorio,* Dante makes the transition from extramission to intromission gradually, not abruptly. He still casts his 'occhio' across distances, his whole body seeming to move wherever the eye goes as though by synecdoche (4.55–6; 10.25). Although Dante continues to use the verb 'ficcare' to describe the act of seeing, he no longer uses it with the object 'viso,' which we saw in the *Inferno* often referred to the visual beam (13.43, 23.2). Eyes are still called 'luci' ('lights'; 15.84, 31.80); but it is not specified whether their light comes from within or is reflected from an external source. By retaining some of the vocabulary of extramission, Dante leaves a certain ambiguity in the early cantos of the *Purgatorio* regarding what model of vision is being used: there could well be a Platonic beam of fire, but in the light of the sun it would be invisible, since (as Dante informs us in the *Convivio*) a greater light causes a lesser light to become invisible (Conv.3.9.12). Once night falls for the first time in purgatory, however, Dante begins to stress the role of the diaphanous medium:

> [io] vidi un che mirava
> pur me, come conoscer mi volesse.
> Temp'era già che l'aere s'annerava,
> ma non sì che tra li occhi suoi e 'miei
> non dichiarisse ciò che pria serrava.

[I] saw one who was gazing only at me, as if he would recognize me. It was now the time when the air was darkening, yet not so dark that it did not make plain between his eyes and mine what it had shut off before. (Pg.8. 47–51)

As Aristotle originally argued in the *De anima,* and as medieval proponents of intromission continued to repeat, vision cannot take place without the light of the sun to activate the diaphanous medium (the 'aere'). By noting that perception improves when he comes nearer to Nino Visconti, Dante illustrates how the medium can interfere with the transmission of visible form: if the medium is not fully transparent, the less of it that the visible form has to traverse, the better.

In order to illustrate the difference between hell and purgatory, Dante replaces a metaphor of vision based on extramission with one based on intromission. But he does much more than that: he magnifies the act of vision so that the wayfarer himself is the visible form being drawn into the eye of God. Dante had prepared the reader for this conceit at the beginning of the *Inferno* by having Virgil refer to Beatrice as the saviour of 'l'umana spezie' (Inf.2.77), that is, of Dante himself.[34] In addition, Dante's entry into the sphere of the eye in purgatory anticipates the inverted image of the celestial spheres he sees in paradise: there, he discovers that his journey only seems to be through increasingly wide spheres; in fact, he is journeying toward a fixed point at the centre of a sphere (Pa.27.16ff.). It makes sense to imagine entering the realm touched by divinity in terms of entering an eye: God is 'the spiritual and intelligible sun' (Conv.3.12.6) and, just as the sun is an eye in the heavens (Pg.20.132), so is He. He is, in the words of the old enigmatic statement found in the *Book of the Twenty-four Philosophers*, 'an intelligible sphere.'[35] Before describing the details of how Dante represents the individual human as an individual visible form, I must emphasize that I do not mean to suggest that the journey of the wayfarer is simply an allegorical representation of the act of vision according to intromission. On the contrary, placing himself in the position of a species allows Dante to use the act of vision as a metaphor for man's knowledge of God: God is not the object man looks upon, but rather man is the object of God's sight.

On leaving the Valley of Princes and beginning his ascent of the mountain, Dante uses a curious metaphor to describe the narrow opening he must pass through: 'Maggiore aperta molte volte impruna / con una forcatella di sue spine / l'uom de la villa quando l'uva imbruna' ['A bigger opening many a time the man of the farm hedges up with a little forkful of his thorns, when the grape is darkening' (4.19–21)]. The passage is often taken as a 'homely simile,' in keeping with the rustic style of the comic mode.[36] But it is interesting to note that in at least three of Dante's rustic similes – one in each of the three *cantiche* – he uses terms which can be applied figuratively to the act of vision. As noted above, in his encounter with Brunetto Latini, Dante likens Fortune's turning wheel to the whirling mattock of a peasant (Inf.15.95–6), recalling the final lines of Guillaume de Lorris's *Roman de la rose*, an allegory (like Dante's) fundamentally concerned with the relation of vision and love. In the *Paradiso*, Bernard of Clairvaux counsels Dante to pause, 'come buon sartore / che com'elle ha del panno fa la gonna' ['like a good tailor that cuts the garment according to his cloth' (Pa.32.140–1)].

This 'humble simile,' as Singleton calls it, refers to the anatomy of the eye: the 'panno' or membrane forms the 'gonna' or tunic of the eye, its outer circumference.[37] The simile underscores the paradox at the heart of Dante's experience of God: even as he is at the innermost point of heaven, he is at its outermost circumference. Interestingly, the image of the tunic also appears in Dante's encounter with Brunetto: the soul tugs at the hem of Dante's garment to get his attention (Inf.15.24).

The homely simile in the *Purgatorio* also refers to the anatomy of the eye. The 'uva' is both the grape and the *uvea* of the eye, what we now call the iris.[38] The *uvea* serves two purposes: it determines the degree to which the aperture of the eye opens and, according to perspectivist accounts, it darkens the visible species entering the eye so that it can be apprehended by the watery humour within.[39] Thus, 'quando l'uva imbruna,' when the *uvea* darkens the species, vision takes place. Having introduced the optical metaphor by referring to the aperture formed by the *uvea*, Dante immediately relates it to his own situation by saying that, to traverse this gap, he must fly through the air. He can pass through this diaphanous medium only when it is actuated by light:

> ... qui convien ch'om voli;
> dico con l'ale snelle e con le piume
> del gran disio, di retro a quel condotto
> che speranza mi dava e facea lume.

... here a man must fly, I mean with the swift wings and the plumes of great desire, behind that leader, who gave me hope and was a light to me. (Pg.4.27–30)

Virgil is Dante's light (6.29), not only showing him what direction to follow, as he did in the *Inferno*, but now facilitating his passage through the medium to God. When the medium becomes more rarified, Beatrice will have to take Virgil's place for, as Virgil puts it, 'Quanto ragion qui vede, / dir ti poss' io; da indi in là t'aspetta / pur a Beatrice' ['As far as reason sees here I can tell you; beyond that wait only for Beatrice' (Pg.18.46–8)]. She will be 'lume ... tra 'l vero e lo 'ntelletto' (6.45), actuating the diaphanous so that Dante may move from intellect (Virgil) to truth (God). Dante will call her 'luce' (33.115).

Dante is not the only species passing into the eye of God; all souls in purgatory are moving in the same direction, each at a rate corresponding to its own potential. It is for this reason that the souls in ante-purgatory cannot travel at night: like Dante, they must have light in order to move through the medium. Sordello explains to Dante that the reason they cannot travel at night is two-fold: they would lose their direction, and also 'la notturna tenebra ... / ... nonpoder la voglia intriga' ['the nocturnal darkness ... hampers the will with impotence' (7.56–7)]. For the souls, like visible forms, without light there is no motion. Having passed through the aperture demarcated by the *uvea*, Dante passes through the gate of purgatory. Dante is carried to that 'intrata aperta' ['open entrance' (9.62)] by none other than Lucy, whose name means light. When Dante is sleeping, he dreams that a golden eagle carries him through the air; when he awakens, Virgil informs him that it was Lucy who carried him to the gate. Before she left, Virgil says, she showed him the entrance with a glance of her beautiful eyes (9.62–3). The angel guarding the gate holds a sword reflecting brilliant rays (9.82–4), recalling the hand mirror held by Oiseuse, the guardian of the garden gate in the *Roman de la rose*: like her, he is a 'cortese portinaio' ['courteous doorkeeper' (9.92)]. In the *Rose*, the garden wall was the first in a series of enclosures to be penetrated by the lover in his effort to reach the object of his desire; in the *Commedia*, the gate of purgatory serves a similar function, for by entering it Dante finds his way to another garden, the earthly paradise.

From the time Dante enters the gate of purgatory, the role of the visual form or imprint is stressed. He sees a white wall covered with carvings, one of which is

> intagliato in un atto soave,
> che non sembiava imagine che tace.
> Giurato si saria ch'el dicesse 'Ave!';
> perché iv' era imaginata quella
> ch'ad aprir l'alto amor volse la chiave;
> e avea in atto impressa esta favella
> 'Ecce ancilla Deï,' propriamente
> come figura in cera si suggella.

so vividly graven in gentle mien that it seemed not a silent image: one would have sworn that he was saying 'Ave,' for there she was imaged who turned the key to open the supreme love, and these words were imprinted in her attitude: 'Ecce ancilla Dei,' as expressly as a figure is stamped on wax. (Pg.10.38–45)

The emphasis in this passage on the impression of form upon matter serves three purposes: it reinforces the mystery of the Incarnation, in which divine imprint was placed on the matter offered by Mary. In this sense, the Incarnation is a reenactment of the creation of Adam, but with an important difference: here, a human being, Mary, wills that this creation take place. At the same time, the stress on the act of impression also reminds the reader that Dante perceives the 'intaglio' before him by means of forms impressed upon the diaphanous medium, then upon the watery matter of his eye, and ultimately upon his imagination.[40]

Finally, just as God's art makes an impression upon the mind of the wayfarer, so too Dante's art makes an impression upon the mind of the reader. The reader's act of understanding is also an act of vision, both the passive experience of taking in the form of Dante's art, and the active experience of interpreting the allegory. Dante urges the reader to carry out such an act of interpretation when he declares in the *Inferno*: 'mirate la dottrina che s'asconde / sotto 'l velame de li versi strani' (Inf.9.62–3): it is an act of intent vision ('mirate') that allows the reader to penetrate the veil of the strange verses and find the truth concealed within. Similarly, in the *Purgatorio* Dante urges 'Aguzza qui, lettor, ben li occhi al vero, / ché 'l velo è ora ben tanto sottile, / certo che 'l trapassar dentro è leggero' ['Reader, here sharpen well your eyes to the truth, for the veil is now indeed so thin that certainly to pass within is easy' (Pg.8:19–21)]. In purgatory, however, there is a difference. The reader must still gaze intently, but now he can do more than just 'mirate la dottrina'; he can 'trapassar dentro,' pass within the veil to be united with the truth. The reader's access to truth deepens as Dante, in his role as Everyman, 'l'umana spezie,' passes deeper into the eye of God. Just as the souls moving through ante-purgatory (including Dante himself) make progress only as long as the medium is prepared for them by light, so too the reader makes progress dependent on the amount of light he has: the intellectual light he possesses, which includes his knowledge of the hymn Dante mentions but does not reproduce, 'Te lucis ante' (8.13). The souls sing it in order to move forward through purgatory; the reader remembers it in order to move forward, interpretively, through the *Purgatorio*.

Dante includes other aspects of intromission theory in the *Purgatorio*. Medieval theories of vision by intromission such as those of Avicenna and Albertus Magnus incorporate galenic descriptions of the anatomy of the eye and the operation of the visual spirit. In keeping with the intromission model prevalent in the *Purgatorio*, Dante alludes to the role of the visual spirit in his account of those being purged of the sin of envy.

Because their sin, *invidia*, is one of vision, they are punished by having their eyes sewn shut. Dante indicates that their moral deficiency can be expressed as a failure of the visual spirit; when it becomes transparent, they will be able to enjoy the sight of God:

> 'O gente sicura,'
> incominciai, 'di veder l'alto lume
> che 'l disio vostro solo ha in sua cura,
> se tosto grazia resolva le schiume
> di vostra coscïenza sì che chiaro
> per essa scenda de la mente il fiume.'

'O people assured of seeing the light on high which alone is the object of your desire, so may grace soon clear the scum of your conscience that the stream of [la mente] may flow down through it pure.' (Pg.13.85–90)

Singleton translates 'mente' here as memory, and this reading makes sense in terms of what the envious souls will experience when Matelda bathes their eyes in Lethe in the Earthly Paradise at the summit of the mountain. Although the narrator knows that the souls will experience Lethe, the wayfarer does not: thus, for the narrator, 'mente' can mean either 'memory' or 'mind'; but, for the wayfarer, it can mean only the latter. He refers to the stream of visual spirit that, as Albertus Magnus tells us, flows from the front part of the brain, through the optic nerve, into the eye.[41] In the invidious, the visual spirit is filled with 'schiume'; only 'long rest in dark, cold places' and 'the application of clear water' will restore the impaired spirit, a remedy that Dante tells us he himself once was obliged to try (Conv.3.9.16). Similarly, damage to the visual spirit occurs in the circle of wrath, where thick smoke is a 'grosso velo' ['thick veil' (17.4)] making it impossible for anyone to see. The 'fummo' or smoke found there is the vapour that, according to Albertus Magnus, rises up from the angry heart, impairing the transparency of the visual spirit. Here, the internal *mezzo* (the visual spirit) becomes the external *mezzo* (the diaphanous air), a transposition in keeping with the transposition of Dante from the role of viewer to the role of visible form.

As Dante progresses through purgatory, the language of optics becomes more sophisticated. Dante begins to describe the phenomenon of reflection, even specifying the direction of angles of reflection when he is almost blinded by the sight of an angel:

Come quando da l'acqua o da lo specchio
salta lo raggio a l'opposita parte,
salendo su per lo modo parecchio
a quel che scende, e tanto si diparte
dal cader de la pietra in igual tratta,
sì come mostra esperïenza e artc;
così mi parve da luce rifratta
quivi dinanzi a me esser percosso;
per che a fuggir la mia vista fu ratta.

As when the beam leaps from the water or the mirror to the opposite
quarter, rising at the same angle as it descends, and at equal distance
departs as much from the line of the falling stone, even as experiment and
science show; so it seemed to me that I was struck by light reflected there in
front of me, from which my sight was quick to flee. (Pg.15.16–24)

By noting how 'esperïenza' can be a source of knowledge, Dante anticipates Beatrice's experiment in optics, as does Virgil a few lines later. He
attempts to explain to Dante how God's 'bene' ['Good' (15.67)] can fill
so many souls and yet remain undepleted by drawing an analogy with
light that fills transparent bodies ['lucido corpo' (15.69)] and yet continues to multiply itself; when Dante still does not understand, Virgil tells
him to wait until he meets Beatrice, who will satisfy his longing to know.

When Statius explains to Dante how souls and bodies develop, he too
uses optical phenomena to illustrate his meaning. He begins by likening
the astral bodies of the souls in purgatory to images in a mirror ['a lo
specchio vostra image' (25.26)], and concludes by likening them to
rainbows:

E come l'aere, quand'è ben pïorno,
per l'altrui raggio che 'n sé si reflette,
di diversi color diventa addorno;
così l'aere vicin quivi si mette
e in quella forma ch'è in lui suggella
virtüalmente l'alma che ristette.

And as the air, when it is full of moisture, becomes adorned with various
colors by another's rays which are reflected in it, so here the neighboring
air shapes itself in that form which is virtually imprinted on it by the soul
that stopped there. (Pg.25.91–6)

By at first likening the souls to a mirror and then to a three-dimensional form, Statius anticipates Beatrice's explanation of how souls are illuminated by divine 'bontate': she too begins by drawing an analogy with a mirror, and then demonstrates how bodies can be filled with physical or divine light. The astral bodies, like the moon, reflect light not because they are backed with any opaque substance (as a mirror is), but rather because they are filled with light.

Shortly before he enters the Earthly Paradise, Dante experiences a dream during an hour under the influence of Venus (27.94–5). He seems to see a woman gathering flowers, who identifies herself as Leah. In biblical exegesis, Leah and Rachel are often identified as types of the active and contemplative lives. Leah's garland of flowers ['ghirlanda' (27.102)] becomes an emblem of the active life; the mirror in which her sister Rachel gazes at herself all day becomes an emblem of the contemplative life. Yet the dichotomy is not absolute: Leah indicates that she, too, looks in the mirror sometimes, for she adorns herself with flowers 'per piacerme a lo specchio' ['to please me at the glass' (27.103)]. Dante's dream of Leah and Rachel, experienced under the influence of Venus, has its fulfilment when he enters the Earthly Paradise. There, he sees Matelda, whose eyes are more brilliant than those of Venus in love (28.65). She represents the active life for, like Leah, she is gathering flowers. But again like Leah, in the realm of the blessed, those who are active are not deprived of the experience of the contemplation of God. Just as Leah (like Rachel) looks in the mirror, so Matelda (like Beatrice) has eyes so full of light that Dante is grateful when she gives him the gift ['dono' (28.63)] of lifting them to gaze upon him.[42]

It is significant that, as soon as he glimpses Matelda, Dante begins to experience both the light and the heat of the sun: he greets Matelda as a 'bella donna' who, Dante says, 'a' raggi d'amore / ti scaldi' ['do warm youself at love's beams' (28.43–4)]. Dante aligns the two manifestations of the sensible sun – light and heat – with the two manifestations of God – intellectual illumination and love (e.g., Pa.15.77). But, in addition, he aligns this dichotomy with the contemplative and active lives, one dominated by vision, one dominated by the manifestation of love, one epitomized by Rachel and Beatrice, one by Leah and Matelda. Matelda's role in relation to Beatrice is highlighted in the Earthly Paradise when Dante likens her to Proserpine or, as he calls her, 'primavera' (28.51). In the *Vita nuova*, Dante had called a lady 'Primavera' (24.3) in order to identify Beatrice with Christ; he does the same here, so that Matelda heralds Beatrice's advent just as John the Baptist did Jesus'. The figure of Matelda makes inevitable the praise welcoming the advent of Beatrice: 'Benedictus qui venis!' (30.19).

Beatrice is a reflection of Christ; but she is also a reflection of the beloved in Guillaume de Lorris's *Roman de la rose*. I have already noted some similarities between the two allegories, both concerned with the relation of love and vision, and other similarities appear at this point. Like the 'fontaine' in Deduit, this 'fontana' has two streams ['due parti aperta' (28.126)]; but while the two streams in the *Rose* feed the fountain, the fountain of Dante's Earthly Paradise gives forth two streams, underlining the fertility of this garden. Dante will see persons there crowned with roses (29.148), like Deduit and his band; but while those in Deduit's garden dance in an aimless *carole*, those Dante sees move forward purposefully. The Earthly Paradise is Eden, but in some respects it is also the garden of Deduit. Yet there is one very important difference: Guillaume's lover never reaches his beloved; Dante does find Beatrice.

Because the Earthly Paradise on one level is an improved version of Deduit, the place of reflection and refraction,[43] it is appropriate that in this garden Dante introduces a new optical model for the allegory of vision: with the arrival of Beatrice, perspectivist vision begins.[44] The first manifestation of a new model of vision is Dante's failure to make out exactly what he sees approaching him. At first, he believes he sees seven golden trees; but as they come nearer, he discovers that they are really giant candlesticks. Their false appearance ['falsava nel parere'] is caused by 'il lungo tratto / del mezzo ch'era ancor tra noi e loro' ['the long space still intervening between us and them' (29.44–5)]. So far, this is in keeping with the intromission theory found throughout the *Purgatorio*, where a long distance causes any lack of transparency in the medium to be amplified. Dante further stresses the role of the *mezzo* by declaring that the fire above the candlesticks was 'più chiaro assai che luna per sereno / di mezza notte nel suo mezzo mese' ['brighter by far than the moon in a clear midnight sky in her mid-month' (29.53–4)].

As the procession continues to approach, Dante experiences other changes in vision: first reflection, and then refraction. He notices that the smooth surface of the stream on his left reflects his image like a mirror ['come specchio' (29.69)]. Then he discovers that the candlesticks are in fact yet another manifestation of God's art: they are paintbrushes, painting the air with seven bands of colour (29.74–5). Dante likens the pattern both to a rainbow and to the iridescent circle that sometimes appears around the moon (29.78), thus anticipating the double rainbow he will see at the summit of paradise. Yet another example of refraction appears in the procession: the personified gospels are adorned with iridescent peacock feathers, each feather adorned with one of Argus's eyes (29.95). The eyes of Argus, the eyes of the sun and moon (cf. 20.132), and the three eyes of Providentia (29.132) all point toward the moment of vision Dante will experience when Beatrice appears at last, her face appearing through a cloud of flowers like 'la faccia del sol' (30.25) rising through the rosy dawn ['rosata' (30.23)]. While Dante at first believes that Beatrice is in herself a source of light, he soon discovers that she is a mirror.[45] By looking into the 'smeraldi' ['emeralds' (31.116)] of her eyes, Dante is able to see the image or 'idolo' (31.126) of the griffin, whose dual nature makes him a model of Christ, both man and God. Through Beatrice, he is able to see the radiance of Christ 'come in lo specchio il sol' ['as the sun in the mirror' (31.121)]. Though it was his greatest desire, Dante cannot bear the sight of Beatrice's smile, which momentarily blinds him just as the sun blinds the viewer who gazes upon it too long (32.12). He falls asleep, and likens his sleeping self to Argus (31.66), lulled to sleep by the songs and stories of Mercury.[46] Looking at Beatrice, he had been, like Argus, all eye; now he is like Argus asleep and vulnerable.

A 'splendor' awakens him (32.71); this is not just a flash of light, but specifically a flash of reflected light, as Dante makes clear in the *Convivio*: 'It is the custom of philosophers to use the word "light" for luminosity as it exists in its original source; the word "ray" for this as it exists in the medium, lying between the source and the first body which it strikes; and the word "splendour"; for this as it exists when reflected in some other being which has been illuminated'[47] (Conv.3.14.5). Dante's explanation of this optical terminology is meant to gloss his explanation of how God affects the universe: '[He] imparts His power to some things by way of a direct ray, and to others by way of a reflected splendour; specifically, the divine light shines into the Intelligences without any intermediary, and is reflected onto other things from these Intelligences who have first been illuminated. ... [T]he divine power without any intermediary draws this love into similarity with itself' (Conv.3.14.4). Thus the 'splendor' that awakens Dante at the end of the *Purgatorio* signifies Beatrice's role for Dante. Virgil was at first a beam to light the way, and then light to activate the diaphanous medium; Beatrice is reflected light, a *mezzo* through whom Dante approaches God, and who teaches Dante to be a *mezzo* himself. From this point on in the *Commedia*, Dante emphasizes the role of the visual ray as it is absorbed or reflected, and how that ray can be transmitted from one body to another. To that end, he relies upon the principles of geometrical optics, a field slightly explored by Albertus Magnus but fully developed only by the perspectivists Bacon, Pecham, and Witelo. This is not to suggest that Dante is an expert in perspectivist theory; on the contrary, he uses it insofar as it is a useful model for the metaphysical truths he wishes to convey.[48]

Vision in the *Commedia*: Paradiso

In the *Paradiso*, Dante uses the vocabulary of perspectivist optics to describe the movement of the light ray, and alludes to phenomena exhaustively discussed in perspectivist treatises such as burning mirrors, rainbows, and the multiplication of species. Yet the limitations of human experimental science (to which Beatrice so feelingly alludes [Pa.2.94–6]) are soon exposed in the *Paradiso*. Dante revives the notion of the visual ray, giving souls in paradise a penetrating, luminescent beam that suspiciously resembles the old Platonic fire. As in the *Purgatorio*, the transition in the optical model is gradual. This change seems at first to be a regression; in fact, it proves to be the only way to reassert the omnipotence of God within a system of optics which places the seeing subject at the apex of the visual cone. The visual beam reappears only in the latter cantos of the *Paradiso*, as when Dante says that Beatrice's eyes 'rifulgea da più di mille milia' ['shone more than a thousand miles' (26.78)]. Similarly, Dante refers to one soul as a 'sacra lucerna' ['holy lamp' (21.73)], using the same analogy he had used in the *Inferno*, where extramission was the norm.

This visual beam differs from that in Platonic theory, however, in one crucial respect: Plato said that visual fire comes from within, but Dante makes it clear that the souls' visual fire comes from without. Their light is merely a reflection of the divine, for even their aerial bodies, which Dante says will shine even more in their fleshly form after the resurrection (14.55–7), shine with a brilliant light whose source is God. Precedents for such reflection can be found both in the scientific writings of Albertus Magnus and in the theological writings of Aquinas: the master had suggested that bodies can become luminescent by means of the multiplication of a faint light ray from an external source, while his student had (following pseudo-Dionysius) argued that created things receive their being from God just as the air receives its light from the sun.[49] In her exposition at the opening of the *Paradiso*, Beatrice had used just such an analogy: but where Aquinas uses it to account for creation, Beatrice uses it to explain the diffusion of grace in human souls. In each case, the relationship of God and man is represented not in terms of subject and object, but in terms of the intermediary which is transformed: the illumination of the diaphanous medium that is the human soul. What appears to be a visual beam extramitted from the eyes of the souls in heaven proves in fact to have its source in the divine light above. The beam that comes from their eyes is itself a visible manifestation of the sight of God which stretches out and, illuminating the medium of the soul, returns back again. Each soul is thus linked to God, locked in a mutual gaze. The vision experienced by souls in heaven is always figured by Dante in terms of mutuality, for while the blessed soul gazes at God, God always gazes back at the blessed soul (Pa.10.12) in a look which is simultaneously self-reflexive. Dante shows that the blessed soul communicates with God through vision when he says to Folco of Marseilles, 'Dio vede tutto, e tuo veder s'inluia' ['God sees all, and into Him your vision sinks' (Pa.9.73)], coining the neologism 'inluia' in order to stress the union of subject and object in paradise.[50]

For Dante himself, however, passage through heaven takes place through reflected vision; not until the end of the *Paradiso* does he ascend through a direct ray. He ascends from circle to circle by means of the reflected ray seen in Beatrice's eyes (e.g., 4.142): the ray is a kind of rope or ladder allowing Dante to ascend. But, as Beatrice herself warns, 'non pur ne' miei occhi è paradiso' ['not only in my eyes is paradise' (18.21)]; Dante's ascent can be mediated by any of the blessed mirrors in heaven, whether a soul (17.123; 18.2) or an angel (9.61). In addition to the passages where Dante explicitly refers to souls as mirrors, he often calls them 'splendori,' i.e. reflected lights (21.32; 9.13). It would probably be most accurate to say that the souls *appear* to be 'splendori'; in their essence, they are mirrors. But at the same time, of course, they are not mirrors like those on earth; as Beatrice explains in the second canto of the *Paradiso*, they are mirrors like the moon. Filled with light, they reflect the superfluous light according to their potential. The same is true of human beings on earth, for they can be mirrors too. They are born with the capacity to reflect divine light; only sin makes them less able to be illuminated (7.79–81).

In describing how souls reflect the divine light, Dante draws upon perspectivist vocabulary and describes phenomena discussed at length in perspectivist texts. One of the most prevalent examples of such phenomena is the 'speculum comburentibus,' the burning mirror. In paradise, Dante's vision becomes better and better because of the increasing transparency of the *mezzo* within, his improving capacity to receive divine light. At the same time, the souls he meets are also each in turn more perfect mediators of the divine vision. Therefore, the light Dante perceives in paradise gradually increases. The souls reflect literal light and heat as well as figurative light and heat, that is, wisdom and love. But just as they magnify God's light, they also magnify his heat.

In his *De anima*, Albertus Magnus mentions that burning mirrors exist, but he does not explain how they work.[51] Witelo, on the other hand, discusses them at length: 'When a concave mirror ... is placed opposite the sun so that its axis points directly toward the body of the sun, all rays parallel to the axis and incident on the mirror are reflected to one point of the axis. ... [I]t is possible to kindle fire by means of the surface of such mirrors. ... [I]f combustible material should be placed at [a given point at a certain distance from the focal point of the rays], fire can be kindled. And this is the best and strongest of all figures for assembling solar rays at one point, since solar rays are assembled at one point by its whole surface and its every point.[52] Beatrice's smile is such a burning mirror. For this reason, when they rise to the sphere of Saturn, she forebears to smile at Dante:

> 'S'io ridessi,'
> mi cominciò, 'tu ti faresti quale
> fu Semelè quando di cener fessi:
> ché la bellezza mia, che per le scale
> de l'etterno palazzo più s'accende,
> com' hai veduto, quanto più si sale,
> se non si temperasse, tanto splende,
> che 'l tuo mortal podere, al suo fulgore,
> sarebbe fronda che trono scoscende.'

'Were I to smile,' she began to me, 'you would become such as was Semele when she turned to ashes; for my beauty which, along the steps of the eternal palace, is kindled the more, as you have seen, the higher the ascent, were it not tempered, is so resplendent that your mortal powers at its flash would be like the bough shattered by a thunderbolt.' (Pa.21.4–12)

Beatrice's beauty is the manifestation of love, that is, divine heat. Her smile becomes more inflaming as her eyes become more brilliant.[53] Just as the brightness of divine light is reflected in her eyes, so the heat of divine light is reflected ['tanto splende'] in her smile, so that, if Beatrice were to smile at him at this lofty height, Dante would be reduced to a heap of ashes.

Dante already had been blinded by Beatrice's radiance in the Earthly Paradise (Pg.32.10–12); the same thing happens to him again in paradise when he looks too fixedly upon John (Pa.25.118–21). But in this passage, the other property of divine light – heat – poses a danger to him. The only solution, clearly, is for Dante to become a mirror himself, so that the concentrated ray does not terminate in him. This is exactly what Beatrice goes on to tell him to do: 'Ficca di retro a li occhi tuoi la mente, / e fa di quelli specchi a la figura / che 'n questo specchio ti sarà parvente' ['Fix your mind after your eyes, and make of them mirrors to the figure which in this mirror shall be shown to you' (21.16–18)]. By making himself into a mirror, so that his eyes ['luci'] become 'chiare e acute' ['clear and keen' (22.126)], Dante now can bear the inflaming sight of Beatrice's smile (23.48; cf. 27.96). Later, near the end of the *Paradiso*, Dante himself will be a burning mirror: the ray of light generated by his ardent gaze at Mary grows hotter as Bernard's fiery beam joins it (31.139–42). Other allusions to *perspectiva* appear in the *Paradiso* as well. Thomas Aquinas commends Dante for having been such a good *mezzo* of divine grace using the model of the multiplication of light, familiar from the writings of Grosseteste and the perspectivists who adapt his theory: 'lo raggio de la grazia, onde s'accende / verace amore e che poi cresce amando, / multiplicato in te tanto resplende' ['the ray of grace, by which true love is kindled and which then grows by loving, shines so multiplied in you' (10.83–5)]. Dante is filled with the ray of divine light, multiplied in the diaphanous medium he has so willingly become.

When he encounters the eagle that represents the secular authority of empire, Dante uses perspectivist language to show exactly in what sense the eagle represents an aspect of divinity. He calls it 'l'imago de la 'mprenta / de l'etterno piacere' ['the image of the imprint of the Eternal Pleasure' (20.76–7)], thus emphasizing its relative distance from God. It is, as it were, a second-hand image, the copy of a copy. On one level, the expression affirms the eagle's status as a work of art; that is, one of God's works of art. Because the eagle is a depiction and not a living reflection of divinity, its pupil contains, not a reflected image, but rather the figure of King David (20.37). At the same time, the phrase 'l'imago de la 'mprenta' also recalls Albertus Magnus's description of the species, which he carefully distinguishes from the actual form of an object:

'specie[s] ... est enim non proprie imago vel forma, sed species imaginis vel formae.'[54] Certainly, in heaven, where the medium is so refined, the degeneration of the visible form as it is reproduced would be minimal; but there does seem to be some degeneration, for in his description of the emblem of the microcosm of the celestial spheres, Dante suggests that the circle nearest the central point is the most similar to it, 'però che più di lei s'invera' ['because ... it partakes most of its truth' (28.39)]. Evidently, at this level, there is still a danger that form degenerates over distance. Dante had already implied this deficiency a few lines earlier in the account of his first glimpse of the emblem reflected in Beatrice's eyes. He likens it to an image seen in a mirror: just as the person looking into the mirror turns to discover that the image matches the original 'come nota con suo metro' ['as a song with its measure' (28.9)], so Dante turned from Beatrice's eyes to see the image of the spheres. 'Nota' and 'metro' are similar, though not the same; just so, the image of the emblem in Beatrice's eyes resembles the emblem itself, but is not identical with it. Not until later in the *Paradiso* does Dante indicate that the *mezzo* is truly imperceptable.

As I have shown above, in the *Paradiso* Dante relies upon the principles and language of perspectiva in order to reveal how God's light is transmitted through the universe and how each soul reflects that light. The souls appear to manifest a platonic fiery beam; but the light cast by their eyes comes, not from an internal source, but from the divine light. Dante emblematizes this unification of extramission (the fiery beam) and intromission (the watery eye) in the 'miro gurge' (30.68), the river of light that casts off drops of water that are also sparks of fire. Dante had earlier likened souls to 'faville' or sparks of the divine flame (7.8; 8.16), little lights 'scintillando forte' (14.110). Here, other souls in the form of 'faville vive' (30.64) issue from 'lume in forma di rivera' ['light in form of a river' (30.61)].[55] One could argue that the 'miro gurge' is only figuratively a river, for Dante explicitly states that it is made of light. But Dante continues to stress its watery nature: Beatrice tells him, 'di quest' acqua convien che tu bei' ['you must needs drink of this water' (30.73)], and Dante willingly does.

> Non è fantin che sì sùbito rua
> col volto verso il latte, se si svegli
> molto tardato da l'usanza sua,
> come fec' io, per far migliori spegli
> ancor de li occhi, chinandomi a l'onda

> che si deriva perché vi s'immegli;
> e ... de lei bevve la gronda
> de le palpebre mie ...

No infant, on waking far after its hour, so suddenly rushes with face toward the milk, as then did I, to make yet better mirrors of my eyes, stooping to the wave which flows there so that we may be bettered in it. And ... the eaves of my eyelids drank of it ... (Pa.30.82–9)

The light is fluid, like milk; it flows in waves, and can be drunk, like water. Dante anticipates the fusion of fire and water of the 'miro gurge' earlier in the *Paradiso* when he sees Rahab, a 'lumera / che ... così scintilla / come raggio di sole in acqua mera' ['light which so sparkles ... as a sunbeam on clear water' (9.112–14)]. More specifically, Dante associates light and rivers when, in the very first canto of the *Commedia*, he calls Virgil a 'lume' that pours out a great 'fiume' of language.[56] The three rhyme words in this *terzina* are 'lume,' 'fiume,' and 'volume': light, river, and book (Inf.1.79–84).[57] Just as Virgil's flowing light of wisdom generated his *Aeneid*, so the sight of the river of light in paradise generates Dante's *Commedia*: its 'splendor' not only prepares him for the final vision of God, but also gives him the ... 'virtù a dir com' ïo il vidi' ['power to tell how I beheld it' (Pa.30.97, 99)].

After his eyes drink of the river of light, Dante partakes in perfect vision, seeing by means of a light 'che visibile face / lo creatore' ['that makes the Creator visible' (30.100–1)].[58] It is a single ray ['raggio'] in the form of a great circle, and its reflected image is the celestial rose, whose every petal is a soul in heaven. Here, at last, the visual beam of extramission has a single source, the diaphanous *mezzo* of intromission is unnecessary, and the geometry of *perspectiva* is irrelevant: 'Presso e lontano, lì, né pon né leva: / ché dove Dio sanza mezzo governa, / la legge natural nulla rileva' ['There, near and far neither add nor take away, for where God governs without intermediary, the law of nature in no way prevails' (30.121–3)]. Because vision is direct, there is no visual species to degenerate as a result of multiplication, or to be dimmed by the intervening medium. This is why, when Dante sees Beatrice far above in the rose, she is right next to him:

> vidi lei che si facea corona
> reflettendo da sé li etterni rai.
> Da quella regïon che più sù tona

occhio mortale alcun tanto non dista,
qualunque in mare più giù s'abbandona,
quanto lì da Beatrice la mia vista;
ma nulla mi facea, ché süa effige
non discendëa a me per mezzo mista.

[I] saw her where she made for herself a crown as she reflected the eternal rays. From the region which thunders most high no mortal eye is so far distant, were it plunged most deep within the sea, as there from Beatrice was my sight. But to me it made no difference, for her image came down to me unblurred by aught between. (Pa.31.71–8)

In the *Convivio*, Dante specified that 'raggio' properly refers to the beam of light as it comes from its source (as opposed to the beam that is generated by reflection). Thus, after Beatrice leaves Dante to take her place in the rose, he must ascend not through reflection, but directly through the divine ray. In the *Inferno*, he moved toward God guided by Virgil's light of reason; in the *Purgatorio*, he moved toward God through a medium whose potential was made active by Virgil's light; in the *Paradiso*, he moved toward God by moving toward his reflection in Beatrice. Bernard helps Dante to prepare his sight to 'montar per lo raggio divino' ['mount through the divine ray' (31.99; cf.33.52–4)]. Once he passes through the ray, Dante experiences mutual vision: 'i' giunsi / l'aspetto mio col valore infinito' ['I united my gaze with the Infinite Goodness' (33.80–1)]. Dante uses the language of extramission to describe his gaze, saying that 'io presunsi / ficcar lo viso per la luce etterna' (33.82–3), presuming to 'thrust' his gaze toward God. But God's eye swallows him up, for 'la veduta vi consunsi' ['my sight was spent therein' (33.84)]. The beam comes from a single source, but the soul must have the will to return the divine gaze and thus produce the beam of light that connects man to God. The gaze is mutual, a reflection of the desire of the soul for God and that of God for the soul, expressed as grace.

In the *Commedia*, Dante establishes a relationship between vision and love, a relationship that Guido Cavalcanti, in 'Donna me prega,' had suggested was impossible. But Dante's was not the first allegory to try to reconcile the two: the *Roman de la rose* is centered on the same concern. For that reason, the figure of Narcissus, so important in the *Rose*, is also important in Dante's *Commedia*. In the earlier poem, Narcissus was a figure of the lover, illustrating how love of another can turn out to be merely veiled love of the self. Jean de Meun's solution to this problem is to return love to a carnal basis: desire is transmuted into consummated lust. Dante's solution is similar in that he, like Jean, redirects narcissistic love toward an external object; but his solution is dissimilar in that, for Dante, the object is God.[59]

As I have just shown in detail, however, it is ultimately inappropriate for God to be the object of anything: He is always the subject, never the passive recipient. This is why neither extramission nor intromission are in the end satisfactory models for the relationship of God and man as it is expressed through the metaphor of vision. Just as Dante explores the limitations and the potential of vision as a metaphor for knowledge, he explores the limitations and the potential of Narcissus as an emblem for the lover. In the *Rose*, Narcissus's sterility expresses itself both in his love, which cannot be consummated, and his vision, which is merely a re-doubled image of the self. Thus it would seem that Narcissus could not be an image for the true lover as Dante conceives of him; but, character-istically, Dante does find a way to reform the figure of Narcissus and incorporate him into the vision of God.[60] Dante first alludes to Narcissus in the *Inferno*, in the last pouch of the Malebolge. But even before that, Dante makes evident the danger of becoming absorbed in a reflected image. Virgil likens himself to 'piombato vetro' ['leaded glass' (Inf.23.25)], so immediately does he apprehend Dante's every wish; thus Virgil is a mirror in which Dante's desire is reflected. The metaphor is related to the description of the hypocrites they are about to meet, who are shiny on the outside, but dull lead ['piombo' (23.65)] on the inside. For Virgil and Dante, the metaphor shows how they correspond to one another: the two are becoming one. Conversely, for the hypocrites, the metaphor shows how their identity is split: one becomes two.

Images of doubling abound in the lower parts of hell, as when frost is said to be the sister to snow, copying her image ['assempra / l'imagine' (24.4–5)]. One soul exchanges its form with another (25.100ff.); in another case, two forms are fused into one (25.68–78). In keeping with this doubling, poor rhymes are plentiful ['piglio' (24.20, 24); 'porta' (24.37, 39)]. Here, too, we find the first example of a simile in which the two terms are nearly indistinguishable (24.25–30), their resemblance evoking the error of Narcissus who mistook a simulacrum for the real thing. By using vocabulary drawn from the closing lines of Guillaume de Lorris's *Rose*, Dante further emphasizes the hopelessness of Narcissus's desire ['villanello' (24.7), 'speranza' (24.12)]. The sterility and the dou-bling so prevalent in the lower parts of hell recall Guillaume de Lorris's depiction of the plight of Narcissus, of the lover who loves himself. Thus it is no surprise that the figure of Narcissus appears at the base of hell: Dante hears one sinner say to another, the Greek Sinon, 'per leccar lo specchio di Narcisso, / non vorresti a 'nvitar molte parole' ['to lick the mirror of Narcissus you would not want many words of invitation' (30.128–9)]. In other words, placed before Narcissus's mirror, Sinon would not hesitate to embrace the image before him.

Sinon and his companion are the last souls Dante encounters in the eighth circle of hell; when Dante moves on the ninth circle, he discovers that it is an icy lake that appears more like 'vetro e non d'acqua' ['glass and not ... water' (32.24)]. This glassy surface recalls the mirror of Narcissus. But Dante does not fall prey to Narcissus's error: when he sees two damned brothers, their bowed faces bearing tears frozen into ice (32.46–51), one of them asks him, 'Perché cotanto in noi ti specchi?' ['Why do you gaze so much on us?' (32.54)]. Because their faces are bent downwards ['col viso in giùe' (32.53)], Dante can see them only as they are mirrored in the lake. If Dante were to look into its glassy surface and see himself, he would commit the error of Narcissus. Instead, he sees rightly: he sees the reflection of another. Dante's rectitude becomes even more evident when, as he passes further along the surface of the lake, he kicks one of the sinners in the face ['viso' (32.78)]. Dante says that he himself does not know if he moved his foot by 'voler ... o destino o fortuna' ['will or fate or chance' (32.76)]; but in fact, as Dante will understand later, he moved it by all of those. His will is one with God's will, which sometimes appears to be destiny or chance.[61]

Dante alludes to Narcissus again in the *Commedia*, but this time he is not in the depths of hell, but rather at the entrance to paradise. After Beatrice explains the reflective properties of the moon to Dante in a disquisition that recalls Nature's lecture on mirrors, the *Roman de la rose* is again called to mind when Dante sees faces in the moon and falls, as he puts it, into 'l'error contrario ... / a quel ch'accese amor tra l'omo e 'l fonte' ['the contrary error to that which kindled love between the man and the fountain' (Pa.3.17–18)]. Dante looks behind him to find the source of what he believes is a reflected image, but what in fact turns out to be the faces of souls in bliss. Beatrice rebukes him, as did Virgil immediately after the allusion to Narcissus in the *Inferno* (Inf.30.130–48); but her rebuke is not harsh, for though Dante has made an error, it is not a fatal error like that of Narcissus.

If Dante had alluded twice in the *Commedia* to the myth of Narcissus, he would have emphasized the sterility of the figure, endlessly gazing upon an inaccessible reflection of the self. Instead, Dante alludes to Narcissus three times and thus, by recalling the Trinity, hints that the myth will be interpreted here, not *in malo*, but *in bono*. In the *Commedia*, Narcissus does not appear only as a negative image of love, but also as a model of the highest form of love, God's love for Himself.[62] In the *Roman de la rose*, the phenomenon of refraction figuratively represented the deceptive vision of Narcissus; in the *Commedia*, refraction figuratively represents God's vision of himself:

> Ne la profonda e chiara sussistenza
> de l'alto lume parvermi tre giri
> di tre colori e d'una contenenza;
> e l'un da l'altro come iri da iri
> parea reflesso, e 'l terzo parea foco
> che quinci e quindi igualmente si spiri.

Within the profound and shining subsistence of the lofty Light appeared to me three circles of three colors and one magnitude; and one seemed reflected by the other, as rainbow by rainbow, and the third seemed fire breathed forth equally from the one and the other. (Pa.33.115–20)

The three persons of the Trinity are represented by three rings: two rainbows, one the image of the other, are the Father and Son, and the ring of fire is the Holy Spirit. That circle of fire is the platonic beam, generated by the mutual gaze of Father and Son. This representation of God, seen in the last sphere, was itself mirrored earlier in the *Paradiso*, in the sphere of the sun. There, the Dominicans and the Franciscans form two rings that resemble 'due archi paralelli e concolori' ['two bows, parallel and like in color' (12.11)]. Almost as an afterthought, as he leaves the circle of the sun, Dante mentions that there was also a third circle, which was none other than the 'vero sfavillar del Santo Spiro' ['true sparkling of the Holy Spirit' (14.76)].

Already in the *Convivio*, Dante had demonstrated that the highest form of intelligence finds its greatest pleasure in contemplating its own act of knowing, which is thus also an act of love:

[L]'anima filosofante non solamente contempla essa veritate, ma ancora contempla lo suo contemplare medesimo e la bellezza di quello, rivolgendosi sovra se stessa e di se stessa innamorando per la bellezza del suo primo guardare.

[T]he soul, when philosophizing, not only contemplates the truth itself; she also contemplates her own contemplation and its particular beauty, consciously reflecting on herself and loving herself on account of the beauty of her first gaze. (Conv.4.2.18)

In the *Purgatorio*, Dante uses Rachel to represent the contemplative soul, perpetually gazing into the mirror to see her own eyes. In this she is a reflection of God, who at the summit of paradise gazes upon Himself. God's contemplation differs, however, in that He is not only the contemplative viewer, but also the 'verace speglio' (Pa.26.106): He contemplates Himself in Himself. But God, like man, is both contemplative and active: He is both Rachel, gazing endlessly upon Himself, and Leah, gathering souls like flowers (Pg.27.100ff.). In Guillaume de Lorris's *Roman de la rose*, the lover's desire for the rose was actually a veil concealing his desire for the image reflected in the mirror of Narcissus. There, knowledge meant despair, the certainty that the lover's desire could not be fulfilled. In Dante's *Commedia*, the rose is itself a mirror, leading him who sees it back to the source of the image, justifying hope and fulfilling every desire. Dante uses the rose and the mirror as emblems in part to connect his work to the *Roman de la rose*, but it is not necessary to note the intertextuality of the *Rose* and the *Commedia* in order to understand the meaning of these emblems, for they already have a rich accretion of associations from philosophical and theological writings, as well as other literature and, of course, art.[63]

Dante uses many other emblems as well, such as the cross (Pa.14.104), the eagle (18.107), and the ladder (21.29). Like the rose and the mirror, each is a visible form showing some aspect of divinity in a way that can be apprehended, not gradually, but in an instant; not through a process of cognition, but through a flash of recognition.[64] The numerical structure of the *Commedia* – both the obvious patterns and the more subtle ones that continue to be discovered – has a similar effect, giving a form to the work that the reader grasps in a moment of vision. The 'fulgore' that flashes through Dante's mind in the last moments of his vision is an instance of such instantaneous illumination (Pa.33.141). These images are a means to ascent, which is why Dante advises the reader to 'ritegna l'image ... come ferma rupe' ['hold the image firm as a rock' (13.2–3)]. They also fix meaning in memory, which is why they are so prominent in the memory systems of antiquity.[65] Dante's similes, which appear throughout the three *cantiche*, serve a similar purpose, making what is unfamiliar accessible by likening it to what is familiar.

Precisely because this is the purpose of a simile, some readers have been troubled by the nature of the some of the images in Dante's *Paradiso*. Freccero has gone so far as to refer to them as 'anti-images,' citing the example of the spirits in the sphere of the moon who resemble 'perla in bianca fronte' ['a pearl on a white brow' (Pa.3.14)]. Freccero sees this image as 'self-defeating,' demonstrating that the wayfarer's experience 'remain[s] out of reach to mortal minds';[66] but I would argue instead that the image is difficult to make out not because of language's inability to convey experience, but rather because at that point Dante's power of vision is still relatively weak. The image is difficult for the reader to make out because it is difficult for Dante, as he demonstrates by immediately making the error Beatrice smiles at (3.17): his vision is our vision. Yet vision does not end in the eye: the image is transported into the mind. To put it another way, the sensible species transmits its form to the mind, where it becomes known as the intelligible species. The medium of that change is the imagination. There has been some tendency to make it seem that Dante values one of the mental faculties above all the others,[67] but I think such a reading is misleading since Dante explores the limitations and potential of each of the faculties, as well as their interrelation.[68] Thus while I will discuss Dante's representation of imagination, I do not suggest that he considers it to be more (or less) important than will, memory, or understanding. Unlike those faculties, however, imagination has a fascinating and awful power: it is both the passive recipient of forms from without, and the womb of the artist's creation.

In the central canto of the *Commedia*, Dante attests to the power of the imagination:

> O imaginativa che ne rube
> talvolta sì di fuor, ch'om non s'accorge
> perché dintorno suonin mille tube,
> chi move te, se 'l senso non ti porge?
> Moveti lume che nel ciel s'informa,
> per sé o per voler che giù lo scorge.

O imagination, that do sometimes so snatch us from outward things that we give no heed, though a thousand trumpets sound around us, who moves you if the sense affords you naught? A light moves you which takes form in heaven, of itself, or by a will that downward guides it. (Pg.17.13–18)

Here, the imagination is clearly passive: it receives forms either from the sense organs (via the *sensus communis*) or from an external source. If it is affected by the latter, then the person experiences dreams, either prophetic *somnia* or meaningless *insomnia*. Dante experiences such a dream in that canto, receiving an 'orma' (literally, a footprint) upon his imagination (17.21). He refers to this dream as a 'visïone' (17.34) and an 'imagine' (17.31; 'imaginar,' 17.43). Dante also refers to the 'alta fantasia' (17.25), which can be taken to refer to the dream itself or to the imaginative faculty. If it refers to the dream, it is 'alta' in the sense that the matter of the vision is lofty; if it refers to the mental faculty, it is 'alta' in a physical sense, for the impression raining down from above strikes the upper part of the imagination, rather than the lower part that is struck by the *sensus communis*. But either way, the imagination is a passive faculty, without the power to create independently.

Elsewhere, however, Dante seems to suggest that the imagination does have the capacity to create. He has another dream, one of a hideously ugly woman; but by gazing at her, Dante transforms her into a beautiful object of desire.

> Io la mirava; e come 'l sol conforta
> le fredde membra che la notte aggrava,
> così lo sguardo mio le facea scorta
> la lingua, e poscia tutta la drizzava
> in poco d'ora, e lo smarrito volto,
> com' amor vuol, così le colorava.

I gazed upon her: and even as the sun revives cold limbs benumbed by night, so my look made ready her tongue, and then in but little time set her full straight, and colored her pallid face even as love requires. (Pg.19. 10–15)

Dante's transformation of the 'femmina balba' is an act of the creative imagination, in the most literal sense. His eyes tell him that she is ugly; but his imagination alters that impression, informing the intellect that she is beautiful. Virgil saves him from this pernicious vision, rending the woman's garments to show the decay and filth within. She is the 'dolce serena,' the sweet siren who captivates men and destroys their ships upon the rocks.[69] Figuratively, she is the creative imagination, the impulse to create art.

In the episode of the siren, Dante seems to be saying that the creative imagination is a path leading toward damnation.[70] Creation is reserved for God; the poet should be only His instrument.[71] When a soul in purgatory asks whether the wayfarer is indeed that Dante who wrote 'Donne ch'avete intelletto d'amore,' Dante humbly replies: 'I' mi son un che, quando / Amor mi spira, noto, e a quel modo / ch'e' ditta dentro vo significando' ['I am one who, when Love inspires me, takes note, and goes setting it forth after the fashion which he dictates within me' (24.52–4)]. Since Amor is another one of God's names, Dante is merely God's scribe. Or is he? When Dante concludes the *Commedia*, after the final flash of brilliance ['folgore' (Pa.33.141)] strikes his mind, he says 'A l'alta fantasia qui mancò possa' ['here power failed the lofty fantasy' (33.142)]. Does the divine vision descend from the heavens, making an impression upon the upper part of the imaginative faculty, or is Dante's creative imagination exhausted? Did Dante really go to the otherworld, or is it just a story?[72] In the ambiguity is the pleasure of the text.

Chapter Seven

CHAUCER'S DREAM VISIONS

In 1981 David Aers concluded an article with a footnote appealing for greater attention to the philosophical milieu which undoubtedly, if only indirectly, informed Chaucer's writings.[1] Since then, several attempts have been made to answer his challenge, most notably by Kathryn Lynch.[2] Recent interpretations of Chaucer's writings have even begun to take into account not only the significance of contemporary changes in philosophical thought but also writings on faculty psychology and perception current in the fourteenth century. Because many of these interpretations centre on a single work by Chaucer, it is possible to come away from them with the impression that his use of faculty psychology appears in essentially the same form throughout his works.[3] Even those studies that consider several of Chaucer's works tend to assume that his presentation of vision and knowledge remains static.[4] I would argue instead that there is a distinct progression in Chaucer's use of faculty psychology, particularly in his use of vision as a metaphor for knowing.[5] In several of his early works, especially the *Book of the Duchess* and the life of St Cecilia which formed the basis of the *Second Nun's Tale* (and, of course, *Boece*), Chaucer represents vision as the highest of the senses, one which accurately conveys reality, seamlessly mediating between the seer and the object. In these works, Chaucer follows the Platonic extramission theory of vision, where the eye emits a fiery beam by means of which the soul knows the object. Thus Blanche the Duchess, for example, both sees by means of light and gives off light: she sees clearly, and she radiates clarity upon those who look at her.

In his subsequent allegories, however, the *Parlement of Fowls* and the *House of Fame*, Chaucer abandons vision as a potential mediator between subject and object, and instead turns to the role of hearing. This shift is, without a doubt, related to Chaucer's movement away from French poetic models after he gained access to the Italian poetry of Dante, Petrarch, and Boccaccio during the late 1370s. The conventions of idealized beauty so central to the *Book of the Duchess* came to be subordinated to a more complex exploration of how knowledge is mediated through the other senses and, especially, through language. While in the *Book of the Duchess* Chaucer uses the Platonic extramission theory of vision found in his French poetic models, in the *House of Fame* he draws upon the perspectivist theory of the multiplication of species to illustrate how sound is transmitted, a theory which by the late fourteenth century could be found in a wide variety of texts. Chaucer's increasingly sophisticated treatment of sense perception is accompanied by changes in how he depicts the process of how language signifies meaning, and results in his abandonment of the genre of allegory after his three major dream visions and – in a last nod to the power of allegory – the prologue to the *Legend of Good Women,* usually dated to 1386–7, shortly before he composed the *Canterbury Tales.* In that prologue, Chaucer marks a transitional moment in which he categorizes his own works based on the very different intentions which governed their composition. Here too he explores for the last time the problem of vision and its relationship to knowledge, illustrating both its deceptive nature and its capacity for redemption.

Chaucer's prologue to the *Legend of Good Women* centres on the image of the daisy. Although the narrator's adoration of the daisy is partly an imitation of the French courtly tradition of 'marguerite' poetry,[6] it simultaneously serves as the vehicle for a meditation on vision and its ability to accurately represent reality, based on the flower's English name. It is, as Chaucer writes, 'the "dayesye," or elles the "ye of day"' (F184).[7] The flower resembles an eye not only because of its appearance, but because it actively projects its own beam of 'clernesse and ... verray lyght' (F84). It acts in the same way as the eye in the Platonic extramission theory of vision, dominant until about the late thirteenth century, which states that the eye emits a sensitive beam that, as it were, reaches out and apprehends the form of the object.[8] The daisy spends the day locked in a mutual gaze with the sun, which like the flower resembles an eye, emitting beams as does the eye: it is, as Ovid puts it, 'mundi oculus.'[9] Chaucer emphasizes the link between the flower and the sun by describing them in similar terms: the 'floures white and rede' (F42) are echoed in his description of 'the sonne, that roos as red as rose' (F112). Their pairing and their common role as the 'eye of day' is just one of several allusions to vision in this prologue. The reliability of vision is discussed in the poem's opening lines, where Chaucer questions the possibility of knowing even those things which we most take for granted, whether 'ther ys joy in hevene and peyne in helle' (F2). He suggests that we can never truly know, since it is impossible to know of either place 'by assay,' but only as we have 'herd seyd, or founde it writen' (F8–9).

At the same time, Chaucer warns, it is wrong to believe only in that which 'men han seen with ye' (F11), and says that even 'Bernard the monk ne saugh nat all' (F16). Bernard of Clairvaux addresses this dilemma in *De consideratione*, where he replies to the demand that he describe the ascent to heaven: 'If you hope for this you will be disappointed; it is beyond description. Do you think I am going to tell you what the eye has not seen and the ear has not heard?' Bernard indicates that what is above can be revealed only by the Holy Spirit: 'What words cannot explain, consideration seeks, prayer asks for, a well-led life deserves, purity attains.'[10] The narrator of Chaucer's prologue attempts to fulfil these requirements, for to him the daisy is the central object of adoration in a religion of love. He rises each morning to see the daisy open, retires at night when it closes, and finally takes to sleeping on the grassy turf both to be nearer to and to resemble the flower he adores.

The narrator's faithful adherence to the flower ultimately brings him a reward: a dream in which he meets a man and woman who personify the qualities emblematized by the sun and the daisy. The man, who is immediately identified as the God of Love, is 'corowned with a sonne' (F230) and, like the sun, he radiates a blinding beam: 'me thoghte his face shoon so bryghte / That wel unnethes myghte I him beholde' (F232–3). Although the narrator is compelled to avert his look, Love gazes at him intently: 'al be that men seyn that blynd ys he, / Algate me thoghte that he myghte se; / For sternely on me he gan byholde, / So that his loking dooth myn herte colde' (F237–40). In each case, the lines of vision do not follow their usual course: the narrator should be intently absorbed in gazing at Love, but instead he is dazzled; Love should be blind, but instead he stares at the narrator. In addition, Love's beams should inflame the narrator's heart; instead, they make it 'colde.' The God of Love's look is not passive but active: though the narrator can scarcely describe how Love appears, the god's stare freezes the man's heart. Conversely, the woman who accompanies him is described in terms of her passive look, not how she aggressively gazes but how she appears. She is clad in a 'habit grene' and wears on her head a 'fret of gold' surrounded by a 'whit corowne ... / With flourouns smale' (F214–17). The significance of her appearance is made abundantly, even redundantly, clear: 'For al the world, ryght as a dayseye / Ycorouned ys with white leves lyte, / So were the flowrouns of hire coroune white' (F218–20); 'the white coroune above the grene / Made hire lyk a daysie for to sene, / Considered eke hir fret of gold above' (F223–5). But Chaucer implies that the reader is incapable of interpreting the meaning of these out-ward signs, for he explicitly states the significance of the woman's ap-pearance despite the fact that her likeness to the daisy should be apparent. By doing so, Chaucer suggests that appearance and inference do not lead to certitude.

In the prologue's opening lines, Chaucer asserts that man is at best limited in what he can know certainly. Even 'that ther ys joy in hevene and peyne in helle' (F2) can be known only 'as he hath herd seyde or founde it writen' (F8). Seeing it 'with ye' (F11) is evidently the only way to know something certainly; yet as the prologue continues, Chaucer shows that even knowing 'with eye' can be deceptive. Love's gaze proves to be blinding; the woman's appearance alone is not enough to indicate her relation to the daisy; and the narrator cannot recognize who she is until Love tells him explicitly. But when her identity is revealed, she implicitly replies to the demand of the prologue's opening lines. She is, Love says:

> the quene Alceste,
> That turned was into a dayseye;
> She that for hire housbonde chees to dye,
> And eke to goon to helle, rather than he,
> And Ercules rescowed hire, parde,
> And broght hir out of helle agayn to blys.   (F511–16)

Alceste is an exception to Chaucer's claim that there is no one who has seen hell and emerged from it (F5–6). Presumably she could provide a first-hand account of the 'peyne in helle' (F2), providing the basis of certain, rational knowledge of matters that are usually known only by faith. Yet Alceste does not describe her experience; instead, she speaks by her action, for her descent and return testify to the redemptive power of love. Like Christ, who harrowed hell for love of man, Alceste descends to hell for love of her husband. In this sense, she is a figure of Christ, and correspondingly the centre of the religion or cult of love followed by Chaucer's narrator, leading him to venerate the flower which 'in figurynge' (F298) is Alceste. Her appearance in his dream causes the narrator to vow to 'serve alwey the fresshe dayseye' (F565), writing stories of exemplary women as Love commands. Thereafter he will see a deeper significance in the flower inclining towards the sun: he will see Alceste contemplating Love. In a sense, the prologue affirms vision's unique role as a sensory mediator of higher meaning: the narrator's meditation on the daisy generates an oracular dream bridging the transcendent realm inhabited by Love and Alceste and the material world in which the narrator lives. The mutual gaze of the sun and the daisy – or Love and Alceste – signifies an absolute communion and even a shared identity, leading the narrator to compare Alceste with not only the daisy, but the sun: 'as the sonne wole the fyr disteyne, / So passeth al my lady sovereyne' (F274–5).[11]

Paradoxically, however, it is just this merging of identities that generates the instability of the allegory operating in the prologue. Alceste is figured by the daisy and Love by the sun, yet Alceste is at the same time 'as the sonne'; the daisy is the 'ye of day' (F184), but so also is the sun. The narrator also identifies himself with the daisy, for his devotion motivates him to imitate it, rising and sleeping in unison with it and even reposing on the same turf. Because of this blurring of the boundaries of identity, the allegorical system coalesces into a reductive unity, where Love, Alceste, the sun, the daisy, and even the narrator are indistinguishable.[12] Simultaneously, the answer only Alceste could provide regarding the nature of hell remains unspoken, for as Bernard of Clairvaux writes in *De consideratione*, such knowledge is available not 'by words' but only as 'revealed by the Spirit.'[13] Bernard goes on to describe three paths by which hidden things can be known: faith, knowledge, and opinion. He states that 'Two of these are certain of the truth, but to faith truth is hidden and obscure; to knowledge it is bare and plain. Yet opinion claims no certainty. It seeks truth by trying to discover what is like truth, rather than by grasping it directly.'[14] In the opening lines of the prologue to the *Legend to Good Women*, Chaucer indicates that knowledge of heaven and hell through hearsay or writings is unreliable (F8), and that first-hand knowledge is impossible (F9). Alceste, who could at least potentially offer a first-hand account, does not tell of her experience in hell, but communicates only through the medium of faith, her act bearing witness to the power of love. Love's redemptive, mediating capacity should be manifest in the relationship of Alceste and Love, of the daisy and the sun, but instead their mutuality disintegrates into homogeneity.

In this sense, the prologue to the *Legend of Good Women* marks a turning point in Chaucer's use of allegory, for in it he at least provisionally dismisses the possibility of knowing truth, whether by knowledge or by faith, and simultaneously dismisses the possibility of articulating certain knowledge through language.[15] For this reason, in the prologue to the *Legend of Good Women* Chaucer abandons the genre of allegory, a mode of writing that presupposes the existence of a coherent world that can be figuratively represented by language. Chaucer's use of the figure of the daisy at the centre of the poem's prologue illustrates the inadequacy of allegory for the description of a world that cannot be known with certainty either by knowledge or by faith.[16] It comes as no surprise, then, that in the work which occupied him during the later years of his career, the *Canterbury Tales*, Chaucer invites his reader to pursue Bernard of Clairvaux's third and last form of inquiry, opinion. The reader never certainly knows the truth which underlies the tale, but can only '[seek] truth by trying to discover what is like truth, rather than by grasping it directly.'

The prologue to the *Legend of Good Women* marks a significant point in Chaucer's works in another sense as well, for in the midst of Love and Alceste's debate on the relative merits of his writings, Chaucer inserts a catalogue of his works to date. This list, like those that appear in the Introduction to the *Man of Law's Tale* and the Retraction, is meant to ensure the fame of his works, especially those that, like the tale of Palamon and Arcite soon to be better known as the *Knight's Tale*, are as yet 'knowen lyte' (F421). It also implicitly divides his works into two groups, those already completed prior to Love's admonition of the narrator and those yet to be written following the oracular dream. These works are divided in another way when, in the course of his catalogue, the narrator distinguishes between Chaucer's translations and his original works, beginning with the two works named first in the prologue which are those most harshly condemned by Love. Love asserts 'Thou maist yt nat denye / ... / Thou hast translated the Romaunce of the Rose / ... / And of Creseyde thou hast seyde as the lyste' (F327–32). Speaking in defence of Chaucer, Alceste says

> He made the book that hight the Hous of Fame,
> And eke the Deeth of Blaunche the Duchesse,
> And the Parlement of Foules, as I gesse,
> And al the love of Palamon and Arcite.
>
> ...
>
> He hath in prose translated Boece ...
> And maad the lyf also of Seynt Cecile.
> He made also, goon ys a gret while,
> Origenes upon the Maudeleyne.   (F417–28)

Like Love, Alceste distinguishes between translations (which in the case of the *Romaunt of the Rose* and *Boece* are very close, line by line translations, close to the modern sense of the word) and works that are 'maad,' that is, where Chaucer is not translator but adaptor, taking on both the fame and the responsibility that follow the claim of authority.[17]

This distinction becomes even more apparent when, defending himself against Love's accusations, Chaucer's narrator disclaims responsibility for the meaning not only of his translation of the *Roman de la rose*, but also of the works he 'maad' himself: he states that true lovers 'oghte rather with me for to holde / For that I of Creseyde wroot or tolde, / Or of the Rose, what so myn auctour mente' (F468–70).[18] In the case of the *Rose*, Chaucer's abnegation of responsibility for what his 'auctour mente' is plausible, for it is possible to argue that a close, literal translation simply reproduces the meaning of the original. (I do not think that Chaucer necessarily believed this to be true, merely that he was aware that the case could be made.) But in the case of *Troilus and Criseyde*, such a claim suggests that the author – in this case Chaucer – does not ultimately control the meaning of his text. For this reason, the narrator goes on to assert that the meaning readers find in his books does not conform to his intention: 'yt was myn entente / To forthren trouthe in love ... this was my menynge' (F471–4). By claiming that the received meaning of his text does not correspond to his intended meaning, Chaucer's narrator calls into question the ability not merely of allegory but of language itself to convey meaning accurately. The inadequacy of language is first illustrated in the allegory of the prologue (where daisy, sun, Alceste, Love, and narrator collapse into a self-referential singularity) and then stated explicitly in the narrator's rather helpless defence of his literary progeny.

Like Jean de Meun, Chaucer articulates the limitations of language and, again like Jean, Chaucer turns to translation as an alternative model for literary creation.[19] Chaucer's contemporary Eustache Deschamps refers to him as 'Grant translateur,' referring not only to Chaucer's translation of the *Rose* but also to his adaptations of Machaut and Froissart in his early writings. Deschamps uses the metaphor of insemination to describe Chaucer's accomplishment, writing that he has 'semé les fleurs et planté le rosier.'[20] This metaphor, as Deschamps was doubtless aware, is crucial to the conclusion of the *Roman de la rose* and reappears throughout Jean de Meun's writings, including his translations and *Testament*, as a metaphor of poetic fertility and creation.

Modern readers are aware of the difficulty of categorizing Chaucer's works as translations and 'original' works. Windeatt, for example, refers to Chaucer's dream poetry, including the *Book of the Duchess*, *Parlement of Fowls*, *House of Fame*, and Prologue to the *Legend of Good Women*, as translations, noting that 'The ways in which Chaucer transforms as he translates these French poems reveals his greatness as a re-creating translator.'[21] Similarly, Edwards argues that 'The techniques of poetic invention that Chaucer follows in his early narrative are closely connected to the practice of translation.'[22] Windeatt's and Edwards's observations are also to some extent true of Chaucer's later works, including many of the *Canterbury Tales*, as is evident from the number of source studies of individual tales that continue to appear. Yet, as Copeland has shown, the division between translation and creation is ultimately a false dichotomy: even works that the author carefully distances from source texts 'may use many of the procedures associated with exegesis and the primary form of translation'; yet, at the same time, 'they also tend to suppress the exegetical character of these moves by integrating them into a larger program of textual reinvention.'[23]

In a sensitively nuanced reading, Copeland goes on to show how the prologue to the *Legend of Good Women* functions like the *accessus ad auctorem* found in the medieval commentary tradition. The exposition of the *intentio auctoris*, however, is directed not toward the intentions of Ovid, whose *Heroides* provided the material for several of the following legends, but towards the intentions of the narrator in producing a redaction of the Ovidian text.[24] His intention, if one can call it that, is coercion: he is ordered to produce the legends as penance for earlier literary transgressions. Ironically, a failure of authorial 'entente' is what has necessitated the penance, for the 'menyng' that the narrator retrospectively (and desperately) avows does not correspond to the readerly responses that led to his condemnation. While on the one hand, as Copeland argues, this strategy is one of authorial empowerment (especially in the G-text of the prologue), on the other hand it is an abdication of authorial responsibility – the consequences, if you like, of authority. This abdication is also in evidence in the retraction which concludes the *Parson's Tale*, where a number of 'translacions and endityngs' are disavowed, including *Troilus and Criseyde*, the *House of Fame*, the *Legend of Good Women*, the *Book of the Duchess*, the *Parlement of Fowls*, and the *Canterbury Tales* (X.1085–6). Even though none of these is a translation in the modern sense, Chaucer refers to them as translations in part to help him renounce his responsibility for them in the retraction.

In the *Canterbury Tales*, Chaucer inhabits the role of the translator in two senses: as Chaucer the writer, he adapts other works at least in part for inclusion in the complete book; as Chaucer the fictional persona, he recounts tales he heard told by someone else. This strategy places the tales told by 'Chaucer' (as pilgrim), the *Tale of Sir Thopas* and the *Tale of Melibee*, in an unusual light, for in some sense Chaucer must claim authority and responsibility for their meaning, unless he is prepared to disavow them as he does in the prologue to the *Legend of Good Women*. In the *Tale of Melibee*, Chaucer articulates the problem of allegory and, more generally, the problem of interpreting language through the metaphor of vision and its privation, blindness. Before approaching the *Melibee*, I will offer a close examination of sensory, particularly visual, terminology and imagery as it appears in Chaucer's *Book of the Duchess, Parlement of Fowls*, and *House of Fame*.[25] In the first of the two chapters on Chaucer, I describe the poems' relation to the allegorical genre, particularly with regard to the enigmatic or oracular dream. In the second chapter, I survey Chaucer's use of personification in the dream visions and use that overview as a point of entry into the *Tale of Melibee*, focusing on the role of Prudence. I then turn to the *Merchant's Tale*, which is in some ways a retelling of the themes of the *Melibee* and which shares its focus on visual deception and blindness. Finally, Januarie's inability to believe the evidence of his own eyes is contrasted with the relationship of visual experience to hesitation and belief as seen in the *Parlement of Fowls*.

The Book of the Duchess

The *Book of the Duchess* is at once an example of both courtly love poetry and allegory. Thus in terms of its genre its ultimate ancestor is the *Roman de la rose*, which at least in its first part, written in the early thirteenth century by Guillaume de Lorris, depicts the unfulfilled longing of the courtly lover in the terms of an allegorical dream vision. The more immediate literary antecedents of the *Book of the Duchess* include Froissart's *Paradis d'amour* and Machaut's *Fonteinne amoureuse*; Chaucer's poem departs from these works, however, in its focus on the love of a woman who is absent, not as the result of a coy romantic game, but because she no longer exists on earth.[26] The poem's purpose is sometimes said to be consolatory, but more specifically, its purpose is to make the absent present, to restore Blanche's corporeal presence on earth by creating a monument with 'walles white' (1318) as her own body. It is not just an elegy, but a memorial.[27]

As I have noted, Chaucer borrows from Machaut's *Fonteinne amoureuse*, particularly in the earlier part of the *Book of the Duchess* for his account of Ceyx and Alcione. The relevance of this Ovidian story is not obvious to most readers: Fisher, for example, refers to it as a 'distraction.'[28] The importance of the story lies not in what Chaucer says, but what he ostentatiously leaves out: in keeping with the poet's usual practice, the metamorphosis – the event that is the climax of the narrative in the Ovidian text – is omitted.[29] At this point in the poem, neither the melancholic narrator nor his counterpart, the Man in Black, can even imagine the redemptive love that causes Ceyx and Alcione to be changed into eternally faithful birds. When the narrator declares that Alcione looked up 'and saugh noght' (213), he describes both her failure to see and his own failure to understand.

The metamorphosis omitted from the Ovidian story reappears elsewhere in the text: in Blanche's transition from the living to the dead, in the Man in Black's movement away from melancholic, potentially fatal despair toward renewed hope,[30] and the transformation of Blanche's absent body into the white castle on a hill. In his adaptation of Ovid, Machaut emphasizes the role of the god Morpheus in the story of Ceyx and Alcione, originally titling the *Fonteinne amoureuse* simply *Morpheus*.[31] Morpheus, whose name evokes the word 'metamorphosis,' is the indirect cause of Ceyx and Alcione's metamorphosis into birds, for he takes on the appearance of Ceyx in order to let Alcione know that her husband is dead, thus precipitating her grief and ultimately her transformation. Similarly, Morpheus precipitates the metamorphosis of the Man in Black from despair to hope, and of the Duchess from absence to presence. He does this by effecting another transformation: changing the daily residue already in the narrator's mind into the stuff of dreams.

Chaucer's narrator is restless at the poem's opening; he considers whether to 'playe ... at ches or tables' (51), but instead picks up a 'bok [which] ne spak but of such thinges, / Of quenes lives, and of kinges' (57–8). The black and white kings and queens of the game he did not play and the kings and queens of the book he did read are transformed by sleep (that is, by the metamorphosing power of Morpheus) into a dream. His dream is apparently significant and even prophetic, generated by the intervention of the god Morpheus; but in fact, according to medieval categorizations of dreams, the dream cannot be meaningful, for it is generated by internal causes, the daily residue contained in the dreamer's mind.[32]

Russell has suggested that, for late medieval authors of dream visions, such ambiguity regarding the source of the dream and the degree to which it is meaningful is quite deliberate. He argues that, in the late Middle Ages, no one really believed in the existence of dreams 'that come from God and ... contain wisdom.'[33] Russell's assertion is difficult to accept for, as Steven Kruger has convincingly demonstrated, the arrival of Aristotelian theories proposing that dreams have a primarily somatic origin did not in fact displace earlier theories that dreams could serve as the media by means of which external truths were conveyed to man. On the contrary, late medieval theories such as those of Holkot integrate the two approaches: 'For Holkot, dreaming remains a complex experience: often explicable as a natural process, it can nonetheless also be inspired by supernatural forces.'[34] Yet Russell's larger thesis is suggestive: he claims that late medieval dream visions such as Chaucer's imply that truth can indeed be found in dreams, not because it is handed down from an external source, but rather because the daily residue itself has a value, generating a dream whose 'content is worthy, its truth universal.'[35]

By indicating that, at least within conventional dream categories, the narrator's dream is not meaningful, Chaucer consequently calls into question the poem's status as allegory, for allegories (at least before the mid-thirteenth century) always claim an external source for the dream vision, one based on a transcendent, irrefutable higher reality. Thus even in his earliest allegory, Chaucer is ambiguous about the potential of the genre, perhaps because its claim to convey truth is too bold; it may be that Chaucer suspected verisimilitude was the most that could be achieved. Nonetheless, the *Book of the Duchess* can be considered within the framework of the allegorical genre because its central purpose is also the central purpose of allegory: it conveys an otherwise inexpressible meaning through a coherent system of figurative language. The dichotomy of black and white, absence and presence, written characters and the blank page permeates the work as a whole, ordering it in a dualism that is focused into an uneasy unity by its conclusion.

The polysemous nature of allegory is manifested both on the level of structure, in the multiple meanings which correspond (often loosely) to the levels of meaning discerned through allegoresis, and on the level of the individual word, in the punning or 'wordplay' which Quilligan identifies as the 'basic component' of allegory.[36] While the entire genre should not be grounded on such a narrow foundation, such wordplay is certainly crucial to the *Book of the Duchess*, where the multiple meanings of the word *hert* percolate through the poem and draw together hunter and hunted, lover and beloved.[37] The term refers to the 'hart,' or deer, and the 'heart'; in addition, the second sense can be further broken down into the corporeal organ of the body and the metaphysical seat of love. There is, further, a secondary figurative sense, for the beloved both possesses and *is* the lover's heart.[38] All of these senses appear in Chaucer's poem, as the literal hunting of the hart observed by the narrator in the first moments of the dream leads to the Man in Black's figurative chase, as he seeks to restore both the integrity of his own feeling heart and the beloved heart whom death has taken away.

Allegory is polysemous not only on the level of the individual word, but also in the work's form. This is frequently expressed in symmetrical structures, such as a theme repeated at the beginning and end of a work to give it a circular (and thus perfect) form. Often the symmetrical structure is numerical, the proportions of the work's form reiterating its essential meaning. Perhaps the most famous example of a meaningful numerical structure is Dante's *Commedia*, which is made up of three *cantichi* each divided into thirty-three *cantos*,[39] affirming the unity of the Trinity. Hart has found a comparable numerical structure in the five parts of Chaucer's *Troilus and Criseyde*,[40] while Peck has described a numerical structure in the *Book of the Duchess*.[41] Just as the numerical form of the poem affirms the unity and perfection of the work's transmission of meaning, so the treatment of vision in the *Book of the Duchess* affirms that sense's capacity to mediate transparently between subject and object. Both aspects partake in a Pythagorean Platonism in which number and light serve to unite contraries in the perfection of form. The Man in Black declares that his beloved's sight penetrated even through his body, and that she communicated her meaning to him through her look: 'I ne tok / No maner counseyl but at hir lok. ... [H]ir eyen / So gladly, I trow, myn herte seyen' (839–42). His description of her beauty stresses her 'look,' both the appearance of her eyes (859–61, 866–8) and the power of her gaze (862–5, 869–75), which has the power to speak wordlessly and thus without mediation: 'ever, me thoght, hir eyen seyde, / "Be God, my wrathe ys al foryive"' (876–7).

Vision is the solution to the problem of mediation between subject and object, and also the key to the resolution of dichotomy. The Man in Black explains his plight in terms of a series of dichotomies, beginning with an erasure of his own identity: 'y am sorwe, and sorwe ys y' (597). He first tells how he has changed, his song turned to plaint, laughter to weeping, 'glade thoghtes' to 'hevynesse' (599–601). But the passage of time implicit in the transition from one state to another disappears as the Man in Black begins to describe his current state using a series of opposing terms:

In travayle ys myn ydelnesse
And eke my reste; my wele is woo,
My good ys harm, and evermoo
In wrathe ys turned my pleynge
And my delyt into sorwynge.
Myn hele ys turned into seknesse,
In drede ys al my sykernesse;
To derke ys turned al my lyght,
My wyt ys foly, my day ys nyght,
My love ys hate, my slep wakynge,
My myrthe and meles ys fastynge.   (602–12)

The metamorphosis implicit in the word *turned* is buried among the repetitions of the word *ys*, stressing that what began as a process of change has become a state of being, one characterized by paradox. But the poem goes on to suggest that, just as paradox is resolved through the application of the reader's wit, so these contraries of bad and good, hate and love, night and day, dark and light, can also be resolved. It will happen through the agency of another dichotomy, that of black and white.

The Man in Black and his lady express this dichotomy both in their appearance and in their names. Her fundamental quality is whiteness, expressed both figuratively in her purity and literally in her body, so that her name expresses her essence: 'good faire White she het; / That was my lady name ryght. / She was bothe fair and bryght; / She hadde not hir name wrong' (948–51). While she still lived, their opposition was resolved into unity by the polysemy of the 'hert,' which was simultaneously his and hers:

Whan I had wrong and she the ryght,
She wolde alway so goodly
Foryeve me ...
Oure hertes wern so evene a payre
That never nas that oon contrayre
To that other for no woo.
For sothe, ylyche they suffred thoo
Oo blysse and eke oo sorwe bothe;
Ylyche they were bothe glad and wrothe.
Al was us oon.   (1282–4, 1289–95)

In one sense, the pairing of white and black generates an irreconcilable opposition; in another sense, it creates contrast, and thus a means of communication. In this context, the Man in Black uses a suggestive metaphor to describe his state as a youth untouched by love, saying that he was 'As a whit wal or a table, / For hit ys redy to cacche and take / Al that men wil theryn make, / Whethir so men wil portreye or peynte' (780–3). The table that can be etched or painted can be taken to mean the receptive page, on which the artist depicts a scene or the writer inscribes his poem. Thus the metaphor suggests that the poem before the reader, composed of black characters on a white 'table,' is a model for the production of meaning out of the seemingly irreconcilable dichotomy of black and white.

On another level, however, the 'white walle or ... table' refers to the receptive imagination in the act of seeing. Boethius describes the intromission theory of species proposed by the Stoics, who were the first to posit the existence of species or visible forms. As Boethius puts it, they believed that 'ymages and sensibilities ... weren enprientid into soules fro bodyes withoute-forth ... ryght as we ben wont somtyme by a swift poyntel to fycchen lettres empreintid in the smothnesse or in the pleynesse of the table of wex or in parchemyn that ne hath no figure ne note in it' (*Boece* 5.m.4.6–20). Boethius goes on to correct the Stoic notion of passive vision by declaring that the beam emitted by the seer is necessary to convey the species to the eye. Thus Boethius keeps the species proposed by the Stoics, but denies intromission, instead stating that the species is conveyed to the eye by the agency of the fiery beam emanating from the eye of the viewer. According to Boethius, the species can be transmitted only in the presence of the beam of light. In the *Book of the Duchess*, the lady herself fulfils that role, for she is not only White but also light: 'she was lyk to torche bryght / That every man may take of lyght / Ynogh, and hyt hath never the lesse' (963–5). Walker rightly notes the significance of light in the description of the lady: 'White's place in the overall design of the *Book of the Duchess*, and her symbolic significance, is defined, not in terms of courtly virtues, but in terms of light.'[42] While Walker stresses the significance of light as a medium of consolation by means of its conquest of spiritual darkness, I would suggest instead that here light is significant primarily as a means of visual illumination. The light offered by the lady is the necessary third part which resolves the dichotomy of black and white, enabling their disparity to produce the impression of the visible species on the imagination.[43]

This resolution of contraries was possible while the lady still lived because of the mediating power of light; but since White is absent, her light cannot resolve the paradoxical state the Man in Black is now in. But there is another faculty which can mediate between contraries:

> She had a wyt so general,
> So holl enclyned to alle goode,
> That al hir wyt was set, by the rode,
> Withoute malyce, upon gladness
>
> ...
>
> I sey nat that she ne had knowynge
> What harm was, or elles she
> Had koude no good. ...
>
> ...
>
> So pure suffraunt was hir wyt;[44]
> And reson gladly she understood;
> Hyt folowed wel she koude good.
>
> (990–3; 996–8; 1010–12)

Wit is the faculty of judgment, discerning meaning by a comparison between two contraries. Chaucer's Pandarus articulates this principle in *Troilus and Criseyde*:

> By his contrarie is every thyng declared.
> ... whit by blak, by shame ek worthinesse,
> Ech set by other, more for other semeth,
> As men may se, and so the wyse it demeth.
> ... [T]hus of two contraries is o lore.   (*Troilus* I.637–45)

But wit has a tertiary, mediating part not only in its task of comparing two contrary principles, but also in its role within the structure of the mind. Wit is the mind's third faculty, mediating between imagination and memory.[45] The lady is dead, and so her wit, like her light, is absent. But in another sense her wit remains, due to the polysemy of the hart.

The Man in Black declares that he would have never 'wraththed' his lady: 'For wostow why? She was lady / Of the body; she had the herte, / And who hath that may not asterte' (1152–4). In other words, she had the 'membre principal / Of the body' (495–6), and thus its fundamental, essential part. But because she was linked in love with the Man in Black, their hearts were interchangable: 'Oure hertes wern so evene a payre ... Al was us oon' (1289, 1295). Therefore, the essence of White remains, and in her, 'wytte' or reason.[46] White's wit, living on in the Man in Black, enables him to restore the proper balance of his faculties. When the narrator first meets him, the Man in Black is in a melancholic state, a physiological state (or complexion) dominated by the imagination or fantasy (*phantasia*).[47] He wears black clothing, manifesting his preoccupation with death and mourning; he sings a sorrowful song; his complexion is 'Ful pitous pale and nothyng red' (470). The narrator repeatedly describes him in terms of death, wondering how 'any creature / [could] have such sorwe and be not ded' (468–9; also 488–9). The Man in Black is preoccupied by imagination, unable to properly place the events of the past into the storehouse of memory; his memory and imagination are improperly mingled. As the Man in Black concludes his account of White, he recalls how, as long as she lived, it was possible to mediate between contraries, to resolve dichotomies: 'Al was us oon' (1295). White's wit, living on in the 'hert' of the Man in Black, restores imagination and memory to their proper roles, resolving the situation of paradox that has afflicted him. Memory takes its proper place in this tripartite configuration as the Man in Black leaves behind his dualistic absorption in memory and imagination in favour of a perfectly balanced, triune mind. At the same time, White is memorialized in the white building of the poem's conclusion, which is simultaneously a kind of memory palace[48] and her continuing corporeal presence on earth.

Although the appearance of this memorial may seem abrupt, it is actually the climax of the extended chess analogy which runs through the poem, beginning with its original appearance as the game the sleepless narrator decided not to play, thus providing part of the daily residue instigating his dream. The game first reappears in the Man in Black's account of his losses at the hand of Fortune, who 'hath pleyd a game / Atte ches with me, allas the while!' (618–19). She defeated him by taking his queen:

> She staal on me and tok my fers.
> And whan I sawgh my fers awaye,
> Allas, I kouthe no lenger playe ...
> Therwith Fortune seyde, 'Chek her!'
> And mat in the myd poynt of the chekker,
> With a poun errant ...'   (654–61)

His loss of the 'fers' is devastating, for in chess the loss of the queen often means loss of the match, since in the medieval game the queen was kept proximate to the king until the very end. Yet the missing chess piece reappears not in the form of the queen, but in that of the rook, as the Man in Black remembers the body of White:

> swich a fairnesse of a nekke
> Had that swete that boon nor brekke
> Nas ther non sene that myssat.
> Hyt was whit, smothe, streght, and pure flat,
> Wythouten hole or canel-boon,
> As be semynge had she noon.
> Hyr throte, as I have now memoyre,
> Semed a round tour of yvoyre,
> Of good gretnesse, and noght to gret.
> And goode faire White she het.   (939–48)

On a superficial level, comparison of the lady's body to a tower is quite conventional, found in several examples of courtly poetry including the *Roman de la rose*. But in the context of the chess analogy which has preceded it, this white tower, made (like chess pieces) of ivory, can be identified as the rook, which in the medieval version of the game was the most powerful piece.[49] White's transformation from 'fers' into rook is significant not only because the rook is a powerful piece in the game, but because it is one of an identical pair. As the queen, White is irrevocably absent; as the rook, White endures in her mirror image, the 'hert' she continues to share with the Man in Black.[50]

Therefore it is as the rook that White materializes in the poem's conclusion. After the Man in Black ends his dialogue with the narrator, he rides off 'Unto a place ... there besyde ... A longe castel with walles white, / Be Seynt Johan, on a ryche hil' (1316–19). The white 'longe castel' conceals the name of the woman memorialized in the poem, Blanche of Lancaster, and its location hides the name of her bereaved spouse, John of Gaunt, also Earl of Richmond. Just as the 'ryche hil' is where the white castle is located, so too the Man in Black is, as it were, the location of White: though materially absent in body, her form persists within his memory. Because White is present inside the mind of the man who mourns her, as the remembered image, his sorrow can come to an end. His moment of emotional release is simultaneous with and, apparently, the consequence of the narrator's comprehension (deferred until the poem's last lines) of the nature of the Man in Black's loss. Even though a protracted narrative concerning the courtship precedes the climactic resolution, both the emotional release of the Man in Black and the intellectual comprehension of the narrator are depicted as sudden, instantaneous acts. In his subsequent works, Chaucer teases apart the process of knowing – whether experienced emotionally, as consolation, or intellectually, as insight. In the *Parlement of Fowls*, he illustrates the doubt and hesitation that precedes (and sometimes precludes) the exercise of the will, while in the *House of Fame*, he locates uncertainty not only in the human being who knows and chooses, but also in the object of knowledge itself: that is, the memorable image of 'fals and soth compouned' which appears near the poem's end. The interplay of knowledge and doubt, certainty and hesitation, remains central to Chaucer's later writings, as will be illustrated with reference to the *Tale of Melibee* and the *Merchant's Tale* in chapter 8.

The Parlement of Fowls

In his works written after the *Book of the Duchess*, Chaucer openly questions the adequacy of vision as a paradigm of unmediated perception and begins to move toward a model of audible exchange, a model most explicitly articulated in the *House of Fame*. In the *Parlement of Fowls*, Chaucer ceases to represent vision as a privileged sense that can potentially bridge the gap between subject and object, and that can be used figuratively to represent the correspondence of intention and reception in language. Even in the work's title, the word 'parlement' signals an interest, not merely in language, but in oral, audible exchange. The visual imagery stressed early in the work soon gives way to the birds' speeches and, later, to the cacophonic 'noyse of foules' (491).

While the *Book of the Duchess* ends with the redemptive power of light and the remembered sight of White, the *Parlement of Fowls* opens with darkness: 'the derke nyght ... Berafte me my bok for lak of lyght' (85–7). Although the onset of darkness causes him to abandon his book, the narrator has read enough of '"Tullyus of the Drem of Scipioun"' (31) to experience an oracular dream, visited by Scipio Africanus. Just as this man, Scipio Africanus Minor, was visited in a dream by his ancestor, Scipio Africanus Major, so Chaucer is visited by the former dreamer. Thus Chaucer inscribes himself within a genealogy of dreaming: since the younger Scipio's dream was prophetic, so should his be. But Chaucer simultaneously suggests that the dream may not be prophetic by his careful ambiguity regarding the dream's source. Although Scipio himself testifies to the dream's oracular status, declaring that he has appeared to requite the narrator's laborious reading (112), the narrator suggests that even Scipio's own statement may be merely a fiction: 'Can I nat seyn if that the cause were / For I hadde red of Affrican byforn / That made me to mete that he stod there' (106–8). The narrator himself is not sure if his dream has an external cause, and is therefore prophetic, or an internal one, instigated by his reading, and is therefore not prophetic.[51]

Whether the dream is oracular or meaningless, in either case its source is the narrator's reading of the 'olde bok totorn' (110). In the *Parlement of Fowls,* the book is emphatically the source of the vision, as Chaucer makes clear in the poem's opening:

> For out of old feldes, as men seyth,
> Cometh al this newe corn fro yer to yere,
> And out of olde bokes, in good feyth,
> Cometh al this newe science that men lere.   (22–5)

This analogy recalls another analogy, that of the corn and the chaff, often used to figuratively represent figurative language itself (for example, *Nun's Priest's Tale,* VII.3443; *Parson's Tale,* X.31–6). The chaff is the covering which is unnecessary and can be cast off; the corn is the essential part, representing meaning or knowledge both in its capacity to nourish and in its capacity for growth and reproduction. But instead of glossing the corn as something transcendent and eternal, Chaucer glosses it as 'science,' a kind of knowledge which is specifically accrued and developed by man. He thus implicitly leaves out the possibility of attaining perfect wisdom from a supernatural source, instead creating a model dependent upon human effort and human genius, not divine inspiration. In one sense, this analogy represents the linear action of *translatio studii,* one book generating another as part of an enterprise in which Chaucer certainly participates. But in another sense, the analogy indicates endless cycling, an action as ceaselessly repetitive as the annual sowing of the fields.

Aers notes that, in this passage, Chaucer 'deletes all human agents, their labour and social relations. A great human task ... is transformed into a closed and purely natural cycle,' and rightly suggests that '[t]hese lines exhibit in a compressed, almost emblematic form an attitude to knowledge and authority the poem will discredit.'[52] In this poem, as is also the case in the *House of Fame,* Chaucer indicates that change is introduced by the writer in each instance of the cycle of translation. This endless circle of literary reproduction is evident in the structure of the *Parlement of Fowls*, for it opens with the narrator's discussion of the role of the book, both in general and specifically in relation to his dream, and closes with a promise that he will continue to read books in the hope that he will 'mete som thyng for to fare / The bet, and thus to rede I nyl nat spare' (698–9). The roundel sung by the birds at the end of the poem (675) reinforces the circular nature of the poem's structure. The dream is the transforming process which generates 'newe science' from 'olde bokes,' and its source is not divine inspiration, but the creative mind of man. According to Macrobian guidelines, the narrator's dream is oracular, a type of dream in which 'we are gifted with the powers of divination,' where 'a parent, or a pious or revered man, or a priest, or even a god clearly reveals what will or will not transpire, and what action to take or to avoid.'[53] Thus the oracle's role is twofold: he both reveals the future and helps the dreamer to choose. The latter aspect is particularly significant, for choice and the will of the individual are central to the *Parlement of Fowls*. Although in this poem Chaucer does not explicitly link will and vision, in later works such as the *Franklin's Tale* and the *Merchant's Tale* the state of the will determines the integrity of the subject's ability to see;[54] for this reason, I will discuss at some length the role of will in the *Parlement of Fowls*.

The narrator's inability to choose is implied early in the poem, when he states that the cause of his 'thought and busy hevynesse' was that 'bothe I hadde thyng which that I nolde, / And ek I ne hadde that thyng that I wolde' (90–1). This two-part opposition materializes within the dream as the two-sided sign on the gate of the park, which creates a dilemma for the narrator which he cannot resolve: 'No wit hadde I, for errour, for to chese / To entre or flen, or me to save or lese' (146–7). Fortunately, his dream is an oraculum, so that Scipio is there to make his choice for him, fulfilling his oracular role of revealing, as Macrobius writes, 'what action to take or avoid': the narrator says, 'Affrycan, my gide, / Me hente and shof in at the gates wide' (153–4).[55] The narrator cannot choose by himself because of his 'errour,' which Scipio says 'stondeth writen in thy face' (155). His will is not steadfast but erring, that is, wandering.[56]

Ferster is perhaps the first to have drawn attention to the importance of will in this poem, showing how the narrator's dream 'reveals will's part in interpretation by showing what happens when will is absent.' She concludes that Chaucer poses the following problem: 'if will does not operate, there can be no interpretation, and in fact no participation in any experience. Yet if will does take part, experience may be prejudiced or even completely subjective.[57] Lynch expands on Ferster's more narrow discussion of the importance of will in the poem,[58] reading the poem in the context of late medieval voluntarism. She suggests that 'the *Parliament's* focus on decision and indecision emerges as a timely anatomy of differing possibilities for human choice.'[59] The individual's ability to choose is demonstrated first in the narrator's dilemma, solved by his oracular guide, and second in the formel eagle's choice of a mate, which is the avowed purpose of the birds' debate. The formel's decision is an affirmation of the autonomy of the individual will, for she chooses *not* to choose: 'I axe respit for to avise me, / And after that to have my choys al fre' (648–9). Lynch ably links this crux in the poem to the fourteenth century's 'shift toward various forms of voluntarism in philosophy, which severed the essential link between Nature, reason, and will, granting the will a larger measure of dignity and independence than it had possessed before.'[60]

Lynch stresses the importance of a metaphor used by Chaucer's narrator in describing his own failure of will, standing indecisively before the two-part gate:

> Right as betwixen adamauntes two
> Of evene myght, a pece of yren set
> Ne hath no myght to meve to ne fro –
> For what that oon may hale, that other le t–
> Ferde I, that nyste whether me was bet
> To entre or leve.   (148–53)

Lynch links this metaphor to the well-known philosophical example of Buridan's ass, a hungry animal positioned equidistantly from two bales of hay. Since they are identical, the animal cannot choose between them, and so remains immobile and finally starves. This example was used frequently in the debate over free will, sometimes to support the notion of its existence, other times to refute it.[61] Lynch attempts to connect Buridan's ass to Chaucer's 'pece of yren' caught between two magnets, noting a seventeenth-century discussion of free will which cites both the hungry ass and the piece of iron as examples. She proposes that 'the original physical context of the sophism had never totally been lost and so might have appeared in an example we no longer have.'[62]

There is, however, a very rich fourteenth-century philosophical context in which Chaucer's indecisive narrator can be situated. In his wide-ranging *Summa logicae et philosophiae naturalis,* John Dumbleton sought to integrate phenomena so widely disparate as motion and velocity, sense perception and knowledge, within a framework based on the concept of 'latitudes,' which were used by certain of the Oxford Calculators to quantify motion and other kinds of alteration.[63] Dumbleton was not alone in espousing a theory of physics based on latitudes; but he was unique in extending this system to include intellectual operations as well. Dumbleton's philosophy of language can be described as nominalist in the most basic sense: he states flatly that terms are imposed on things at will.[64] Yet, like many so-called nominalists, he does seem to posit some mode of real existence for universals; Dumbleton's opinion on this subject is unclear, because he apparently died before he could complete the projected tenth book of his *Summa,* which was to address 'universals, which are called "ideas" by Plato.'[65]

Nominalism has often been adduced as an important formative principle in Chaucer's theory of language;[66] yet, too often, modern critics who try to read Chaucer's work in this way define nominalism only very loosely, as a rough synonym for scepticism.[67] Dumbleton's *Summa* offers a more concrete intellectual context for the exercise of the will in the *Parlement of Fowls,* because his work was transmitted both orally and in written form throughout the second half of the fourteenth century, especially at Oxford.[68] Like the works of the other Calculators, Dumbleton's logic was absorbed into the arts curriculum at Oxford.[69] His treatment of insolubles in the first book of his *Summa* engages the positions of Bradwardine, Swyneshed, and Heytesbury; the same views of these same three philosophers are wrestled with by Ralph Strode in *his* treatise on insolubles, written probably during the early 1360s.[70] Too often, arguments seeking to connect Chaucer to movements in fourteenth-century philosophy fall down in the face of the apparent nominalist-realist divide: Delasanta, for example, suggests that Strode was a Thomist 'moderate realist' who straddled the divide separating the 'ferocious ultra-realist' Wycliff from the voluntarist-nominalist Ockham.[71] As Courtenay has recently shown, however, the nominalist position with regard to divine omnipotence was far from homogeneous;[72] the nominalist philosophy of language was perhaps even less so. It may be the case that agreement with a given philosophical position is less to be sought than a shared intellectual environment: Strode and Dumbleton shared not only the physical environment of the University of Oxford, but also the virtual environment of Bradwardine, Swyneshed, and Heytesbury on the subject of logical insolubles. Heytesbury may have provided a more personal link between Dumbleton and Strode: a fellow of Dumbleton's at Merton College, he died at Oxford in 1372/3, long after Strode had come into residence in 1359 and written his own treatise on insolubles in the 1360s.[73] Certain opinions ascribed by Strode to Heytesbury are ascribed by others to Dumbleton, illustrating the extent to which the work of these Oxford logicians overlapped.[74]

It is not necessary, therefore, to argue that Chaucer read a copy of Dumbleton's *Summa* (though that is not impossible). It is clear that Strode, in writing at Oxford on the topic of logical insolubles, would have had to familiarize himself with other Oxford texts on the same topic such as Dumbleton's, written only shortly before and dealing with many of the same arguments and the same authorities. It is not unreasonable that Strode, friend of Chaucer in London during the 1370s, would have discussed with him theories of language and logical argument.[75] Dumbleton alone among the Calculators extended his writings in logic to make room for movement, both on the level of the individual who hesitates and is persuaded, and on the level of the logical proposition that can potentially encompass within it both truth and falsehood. The narrator of the *Parlement of Fowls* inhabits just this awkward space between doubt and certainty. As Lynch has shown, his paralysis is a suspension of the will, torn between two conflicting impulses. Yet the language of the passage attests to the range of the narrator's experience, a range which Lynch's explanation does not fully address:

> [W]ith that oon [vers] encresede ay my fere,
> And with that other gan myn herte bolde;
> That oon me hette, that other dide me colde.   (143–5)

Caught between contraries of heat and cold, eagerness and fear, the narrator cannot choose whether to 'entre or flen,' whether to 'save or lese' himself. No choice – and therefore no motion – is possible. It is not especially unusual to align physical phenomena in the world with emotional changes in the individual; after all, the medical theory of humours assumes that the fourfold system of elements which orders the whole of the natural world is also the underlying structure of man. What *is* unusual is the extension of this system into the intellective realm, addressing the power of the will, certainty, doubt, and hesitation. Dumbleton is alone among fourteenth-century philosophers in making this synthesis; Chaucer, alone among medieval poets. The movement of the magnet, addressed by Dumbleton in the sixth book of his *Summa*, corresponds to the movement of the mind addressed in the first book. Both are governed by the same theory of motion and alteration, and both can be quantified in terms of latitudes.[76]

In the *Parlement of Fowls*, the narrator does not painstakingly pick his way along latitudes of belief and doubt; such detailed treatment of hesitation does not appear in Chaucer's works until book 2 of *Troilus* in the agonized decision-making of Criseyde.[77] Instead, the narrator's guide Africanus cuts the Gordian knot which, as it were, binds the narrator, forcibly making his decision for him. This new focus on the role of the will, based generally in changes in medieval voluntarism and more specifically in the system of motion and alteration outlined by Dumbleton, is just one aspect of the development of the dream vision as seen in Chaucer's poetry. The *Parlement of Fowls* stands in a medial position between the cohesive allegory of the *Book of the Duchess* and the fragmented irresolution of the *House of Fame*, where the transcendence of vision gives way to the cacophony of rumour. Both the discussion which gives the *Parlement of Fowls* its title and the deafening 'noyse of foules' which permeates the poem demonstrate the poet's interest in the audible; at the same time, however, elaborate visual images abound as the narrator makes his way through the garden and temple of Venus. Yet, after exiting the temple, the narrator is decreasingly preoccupied by what he sees and increasingly aware of 'voys,' both that of Nature and of the birds. Significantly, it is when darkness falls that the sound springs forth:

> dounward went the sonne wonder faste.
> The noyse of foules for to ben delyvered
> So loude rong, 'Have don, and lat us wende!'
> That wel wende I the wode hadde al to-shyvered ...
> The goos, the cokkow, and the doke also
> So cryede, 'Kek kek! kokkow! quek quek!' hye,
> That thourgh myne eres the noyse wente tho.   (490–500)

It is significant, too, that the bird whose choice is the crux of the poem is a 'formel egle' (373), for Chaucer begins his category of birds with 'the royal egle ... / That with his sharpe lok perseth the sonne' (330–1). The power of the eagle's gaze should be epitomized in the formel, whose name indicates that she is the exemplar of the species.[78] But her decision is motivated by what she hears, not what she sees, and is expressed, not by her look, but 'with dredful vois' (638).

Time's role in the *Parlement of Fowls* also signals a shift in Chaucer's use of the allegorical genre. In the *Book of the Duchess*, time was represented as cursus, a repetitive cycle emblematized in the closing lines of the poem by the bell which rings twelve at midday and at midnight, uniting the opposites of light and dark in a single moment. Entzminger has suggested that a similar notion of time as cursus appears in the *Parlement of Fowls*, governed by the authority of Nature: 'Nature is primarily concerned with the yearly round, and she convenes the annual session simply to assure that delegates will be available for future parliaments.' He finds significance in the number of lines in the poem (699), and suggests that 'the imperfect brevity is part of the poem's meaning. For while the poem is cyclical, returning to its beginning, it is also ongoing.'[79] It is much more likely that the 'imperfect' number indicates just that: imperfection. The cyclical notion of time, like the naive belief that vision can transparently mediate between subject and object, is tested and found wanting in the *Parlement of Fowls*. The ambiguity of the closure of the numerical cycle of the poem is echoed in the closure of a thematic cycle concerning books and their relation to dreams. Although the cycle is completed in the moment that the dreamer awakens, it questions rather than affirms the existence of the transcendent domain of Nature. The narrator recounts how 'with the shoutyng ... / That foules maden at hire flyght awey, / I wok, and othere bokes tok me to' (693–5). By returning to his book, the narrator recalls the poem's beginning and the book's role in generating his dream. He even hopes that his future reading will generate more such dreams (696–9), continuing the endless process of generating 'newe science' from 'olde bokes' (24–5), a cycle dependent not upon divine inspiration but upon the fecundity of the human mind.

Though the narrator apparently indicates that the sound he hears is that of the birds in the dream, he simultaneously implies that it was the song of real birds; we know it is morning because the narrator begins to read the book he put down last night, when the darkness 'Berafte me my bok for lak of lyght' (87). The reader is thus sharply reminded that the birds' debate and even Nature herself were – or, more accurately, may have been – illusory. The ambiguity is intentional, for even while Chaucer suggests that the transcendent world does not exist, he is not quite ready to let go of it yet. This is why voluntarism of the poem is limited; why the formel, before declaring her decision, first asks Nature's permission to do so for, she says, 'I am evere under youre yerde' (640).

The House of Fame

The *House of Fame* is sometimes read as a commentary, whether homage or parody, on Dante's *Commedia*. This is largely due to the number of explicit echoes of themes or phrases drawn from the earlier work,[80] particularly the image of the eagle which appears in Dante's *Commedia* and which resurfaces in the *House of Fame* to carry the rotund narrator up into the heavens. I have already noted the significance of Chaucer's use of an eagle in the *Parlement of Fowls* to make the decision at the climax of the narrative; in the *House of Fame*, Chaucer uses the eagle once more to deny that vision has the capacity to mediate transparently between subject and object. For Dante, as for Chaucer in the *Parlement of Fowls*, the eagle's significance lies in its capacity for powerful sight.[81] In *Purgatorio*, a golden eagle appears to lift Dante up to the gates of purgatory (9:19–33); Dante faints as he is being carried, but upon his arrival at the gates he is told that he was brought by a lady who declared 'I' son Lucia' (Pg.9:55). She is both the saint, Lucy, and *luce* or light. This passage is often suggested as a source for Chaucer's eagle in the *House of Fame*, for in both cases the eagle carries the narrator physically. But there is another eagle in Dante's *Commedia* that may illuminate Chaucer's use of the figure. In *Paradiso*, the narrator sees small lights which spell out each successive letter of the phrase, 'Diligite iustitiam qui iudicatis terram' (Pa.18.88–93). They hold the shape of the final 'm,' which metamorphoses into the shape of an eagle. In this moment, Dante's narrator sees *res* and *verbum* together, the 'm' for 'monarchia' emblematized by the eagle.[82] This closure of the gap between signified and signifier is subsequently stressed by the blurring of the boundaries separating different kinds of sense perception. Dante's eagle speaks to the narrator, just as Chaucer's eagle does in the *House of Fame*; but while Chaucer's eagle attempts to revive the narrator by means of verbal reassurances, the voice of Dante's eagle affects not only hearing but also vision, making the narrator's 'short sight clear.'[83]

Though Chaucer's eagle is certainly patterned on the eagles of the

Commedia described above, his differs in a very important respect: while Dante's eagle speaks through vision, embodied as light, Chaucer's communicates through the sense of hearing. The first suggestion that Chaucer's eagle is unusual appears in the narrator's description of the bird, which tells how he looks, but not how he sees:

> Myn eyen to the hevene I caste.
> Thoo was I war, lo, at the laste,
> That faste be the sonne, as hye
> As kenne myghte I with myn ye,
> Me thoughte I sawgh an egle sore ...
> Hyt was of gold, and shon so bryghte
> That never sawe men such a syghte,
> But yf the heven had ywonne
> Al newe of gold another sonne;
> So shone the egles fethers bryghte.   (495–507)

The eagle is normally distinguished by his penetrating gaze, powerful enough to look into the sun. But here that capacity is seemingly reserved for the narrator, who looks at the brilliant bird which seems itself 'another sonne.' Yet even the narrator's visual capacity is limited: he cannot look at the real sun, for it blinds him so that he can only declare 'me thoughte I sawgh an egle.' He can gaze only at the sun's simulacrum, the eagle. Finally, the eagle, like the formel of the *Parlement of Fowls*, does not communicate through his gaze but through his voice, awakening the stunned narrator with a loud squawk: 'at the last he to me spak / In mannes vois, and seyde, "Awak!"' (555–6).

Just as Chaucer returns to the symbol of the eagle, seen earlier in the *Parlement of Fowls*, in the opening lines of the *House of Fame*, he also returns to the theme with which he closed the other poem. His determination to continue reading in order to 'mete sumthyng for to fare / The bet' (*Parlement* 698–9) is echoed in the opening invocation of the *House of Fame*, 'God turne us every dreme to goode' (1). Many of the issues raised and only tentatively resolved in the *Parlement of Fowls* find a clearer expression in the *House of Fame*, including the specific question of the source of dreams and the more general question of the origins of language. Chaucer begins by rejecting, not simply Macrobian categories of dreams, but the very idea of classifying them at all. He does so on the basis that one can never know the cause of a dream, an assertion that was implicit in both the *Book of the Duchess* and the *Parlement of Fowls*:

hyt is wonder, be the roode,
To my wyt, what causeth swevenes.

... [B]ut whoso of these miracles
The causes knoweth bet then I,
Devyne he, for I certeinly
Ne kan hem noght, ne never thinke
To besly my wyt to swinke
To knowe of hir signifiaunce
The gendres, neyther the distaunce
Of tymes of hem, ne the causes,
Or why this more then that cause is.   (2–3, 12–20)

The search for causes is central to the *House of Fame*: at the beginning of the poem, the search appears at best comically difficult; at the end, the search is shown to be futile, undercutting the possibility of conveying truth through language. It is in this sense that the *House of Fame* is an allegory about the impossibility of writing allegory, for the allegorical writer must be able to assume that it is possible to convey figuratively an otherwise inexpressible truth through the veil of language.

The search for causes ultimately ends in the search for origins, a quest which is manifested in the rhetorical structure of the *House of Fame* as the narrative repeatedly begins again[84] and concludes with the promise to begin once more: 'Atte laste y saugh a man ... ' (2155). Explicit allusions to the search for causes are found throughout the poem, from the narrator's invocation of 'he that mover ys of al' (81), through the account of Dido and Eneas (283, 369) and the narrator's journey with the eagle, who tells 'what I am, / And whider thou shalt, and why I cam' (601–2; cf. 612, 661, 667). As a search for origins, then, the *House of Fame* is a kind of creation story, seeking the source of fame. Grennen suggests that the *House of Fame* can be read as a retelling of another creation story, that of Plato's *Timaeus*: he suggests that the Middle English text 'substitut[es] a cosmos of unpredictable and jarring confusion for the arithmetical harmonies of the Platonic cosmos.' The *House of Fame* is characterized by 'sound rather than light'; in it, 'the insubstantialities of intuition, mystical glow, and the shimmering light of intellectual day, have given way to the palpabilities of statues, trumpets, and wicker houses.'[85] Grennen's reading provides a literary context for Chaucer's treatment of the senses, exploring the mechanism of fame in its component parts of language, voice, and sound.

Chaucer first describes fame in terms of light in his account of Dido, who declares 'O wikke Fame! – for ther nys / Nothing so swift, lo, as she is! / O, soth ys, every thing ys wyst, / Though hit be kevered with the myst' (349–52). Like light, fame travels virtually instantaneously, penetrating mist. The eagle makes the analogy more explicit in his discussion of the dynamics of sound, which he explains in terms of a theory of light. He states that 'Soune ys noght but eyr ybroken' (765), and that like any element, it must find its proper place, which for sound is the House of Fame. The sound, or broken air, does not fly directly there, but 'thurgh hys multiplicacioun' (784), that is, through the multiplication of species. The eagle explains the phenomenon with the example of a stone thrown into a pond:

> Wel wost thou hyt wol make anoon
> A litel roundell as a sercle ...
> And ryght anoon thow shalt see wel
> That whel wol cause another whel,
> And that the thridde, and so forth ...
> Ech aboute other goynge
> Causeth of othres sterynge
> And multiplyinge ever moo.   (790–801)

In his widely used text on optics, John Pecham uses a similar metaphor to explain the emanation of the visual species: 'all the pyramids in a single body of illumination constitute essentially one light, nevertheless they differ virtually, i.e., in efficacy. In the same way, when a stone is thrown into water, distinct circles are generated, which nevertheless do not differentiate the water.'[86]

In a learned treatment of the grammatical sources of the *House of Fame*, Martin Irvine argues that, in the passage quoted above, Chaucer draws upon late medieval commentaries on Priscian used in the schools. Interestingly, however, the passages that most closely resemble Chaucer's text are themselves informed by contemporary innovations in optics and thus use perspectivist terminology. The late medieval glosses on Priscian quoted by Irvine differ from the earlier ones cited in that they frequently use the term 'species' to refer to the sound transmitted to the air. The late thirteenth-century *Tractatus super Priscianum maiorem*, attributed to Robert Kilwardby, even refers to the phenomenon of the multiplication of species in relation to the generation of sound ['multiplicatio speciei uocis sensibilis'].[87] This vocabulary is drawn from Robert Grosseteste, who in the early thirteenth century illustrated the multiplication of species using the model of light, and Roger Bacon, who in the mid-thirteenth century more fully developed the concept in relation to the act of vision and popularized the term. Thus, whether Chaucer gets his knowledge of species directly from Pecham or indirectly from grammatical texts influenced by perspectiva is of little consequence: he is partaking in a vocabulary that stresses the role of the intermediary, the species, in the act of perception.

It is important not to lose sight of the fact that the eagle's exposition is, above all, funny. To say that 'Soun is noght but eyr ybroken' (765) is to recall, as the Summoner puts it, that 'The rumblynge of a fart, and every soun, / Nis but of eir reverberacioun' (*Summoner's Tale*, III.2233–4).[88] The eagle himself suggests that his technical explanation is not necessarily to be taken entirely seriously, for he concludes his account of the perspectivist theory of the transmission of sound with a caveat:

> Everych ayr another stereth
> More and more, and speche up bereth,
> Or voys, or noyse, or word, or soun,
> Ay through multiplicacioun,
> Til hyt be atte Hous of Fame –
> Take yt in ernest or in game.   (817–22)

That is, you can take it 'in game,' or take it in earnest, believing, with the perspectivists, that sound is transmitted through the multiplication of audible species.

To take it 'in game,' however, does not necessarily mean giving up transmission via species as a metaphor. After he reaches the House of Fame, the narrator hears a 'noyse' which 'ferde / As dooth the rowtynge of the ston / That from th'engyn ys leten gon' (1932–4). The noise emanates from the house like a stone hurled out of a catapult, recalling the eagle's explanation of the multiplication of species using the example of a stone tossed into a pond. The narrator goes on to describe not just sound, but gossip or 'tydynges' in terms of species:

> Whan oon had herd a thing, ywis,
> He com forth ryght to another wight,
> And gan him tellen anon-ryght
> The same that to him was tolde,
> ...

And nat so sone departed nas
Tho fro him, that he ne mette
With the thridde ...

...

Were the tydynge sothe or fals,
Yit wolde he telle hyt natheles,
And evermo with more encres
Than ys was erst. Thus north and south
Wente every tydyng fro mouth to mouth.   (2060–76)

Just as the stone in the pond created circles with 'Every sercle causynge other / Wydder than hymselve was' (796–7), so the 'tydinge' is transmitted 'evermo with more encres' (2074). Yet the transmission of tidings can only be figuratively expressed in terms of species, for the accuracy of describing sound as species has already been challenged, albeit indirectly, by the eagle. He tells the narrator that the 'tydynges' they hear are not uniform and indivisible, as species must be, but are instead composite, 'of fals and soth compouned' (1029). The combination of 'fals and soth' materializes near the poem's end, as the narrator sees two conflicting tidings struggle to escape the whirling house beside the temple of Fame. They swear to 'medle us ech with other, / That no man ... / Shal han that on of us two, but bothe / At ones' (2102–5). 'Thus,' says the narrator, 'saugh I fals and soth compouned / Togeder fle for oo tydynge' (2108–9).

Take it in earnest or in game: the earnest side, as it were, of Chaucer's playful image of the struggling tidings of truth and falsehood can be found in fourteenth-century treatises on insolubles. While Chaucer's basic understanding of insolubles may have come (either orally or in written form) from Strode, Dumbleton's account of insolubles accentuates the quality of motion in the interaction of truth and falsehood in logical propositions. For Dumbleton alone, in his grand theory of latitudes, truth and falsehood are not static, but rather part of a dynamic process. This capacity for change resides both in the knower, whose 'latitude of hesitation' can be assessed and even quantified, and in the object of knowledge, which can be 'true,' 'false,' or 'true and false' ('vera cum falsa').[89] Chaucer's struggling 'tydings' have their 'ernest' progenitor in Dumbleton's theory of motion, the dynamics of change in both the natural world and the logical domain of the mind. Such instability is no foundation for certain knowledge; yet this is just what the narrator seems about to offer in the poem's last lines, when he introduces a 'man of gret auctorite' (2158). But the man does not, and indeed cannot appear, for in a world where the veracity of knowledge can only be weighed and measured, believed or discredited, what good is authority?

The depiction of sense perception in the *House of Fame* underlines the contingent nature of all knowledge, as coherent images give way to noisy confusion. The Platonic notion of vision's capacity to mediate between subject and object, offering both sensible and intellectual illumination, is used to good effect in the *Book of the Duchess*; by the *House of Fame*, however, vision's status has changed markedly. The distinction between vision and the other, lower senses is elided, while the experience of seeing offers the narrator, not a glimpse of Pythagorean unity, but overwhelming, dizzying multiplicity. This can be seen in the poem's numerous examples of ekphrasis, such as that of Venus's 'temple ymad of glas' (120), which is engraved with an account of Troy. It is difficult to tell which parts of the story are pictorially represented and which are inscribed in letters, for Chaucer repeatedly uses the ambiguous term 'grave' (193, 212, 253, 256, 451). Only at one point does the ambiguity seem to be resolved, when the narrator declares that it was terrifying 'To see hyt peynted on the wal' (211). But almost immediately Chaucer undermines the certainty that he refers to a pictorial image by asking 'What shulde I speke more queynte, / Or peyne me my wordes peynte / To speke of love?' (245–7). By again blurring the distinction between pictorial and verbal representation, he allows the ambiguity created by the word 'grave' to stand.[90]

Chaucer concludes his account of the depiction of Trojan history by stating 'Yet sawgh I never such noblesse / Of ymages, ne such richesse, / As I saugh graven in this chirche' (471–3). Yet the nature of the images he saw remains ambiguous: it is not clear whether they were in fact pictures or whether the images he saw were in his mind, conjured up by the story engraved on the wall. Chaucer implies that the latter kind of image is more verisimilar by stressing the deceptive nature of appearances: 'Allas! what harm doth apparence, / Whan hit is fals in existence!' (265–6). The narrator shows that he remains wary of the danger of visual deception when he emerges from the temple of Venus into a wasteland and begs 'O Crist ... Fro fantome and illusion / Me save' (492–4).[91] He is rewarded by the apparition of the eagle, who carries him safely away to Fame's palace. There the narrator finds a kind of memory palace containing statues standing on pillars made of various metals, each of which has 'hy and gret sentence' (1425), that is, allegorical significance. Rowland suggests that both Fame's palace and Venus's temple are 'typical memory hall[s],' where the poet 'reveals the *loci* for various images and explains the significance which the images retain in the memory.'[92] But the ordered system of the memory palace rapidly degenerates into confusion, a failure forecast by the narrator's early warning that he would recount his adventures merely 'as I kan now remembre' (64): he says that the hall was as full of statues 'As ben on trees rokes nestes ... [H]it a ful confus matere / Were alle the gestes for to here' (1516–18). The narrator's attention is then drawn from this disordered 'syghte' (1520) to an approaching 'noyse' (1521), signifying his movement from the unsatisfactory realm of visible, figurative representation to a different realm of audible, mimetic representation.

The *House of Fame* contains a number of foreshadowings of the *Canterbury Tales*, most famously the 'ryght gret companye ... of sondry regiouns, / Of alleskynnes condiciouns' (1528–30), but also the notion of pilgrimage (116), the 'soune [which] ys noght but eyr ybroken' (765) that reappears in the *Summoner's Tale* (III.2233–4), and the 'fundament' of 'Seynt Thomas of Kent' (1131–2) echoed in the *Pardoner's Tale* (VI.948–50).[93] Most importantly, however, it prepares the way for the *Canterbury Tales* by showing the limitations of language. The totalizing system of the allegorical dream vision, with its omniscient narrator and guarantee of divine authority (through the oracular or prophetic dream), is plainly inadequate. Developments in the philosophy of language, especially terminist logic with its knotty insoluble, required that poets find a new way to illustrate the transmission of meaning. For Chaucer, that new way would be based on oral discourse and exchange. Having demonstrated that truth cannot be expressed directly through language, whether literally or figuratively, Chaucer chooses to pursue the next best thing: verisimilitude, fiction that resembles reality.

Chapter Eight

CHAUCER'S PERSONIFICATION AND VESTIGIAL ALLEGORY IN THE *CANTERBURY TALES*

Chaucer's changing use of personification reflects the gradual dissolution of the link between the abstraction and its embodiment, between universal concept and particular expression. This can be seen clearly in the changing status of the personified abstractions of Chaucer's dream visions, as well as in the portrayal of Prudence in the *Tale of Melibee* and the significantly named Januarie and May of the *Merchant's Tale*. While personification and allegory are not synonymous, they are interrelated: as Quilligan puts it, personification is 'one of the most trustworthy signals of allegory.'[1] This is true insofar as the impulse to create personifications from abstractions is allegorical; as Whitman notes, 'the whole history of the word ["personification"] demonstrates a close connection with the creative technique of saying something in fiction and meaning something else in fact.'[2] It is therefore unsurprising that modern readers have sought to interpret late medieval changes in both allegory and personification in terms of nominalist philosophy: in an influential article, Delany has argued that the analogical system fundamental to allegory was shaken by these new currents in philosophy, while Paxson suggests that personification in texts such as *Piers Plowman* 'works as a parody of or satire on ... philosophical Realism.'[3] The dream vision form and personification are both arguably manifestations of a mode of analogical thought that was called into question by late medieval scepticism. Yet, far from vitiating the power of analogy, nominalist philosophy (often loosely equated with scepticism) may have, as Stephen Penn has suggested, actually 'contributed positively to its development.'[4] Debates on what kind of existence universals might have (if any), and their relationship to the terms by which they are signified, undoubtedly had an impact on the practice of allegory and the use of personification. In the follow-

ing pages, I will analyse the use of personification in the dream visions
before turning to its use in the *Canterbury Tales*. I will then show how, in
the *Tale of Melibee* and the *Merchant's Tale*, Chaucer's understanding of at
least some of the implications of terminist theories of language are
manifested both in the use of personification and in the depiction of
how knowledge is constituted and received. The waning effectiveness of
vision, represented both in the wounded body which motivates the
dialogue of the *Melibee* and in the unreliable sight of Januarie in the
Merchant's Tale, is metonymic of man's imperfect apprehension of his
world. Blurred vision is just a symptom of a more general intellectual
malaise.

In the *Book of the Duchess*, Chaucer conspicuously avoids using traditional personifications, which appear only as allusions in the discourse
between the narrator and the Man in Black and not as actors within the
narrative. In the first lines of the poem, Chaucer names abstractions
that, in a dream vision, we would normally expect to find as personifications. His narrator tells how he has 'so many an ydel thoght' (4), corresponding to Idelnesse or Oiseuse of the *Roman de la rose*, and refers to
'kynde' and 'nature' (16, 18), but only as unembodied cosmic principles, not personifications.[5] As a work of consolation, the *Book of the
Duchess* follows the model of Boethius's *De consolatione Philosophiae*, a work
it also relies upon for the Platonic extramission theory of vision espoused in the poem. But the imitation includes a comic inversion, for
instead of the personified abstraction Philosophy, the rather obtuse
narrator offers consolation. Instead of an allegorical personification who
reflects the divine mind, we find a fictional persona who has to struggle
for comprehension. Personifications appear in the *Book of the Duchess*
only indirectly, as when the Man in Black refers to Love (766), Yowthe
(797), Fortune (811), or Nature (871). These are not so much agents in
the world as part of a common vocabulary derived from the literary
heritage shared by the narrator and the Man in Black. The only personification who clearly is an embodiment of an abstract principle and whose
presence permeates the poem is White (948). Her efficacy and validity as
a personification are based on the newness of the term's coinage: 'White'
has not been debased by being currency in the economy of literary and
verbal exchange, an exchange which (as we learn in the *House of Fame*)
always introduces error.

Kathryn Lynch has argued that White's status as a quasi-personification in the poem must be understood within the framework of nominalist discussions of knowledge: 'The quality white and its substantive form

"whiteness" were probably the most frequently invoked terms in late medieval Ockhamist discussion of the problem of how one knows, whether by cognition of singulars or of universals.' This leads her to conclude that, in Chaucer's dream visions, 'the form's epistemological argument becomes a nominalist rather than a realist one, an argument that values the singular over the universal.'[6] To couch the argument in terms of realism versus nominalism, however, is to lose sight of the variety of positions on the status of universals held by philosophers who can loosely be referred to as 'nominalist.' Dumbleton's position on universals is a case in point: in the description of his planned tenth book of his *Summa*, Dumbleton indicates that he will treat 'universals, which Plato calls ideas.' For Dumbleton, the realist and nominalist positions on universals are not to be chosen between, but reconciled. In the first book of his *Summa*, he breaks off a discussion of how terms are imposed, promising to continue the topic in connection with universals in the tenth book.[7] In the first, he confines himself to explaining how terms signify things through acts of memory, both with regard to the initial imposition of the term and with regard to the recognition of the thing by means of the term one knows. The example he uses concerns 'white': the knowledge of whiteness (that is, its 'intention') comes from seeing a white object, and the term 'white' comes from the experience of hearing or seeing the term applied to the white object we see. Thereafter, says Dumbleton, 'when a person who knows "white" either sees it written or hears it spoken, the true intention of white is called forth from the person's memory.' The linking of the term with the thing it signifies does not occur naturally, but arises from human actions: 'the term and the intention have the same name simply because each intention is called forth by a term by means of our own acts.'[8] Here, the universal quality signified by the term 'white' clearly exists; but it exists not in some nebulous realm of Platonic ideas, but in the mind of the person who thinks. It is in all these senses that the Man in Black remembers White.

In the *Parlement of Fowls*, personifications take on a more dynamic role in the narrative. Two goddesses, Nature and Venus, dominate the poem, though several other personifications are mentioned briefly in the narrator's description of Venus's temple.[9] Nature and Venus are frequently linked in allegorical literature, based on their common yet divergent interest in carnal love.[10] Nature promotes intercourse in order to ensure the perpetuation of the species, while Venus promotes intercourse as a manifestation of love. The binary opposition of the two goddesses is implicit in the inscription on the twofold sign above the gate

(120–47), a duality forecast by the narrator's dilemma caught between lacking what he 'wolde' and having what he 'nolde' (90–1).

The narrator finds Venus virtually naked, reclining on a bed with Richesse (260–73). Her domain is interior, within her temple; conversely, Nature's domain is outdoors, her traditional begemmed temple replaced by one made of branches.[11] Venus does not speak, but communicates her advocacy of carnal love by her appearance; Nature speaks openly, presiding over the debate of the birds. The opposition of the two is also articulated in how they are named: while the narrator always uses the same name to refer to Nature, he refers to Venus using several different names ('Cytherea' 113, 'Venus' 261, 'Cypride' 277). Venus is also present in other parts of the poem by proxy, in her allegorical son Cupid, and the 'sparwe, Venus sone' (351). Her reappearance under various names emphasizes the pervasiveness of love, whose importance the narrator avows in the poem's opening two stanzas. At the same time, Venus's different names, like the assembly of personifications who surround her temple,[12] place her opposite to Nature in the traditional dichotomy of art and nature. Her many names attest to the instability of a personification once it has entered into the economy of literary exchange, altered and renamed by each successive author. The abundance of personifications surrounding her attest to the author's power of literary creation, an embodiment which mimics God's creative power expressed through his agent, Nature.

The dichotomy of art and nature is not firmly fixed, for the lush greenery of the garden is described, paradoxically, in terms of art: 'overal ... / Were trees clad with leves that ay shal laste, / Ech in his kynde, of colour fresh and greene / As emeraude' (172–5). Ferster suggests that here Chaucer 'demonstrates that description of nature, no matter how realistic, is a product of art.'[13] Nature's opposition to art is further drawn into question by the discrepancy between her name and the name of the domain she embodies: the goddess is called 'Nature,' but her domain is called by the Germanic synonym 'Kynde' (672; cf. 311). The two names are used interchangably in the *Romaunt of the Rose* (1697–9), but here Chaucer uses two different names to distinguish between the personification and the abstraction she represents. This usage strikingly demonstrates the dissolution of the correspondence between personification and abstraction, between the particular and the universal. This dissolution is also manifested, here as in Jean de Meun's continuation of the *Roman de la rose*, by a tendency for personifications to take on characteristics of other personifications, something inconceivable within the bounds of traditional, vertical allegory. For example,

Ferster notes that 'Pleasaunce is in Cupid's part of the garden, but Nature tells the birds, "I prike yow with plesaunce."'[14] Even more notably, Nature tries to help the formel make a decision, suggesting 'If I were Resoun, thanne wolde I / Conseyle yow the royal tercel take' (632–3). But it is ridiculous to imagine that Nature could be Reason; only human beings (and their fictional counterparts) can imagine inhabiting another point of view. By giving individual perspective to personifications, Chaucer metamorphoses them from embodied abstractions into living persons.

The distinction between nature and art, already blurred in the *Parlement of Fowls*, is dissolved in the *House of Fame* as the narrator likens the poet Orpheus to the goddess Nature: 'smale harpers ... gunne on hem upward to gape, / And countrefete hem as an ape, / Or as craft countrefeteth kynde' (1209–13). He goes on to describe 'magiciens' who practice 'magik naturel' (1260, 1266), an oxymoron which breaks down the conventional opposition of art and nature. Neither Nature nor Venus appears bodily in the *House of Fame*: though the narrator enters Venus's temple, he sees her only 'in portreyture' (131). Other than Fame herself, personifications and gods appear only in invocations ('Cipris' 518, 'Thought' 523) or indirectly, as when the eagle mentions that Jove was the cause of his journey (661) or notes how badly Fortune had treated the narrator (2016).

Only one personification appears bodily in the poem: 'a femynyne creature' described at length and finally identified as the 'Goddesse of Renoun or of Fame' (1365, 1406). She is described in terms of multiplicity:

> [A]s feele eyen hadde she
> As fetheres upon foules be,
>
> ...
>
> ... also she
> Had also fele upstondyng eres
> And tonges, as on bestes heres. (1381–90)

Her 'feele eyen,' ears, and tongues embody the multiplicity promised by the eagle as he bore the narrator upward:[15]

> For truste wel that thou shalt here
>
> ...
>
> Mo wonder thynges, dar I leye,
> And of Loves folk moo tydynges,

Both sothe sawes and lesinges,
And moo loves news begonne,
And longe yserved loves wonne,
And moo loves casuelly
...
Mo discordes, moo jelousies,
Mo murmures and moo novelries,
And moo dissymulacions.   (672–87)

Chaucer stresses the word 'moo' another five times in the next ten lines in a comic passage that lets the narrator 'here' (672) what he 'saugh' (127) earlier in Venus's temple:

                ther were moo ymages
        Of gold, stondynge in sondry stages,
        And moo ryche tabernacles,
        And with perre moo pynacles,
        And moo curiouse portreytures,
        And queynte maner of figures
        Of olde werk, then I saugh ever.   (121–7)

Fame's body expresses the plurality of sensory perception, whether of sight or of hearing. She attests to the unattainability of a singular meaning and to language's capacity to convey truth only approximately, through many mimetic representations of reality.

The *Tale of Melibee*, one of the two *Canterbury Tales* narrated by Chaucer *in propria persona*, is a translation of a Latin prose treatise by way of a French intermediary. It is largely a dialogue between Melibee and Prudence, who as one of the four cardinal virtues, is a personification with a richly elaborated literary and iconographic tradition behind her. She is the beautiful 'heroine,' as Trout puts it, of Alanus de Insulis's *Anticlaudianus*,[16] and stands out as well among her three sister virtues in the iconographic tradition. During the fifteenth century, symbols for the various virtues and vices proliferated, and at that time Prudence came to be represented with a sieve (to denote her ability to winnow out the truth) or a mirror (to illustrate her aspect of *praevidentia* or foresight).[17] Before that, however, Prudence was shown carrying a book in order to distinguish her from the other cardinal virtues, denoting her role as 'scientia scripturarum' ('knowledge of the scriptures'), as Isidore puts it.[18] By the twelfth century, Prudence was often shown with an additional symbol, usually a snake or (less often) a dove, referring to the injunction to be prudent found in the gospel of Matthew: 'Estote ergo prudentes sicut serpentes et simplices sicut columbae' ['Be ye therefore wise {lit. 'prudent'} as serpents, and harmless as doves'; Matt. 10:16].[19] In *Troilus and Criseyde*, Prudence's aspect of foresight is emphasized when Criseyde laments her failure to have anticipated the future: 'Prudence, allas, oon of thyn eyen thre / Me lakked alwey!' (5.744–5).

Thus Prudence makes her appearance in the *Tale of Melibee* with a whole series of associations. Her bookishness is much in evidence, for Prudence's discourse is nothing so much as a series of learned extracts, often proverbs, applied more or less aptly to the subject at hand. David Wallace points out that Prudence resembles no other figure in the *Canterbury Tales* so much as the Wife of Bath: both wield the tools of rhetoric in order to carve out a 'specifically feminine discursive space.'[20] Both Alison and Prudence struggle against husbandly anger, and both do so in the context of textual interpretation: the clerical glossing which is Alison's nemesis, and the faulty understanding of Melibee which Prudence struggles to remedy. One might expect that Prudence in the *Tale of Melibee* would have all the awesome status typical of an allegorical personification, based both on the iconic status of Prudence as one of the cardinal virtues and on the allegorical model for Prudence found in Boethius's *De consolatione Philosophiae*.[21] Like Philosophy, Prudence attempts to restore the impaired faculties of a man by persuading him to fix his sight on the proper object. But Prudence does not display such majesty: she is, as a number of critics have observed, not so much an abstraction as 'a prudent wife.'[22]

Prudence's other iconographic attributes, the snake and the dove, will be discussed below in connection with the *Merchant's Tale*. I want first to turn to her quality of foresight, that threefold vision of past, present, and future to which Criseyde refers. Prudence's extraordinary power of vision would appear to be precisely what Melibee needs, for as we learn at the outset of the Tale, he has suffered (at least vicariously) a terrible blow. Because the *Tale of Melibee* is a translation (perhaps an early work revised for inclusion in the *Canterbury Tales*),[23] differences between the Middle English version and the French translation used by Chaucer along with the Latin original are bound to be revealing. The first of these alterations appears in connection with the wounds suffered by Melibee's daughter and thus, by extension, by Melibee himself. Three of Melibee's enemies entered his house, beat his wife, and 'wounded his doghter with fyve mortal woundes in fyve sondry places – / this is to seyn, in hir feet, in hire handes, in hir erys, in hir nose, and in hire mouth' (VII.971–2). It has been suggested that Chaucer follows the erroneous reading found in a single manuscript that substitutes 'piez' for 'yeux.'[24] But this places Chaucer in the awkward position of not understanding the allegory he is translating, especially since Prudence goes on to helpfully interpret the wounds as afflicting the five senses (VII.1424).[25] I suggest instead that Chaucer retains the erroneous reading 'piez' in order to emphasize, not merely injury to, but the lack of one of the senses. Sophie's eyes, the embodiment of wisdom's capacity to perceive truth, are absent, figuratively representing Melibee's intellectual blindness.

Wallace points out the extent to which Melibee makes himself the focus of attention, so that the suffering of the females in his household is subordinated, even assimilated to the suffering of the male head of the household: 'Melibee's pain becomes the subject of the text.'[26] This is, I would argue, especially true of Melibee's daughter, because Prudence retains at least some autonomy, acting as Melibee's interlocutor throughout the text. The daughter, by contrast, exists only as the mutilated body through which Melibee suffers: her pain is his pain, and her loss is his loss. The nature of that loss is made clear by the alterations made by Chaucer in his Middle English translation. The five wounds, which in the original refer to the five senses, are modified to omit the eyes: the power of sight is not merely injured, but entirely absent. In addition, a simple change instituted by Chaucer in the Tale's opening lines skews the narrative in such a way as to make clear what it is that Melibee lacks: 'A yong man called Melibeus ... bigat upon his wyf, that called was Prudence, a doghter which that called was Sophie' (VII.967). Sophie; that is, philosophy. While his source, Reynaud de Louen's *Livre de Mellibee et Prudence*, refers simply to an anonymous 'fille,' Chaucer gives her a name. The nurse whose very milk Boethius thirstily drinks in is here transformed into a helpless, wounded child, an emblem not of redemption but of loss.

Melibee himself alludes to his impairment in terms of vision when he excuses his inability to comprehend Prudence's counsel by declaring that 'troubled eyen han no cleer sighte' (VII.1702). Prudence attempts to fill the gap left by Sophie's absence, offering him 'sapience' (VII.1071). But prudence and sapience are not synonymous, as Prudence herself acknowledges by distinguishing between the two: 'The book seith that in the olde men is the sapience and in longe tyme the prudence' (VII.1164).[27] Nonetheless, Prudence attempts to offer Melibee wisdom, declaring 'if ye governe yow by sapience put awey sorwe out of your herte' (VII.994). But she can offer it only secondhand, quoting biblical and classical works describing wisdom; Prudentia cannot *be* Sapientia. Her words of counsel cause Melibee to lose his grasp of the distinction between wisdom and prudence, when he asserts that she has proven her 'grete sapience' (VII.1114). Here Prudence steps out of her proper role, much as Nature in the *Parlement of Fowls* does by remarking, 'If I were Resoun ...' (632). Once again, the personification is no longer firmly linked to the abstraction it embodies.

It takes a long time for Melibee to understand at last the teachings of Prudence. This is due to the impairment he has suffered (vicariously through Sophie); but Melibee imagines that the problem lies not in him but in Prudence. He initially discounts her counsel on the basis that 'alle wommen been wikke' (VII.1057). One can hardly imagine Boethius saying this to Philosophy. Like Boethius, however, Melibee needs to be healed and, like Boethius, the cure will be effected through a dialogue. But while Philosophy wields the rapier of logic, Prudence uses the blunt instrument of rhetoric. Her interpretive inadequacy is evident when Prudence attempts to explain Melibee's predicament allegorically:

> The three enemys of mankynde – that is to seyn, the flessh, the feend, and the world – / thou hast suffred hem entre in to thyn herte wilfully by the wyndowes of thy body, / and hast nat defended thyself suffisantly agayns hire assautes and hire temptaciouns, so that they han wounded thy soule in fyve places; / this is to seyn, the deedly synnes that been entred into thyn herte by thy fyve wittes. (VII.1421–4)

Prudence's interpretation is inconsistent with what we already know: the five injuries cannot be to Melibee's soul, for Sophie does not represent something within him, but an abstract principle existing outside him. Prudence's interpretation works only locally, not forming part of an allegorical system.[28]

A related inconsistency appears in Prudence's interpretation of the figurative significance of Melibee's name: 'Thy name is Melibee; this is to seyn, "a man that drynketh hony." / Thou hast ydronke so muchel hony of sweete temporeel richesses, and delices and honours of this world / that thou art dronken' (VII.1410–12). But Melibee has already offered another, implicit interpretation of his own name: he quotes Solomon's statement that 'wordes that been spoken discreetly by ordinaunce been honycombes, for they yeven swetnesse to the soule and hoolsomnesse to the body' (VII.1113). What Prudence offers Melibee throughout the

Tale is precisely these 'sweete wordes' of counsel (VII.1114). But there is a contradiction. It cannot be the case that both Melibee and Prudence are correct, for if the honey is 'sweete temporeel richesses,' it cannot be 'sweete wordes' of counsel. They are contraries, one a source of corruption, the other a source of regeneration.

One might resolve this by concluding that Melibee is simply a poor interpreter: as Patterson puts it, even though 'Prudence's task is to teach Melibee how to interpret ... as soon as Melibee is given the opportunity to make a decision on his own ... he reveals the failure of this effort.'[29] Prudence certainly fosters this impression by continually rebuking Melibee for his faulty interpretation of her 'sentence.' For example, she asks 'I wolde fayn knowe hou ye understonde this text and what is youre sentence' (VII.1278), and follows his reply by exclaiming that 'it sholde nat han been understonden in thys wise' (VII.1384). Yet, as we have seen, Prudence's interpretive ability is also weak, functional perhaps on a local level, but undeniably inconsistent. Logical coherency is clearly not Prudence's strong point. She is, nonetheless, persuasive, as is brought out in the tale's conclusion, where Chaucer once again departs significantly from his source. He amplifies Reynaud de Louens's simple statement that Prudence's words put Melibee in a state of 'grant paix,' adding a detailed description of Melibee's will: 'Whanne Melibee hadde herd the grete skiles and resouns of dame Prudence, and hire wise informaciouns and techynges, / his herte gan enclyne to the wil of his wif, considerynge hir trewe entente, / and conformed hym anon, and assented fully to werken after hire conseil' (VII.1870–2). This passage is striking in two respects: first, in its description of the persuasion of Melibee as a process; second, in its assignation of a passive role to the man who conforms his will not to the omnipotent rule of God, but to the ultra-mundane rule of his wife.

First, the process. Chaucer was clearly interested in the question of how the will operates, as Lynch has shown. The indecisive narrator of the *Parlement of Fowls* is engaged in the exercise of the will; the problem, of course, is that his will pulls him in two directions at once. The will is in a dynamic state, moving now toward greater 'hette,' now toward greater 'colde'; now toward a state of boldness, now toward 'fere' (143–5). The dynamics of motion had been quantified by several of the Oxford Calculators, but only John Dumbleton extended the system to include measurable degrees (or latitudes) of certainty and doubt. Chaucer illustrates how the operation of the will is not instantaneous but part of a process, both in the dynamic suspension (as it were) of the narrator's will in the

Parlement of Fowls and in the successful conforming of the will in the *Tale of Melibee*. Dumbleton does not explicitly extend his account of degrees of hesitation and belief to encompass the operation of the will, probably because this would require that he extend his argument to address the capacity of the human will relative to the omnipotence of God, whether as the *potentia Dei absoluta* or the *potentia Dei ordinata*, a central topic in the philosophy of Ockham and his followers.[30] (The scope of his *Summa* was already ambitious enough.) Yet implicit in Dumbleton's description of the latitudes of belief and hesitation is a notion of the human will which foregrounds its reliance upon the adjudicative faculty of reason. His exposition of the process of weighing evidence, now drawn toward belief, now drawn toward hesitation, is in effect an anatomy of interpretation.

Scholars have occasionally questioned the applicability of Dumbleton's psychology of latitudes, seeing it as rather eccentric and without any practical use;[31] yet surely it would be fascinating to anyone interested in the weighing of evidence, whether in the context of the law or of the letter. In his masterful survey of fourteenth-century English philosophy, Courtenay draws attention to the movement of scholars away from the faculty of theology and toward law near the end of the century.[32] Perhaps Dumbleton's effort to schematize and quantify the process of weighing evidence was a harbinger of this intellectual turn. It certainly was a development that attracted Ralph Strode, Oxford philosopher turned London lawyer; perhaps it appealed to his poetic friend as well.[33]

Having addressed Chaucer's depiction of Melibee's will in terms of process, I will now turn to the object towards which Melibee's will inclines: that is, his wife. In an allegorical sense, of course, it is desirable to turn one's will towards Prudence; that is, if Prudence represents a transcendent abstraction. The *Tale of Melibee*, however, relentlessly invites its reader to look first to the literal level, as Wallace emphasizes;[34] and, after all, what Melibee's will inclines toward is not his wife (that is, Prudence 'herself') but 'the wil of his wif' (VII.1871). The impotence of Melibee's will was earlier implied by his inviting Prudence to 'Seyeth shortly youre wyl' (VII.1712), her avowed intention to take into account the 'wil' of his enemies (VII.1723), and Melibee's concession 'Dame ... dooth youre wil' (VII.1724), which, of course, Prudence takes as evidence of 'the goode wil of hire housbonde' (VII.1726). Thus read as a individual work, separate from the *Canterbury Tales*, the *Tale of Melibee* concludes by questioning the autonomy of Melibee's will and, by extension, the freedom of human will. In adapting the *Tale of Melibee* for inclusion in the *Canterbury Tales*, however, Chaucer denies not the autonomy of human will but the autonomy of husbands' will. When the tale ends, the Host declares that he wishes that 'Goodelief, my wyf, hadde herd this tale' (VII.1894) and compares her unfavourably to 'this Melibeus wyf Prudence' (VII.1896). This link affirms the primacy of the literal level of the tale, both by putting Prudence and Goodelief on the same level of reality and by making Melibee as much of a henpecked husband as the Host is. Patterson notes the importance of the connection: 'just as Melibee has learned nothing, neither has Harry ... [J]ust as Harry Bailly remains dominated by Goodelief ... so does Melibee.'[35] The link following the *Tale of Melibee* momentarily breaks the frame, blurring the distinction not only between characters within the tale and those within the frame, but between personifications and fictional personae.

In the *Merchant's Tale*, Chaucer reexplores the issues of interpretation and persuasion, doubt and certainty, that were so central to the *Tale of Melibee*. Like Melibee, Januarie succumbs to the will of his wife, moved not by the 'sweete wordes' of Prudence (VII.1114) but the 'suffisant answere' of the adulterous May (IV.2266). The 'sweete venym queynte' (IV.2061) that robs Januarie of his bodily sight is apparently that of Fortune; at the end of the Tale, however, Januarie's power of vision is clearly under the control of not divine Fortune, but human (that is, wifely) will. 'Household rhetoric,' in Wallace's phrase, works just as well in the parodic environment of the *Merchant's Tale* as in the didactic world of the *Melibee*. Other, less central themes persist as well:[36] first, the counsel offered to Melibee by Prudence and his advisors reappears as Januarie consults his brothers Placebo and Justinus for advice on whether to marry. Placebo, as his name suggests, wishes only to please; he assures Januarie that 'I holde youre owene conseil is the beste' (IV.1490). He elides the distinction between prudence and wisdom, telling Januarie 'ye been so ful of sapience / That yow ne liketh, for youre heighe prudence, / To weyven fro the word of Salomon' (IV.1481–3). Conversely, Justinus suggests reasonable caution, but Januarie rejects his counsel as emphatically as Melibee rejected Prudence's: 'Straw for thy Senek, and for thy proverbes! / ... Wyser men than thow / ... assenteden right now / To my purpos' (IV.1567–71). Januarie ends up taking no one's counsel but his own, convincing Placebo and Justinus to 'assen[t] fully that he sholde / Be wedded whanne hym liste and where he wolde' (IV.1575–6). This decision to rely upon only internal counsel immediately generates the 'heigh fantasye' (IV.1577) that dominates Januarie's mind throughout the tale.

The marriage of Melibee and Prudence is also echoed in the *Merchant's*

Tale, for Januarie marries in order to procure 'the lusty lyf, the vertuous quyete, / That is in mariage hony-sweete' (IV.1395–6). But while Melibee drinks in Prudence's 'swete wordes' of counsel (VII.1114), Januarie drinks in the sweetness of May's body. He makes this explicit by declaring 'How fairer been thy brestes than is wyn! / The gardyn is enclosed al aboute' (IV.2142–3), echoing the corresponding passage in the Song of Songs: 'I come to the garden, my sister, my bride ... I eat my honeycomb with my honey' (Song of Solomon 5:1). Just as Melibee is 'dronken' with 'sweete temporeel richesses' (VII.1411–12), so Januarie is 'dronken in plesaunce / In mariage' (IV.1788–9). While the nubile May does not outwardly resemble the matronly Prudence, she is characterized in terms of the two symbols of Prudence found in medieval iconography, the dove and the serpent. May is both her husband's 'dowve sweete' (IV.2139; cf. Song of Solomon 2:14) and the venomous 'monstre' that renders Januarie blind (IV.2062). Both tales have as their basis the writings of Solomon, but where *Melibee* relies upon his Proverbs and Wisdom,[37] the *Merchant's Tale* follows his Song of Songs.

By invoking the Song of Songs, January likens May to the bride whose body is 'a garden locked, a fountain sealed' (Song of Solomon 4:12). Thus when January states 'The gardyn is enclosed al aboute' (IV.2143), he refers both to the literal garden they inhabit and to the figurative garden of May's body. Her body becomes the exclusive source of the 'paradys' that the Merchant avows can be found in 'wedlok' (1264–5): 'wyf is mannes helpe and his confort, / His paradys terrestre, and his disport' (IV.1331–2; cf. 1822). Januarie knows that, as an earthly paradise, May's body is a foretaste of the heavenly paradise, for he states that, in her, 'I shal have myn hevene in erthe heere' (IV.1647). Yet Januarie soon discovers that his garden is not paradise, but the inferno, for the engaging garden reminiscent of that presided over by Deduit or Myrthe in the *Roman de la rose* (*Romaunt* 601) is instead ruled by Pluto, god of the underworld.[38] The garden's paradoxical status as simultaneously heaven and hell is finally resolved in its assumption of a third, middle status, forecast by Justinus' warning that 'she may be youre purgatorie' (IV.1670). In this sense, the *Merchant's Tale* is not only a reworking of the theme of the two kinds of honey-sweetness in marriage found in the *Tale of Melibee*, but also a parodic reenactment of the three parts of the *Commedia*, Dante's grand cosmology played out instead on the landscape of the female body.

It is difficult to speak of the *Merchant's Tale* as being of a particular genre, for it incorporates fabliau and romance, biblical story and classical myth. By alluding in the tale to not just *Melibee* but also the *Roman de la rose* and Claudian's *De raptu Proserpinae*, Chaucer places the tale in the context of the genre of allegory.[39] Also, as Neuse remarks, the marriage of May and Januarie 'seems pregnant with allegorical possibilities,'[40] both because their names emblematize their difference in age and because they metonymically evoke the seasons of spring and winter, renewal and dormancy. May's name describes not her essential character but her status at the moment, her present youth and her appearance 'lyk the brighte morwe of May, / Fulfild of alle beautee' (IV.1748–9). January's name does not only refer to his age and hoary appearance; it also evokes the Roman god of the new year and of thresholds, Januarius who looks both backward and forward.

Neuse suggests that Chaucer's marriage of youth and age may be in part based on the *Roman de la rose*, where Elde is grouped with the vices on the wall paintings outside the garden while Yowthe frolics within.[41] Calabrese also stresses the relation of the two works, arguing that the *Merchant's Tale*, like the *Rose*, is 'allegorical,' for both 'contain figures named "Reason" and "Justinus."' I would hesitate, however, to call the *Merchant's Tale* an allegory simply because one could argue that personifications appear in it. More pertinent is Calabrese's comment that 'both works are about faulty judgment, fantasy, illusion, false seeming, and spiritual blindness. ... about false paradises, idolatry, and in many ways, mirrors.'[42] Vision is at the centre of the *Merchant's Tale*, from Januarie's imaginary vision of various women to his blinding and ultimate regaining of something less than 'parfit sighte' (IV.2383).

Patterson rightly states that 'the *Merchant's Tale* is on several levels *about* "fantasye" and self-enclosure.'[43] But it is important to realize that Januarie's 'heigh fantasye' is altogether different from the word's modern meaning: it denotes his imagination, one of the primary faculties of the mind.[44] Chaucer tells how Januarie's 'fantasye ... fro day to day gan in the soule impresse' (IV.1577–8), referring to the imprinting of the species from Januarie's imagination onto the central faculty of judgment. He follows the long tradition referring to the imagination as the mirror of the soul:

> As whoso tooke a mirour, polisshed bryght,
> And sette it in a commune market-place,
> Thanne sholde he se ful many a figure pace
> By his mirour; and in the same wyse
> Gan Januarie inwith his thoght devyse
> Of maydens whiche that dwelten hym bisyde.   (IV.1582–7)

By placing the mirror in the 'commune market-place,' Chaucer stresses the role of the *sensus communis* in passing the sensible species to the imagination. At the same time, the 'commune market-place' also evokes the mercantile ethic which informs the tale.[45]

Januarie concludes by choosing one image 'of his owene auctoritee' (IV.1597), but that decision is framed in terms of blindness: 'For love is blynd alday, and may nat see' (IV.1598). By allowing himself to be governed by imagination rather than by judgment, relying upon no outside source of counsel but only his own desire, Januarie sets the stage for his own downfall. He continues to be preoccupied by the imagination, 'purtrey[ing] in his herte and in his thoght / Hir fresshe beautee' (IV.1600–1). Januarie is certain that his decision is right, thinking 'ech oother mannes wit so badde / That inpossible it were to repplye / Agayn his choys; this was his fantasye' (IV.1608–10). That is, his fantasy or imagination rather than his reason tells him that others' judgment is faulty. At the same time, the concluding phrase suggests an equivalence between 'choys' and 'fantasye' ('his choys [this was his fantasye]'), indicating that Januarie's will is operating, not in the rational faculty, but in the imaginative faculty of his mind. Because Januarie is dominated by his imagination, he falls in love not with a real woman, but with the idea of a woman. He is never able to see the real May because he is too preoccupied with the image he has generated in his own fantasy.[46] After the wedding, Januarie's wilful blindness metamorphoses into physical blindness; but even this apparent 'real' blindness may be caused by a failure of will. Medieval perspectivists such as John Pecham argued that vision is not a passive act, but rather that 'the natural light of the eye contributes to vision by its radiance.' Blindness is the result of that natural light being disturbed by infection or injury; but vision can also be impaired by a failure of the rational faculty. Pecham writes 'Reason operates imperceptibly in the distinction of visible objects. For no visible object is recognized without a distinction of the visible intentions or without a comparison or relation to the universals of known things previously abstracted from sensible things; and this cannot occur without reasoning.'[47] Januarie's rational faculty fails to make the link between the sensible species he receives and the universal form in his mind, making him psychologically (but not physiologically) blind.[48]

Januarie's blindness is miraculously cured by Pluto as he stands below the tree in which Damyan is having intercourse with his wife. In other versions of the tale, the blind man is invariably cured by God, or by St Peter acting in accord with God's will.[49] Chaucer substitutes Pluto in part to transform the earthly paradise of the garden into a hell on earth. But

Pluto's debate with Proserpine regarding whether or not to restore the sight of a cuckolded husband also affords an opportunity to reexplore the question of wifely counsel previously articulated in the *Melibee*. The couple appear first in the poem not as living characters but as literary allusions, dancing in the garden long ago 'as men tolde' (IV.2041). They materialize at the moment before May's deception of Januarie, when Damyan is already concealed in the tree above. Pluto, outraged by May's deception of Januarie, quotes Solomon, 'fulfild of sapience' (IV.2243), to support his claim that women are wicked, and declares 'Now wol I graunten ... / Unto this olde, blynde, worthy knyght / That he shal have ayeyn his eyen syght' (IV.2258–60). Proserpine belittles the importance of 'this Jew, this Salomon' (IV.2277), for 'He was a lecchour and an ydolastre, / And in his elde he verray God forsook' (IV.2298–9). Their debate recalls Melibee's rejection of Prudence's counsel on the basis that 'alle women been wikke' (VII.1057), where he quotes the same passage from Ecclesiastes cited by Pluto (IV.2246–8). Just as Melibee responds to Prudence's 'semblant of wratthe' by finally declaring 'I putte me hoolly in youre disposicioun and ordinaunce' (VII.1706, 1725), Pluto acquiesces 'Dame ... be no lenger wrooth. / I yeve it up!' (IV.2311–12).

While the debate of Pluto and Proserpine alludes to the *Tale of Melibee*, their union also evokes that of the God of Love and Alceste in the prologue to the *Legend of Good Women*. In each case, a male god is accompanied by a woman who has been removed from her rightful place on earth into Hades. There are also a number of very specific parallels between Alceste and Proserpine: one is the flower of good women, the other a defender of 'Wommen ful trewe, ful goode, and vertuous' (IV.2281). They parallel one another not only materially, but also spiritually: because both have descended into hell and returned, they are figures of Christ. But on this level, Proserpine is a parodic version of Alceste, for while Alceste facilitates clear vision, Proserpine resists Pluto's efforts to restore Januarie's sight. Her support of May ensures that Januarie will regain only 'som glymsyng, and no parfit sighte' (IV.2383). Alceste and Proserpine also parallel one another by each being the type of a particular kind of woman. In the *Legend of Good Women*, a group of women 'trewe of love' praise Alceste as the 'trouthe of womanhede' (F290, F297); in the *Merchant's Tale*, Rebekke, Judith, Abigayl, and Ester are named as women who persuaded men by offering 'good conseil' (IV.1362–74), forerunners of Proserpine's persuasion of Pluto. Otten notes the importance of these biblical figures and Alceste, stating that Proserpine, 'like them, is a Deliverer. But there is a contextual difference: she announces herself the saviour of an adulterous woman, May.'[50] At the same time, as deliverers, these biblical women are also forerunners of Mary, who by bearing Jesus played a role in the redemption of man. In this context, May's identification with Mary through the tale's resemblance to the 'Cherry-Tree Carol'[51] takes on new resonance: May delivers her people (that is, women) by conceiving a child whose father is not her husband.

Although Pluto avows his intention to cure Januarie, stating 'I wolde graunten hym his sighte ageyn' (IV.2313), May claims credit for the cure: 'I have maad yow see' (IV.2388).[52] But what May has in fact done is to teach Januarie how to not see without being blind, by telling him that he cannot trust his eyes. Her deception of January is the comic manifestation of Chaucer's determination, most explicitly articulated in the *House of Fame*, that vision is incapable of transparently mediating between the viewer and the world. In the Prologue to the *Legend of Good Women*, Chaucer argues that there are things which the eye cannot see but which nonetheless exist (F10–11); in the *Merchant's Tale*, he suggests that there are things which the eye can see but which do not exist unless the will accepts their existence.

Chaucer expresses the possibility that a thing may be true and not true, seen by the eye but not seen by the mind, in terms of double vision. Januarie's name, that of the Roman god who looks both forward and backward, anticipates his fate. May says that Januarie's vision is not yet 'ysatled' and is only 'glymsyng' (IV.2405, 2383), but Januarie, who 'Up to the tree ... caste his eyen two' (IV.2360), claims instead that he saw her there with Damyan 'With bothe myne eyen two' (IV.2385). By stressing the two eyes, as opposed to the monocular vision of the daisy in the Prologue to the *Legend of Good Women*, Chaucer gives Januarie double vision, able to switch between fantasy and reality at will: if he exercises his will, he will see reality; if he subordinates his reasonable faculty to the imaginative, he will see only fantasy. The doubleness of Januarie's vision is a model for the doubleness of the tale's meaning. Otten suggests that the biblical allusions in the *Merchant's Tale* provide two levels of signification: 'Chaucer could count on the dual vision' from which 'springs the humor, and in the recognition is the instruction.'[53] The *Merchant's Tale* is not an allegory, any more than it is a fabliau;[54] but it does what an allegory does, even as it does what fabliau does. On a literal level, the tale is about the woe that there is in marriage, and wives' tendency to deceive their husbands; on a figurative level, the interaction of May and Januarie depicts the unreliability of sense perception.

This definition of the tale in terms of a conventional allegorical structure, where the literal level exists merely as a means to the figurative, is necessarily reductive. The 'literal' level of the *Merchant's Tale* is also allegorical, in the sense that it requires an informed reader to provide the context for its interpretation, a context found in Solomon's book of Proverbs:

> My child, be attentive to my wisdom;
> incline your ear to my understanding,
> so that you may hold on to prudence,
> and your lips may guard knowledge.
> For the lips of a loose woman drip honey,
> and her speech is smoother than oil;
> but in the end she is as bitter as wormwood,
> sharp as a two-edged sword.
> Her feet go down to death;
> her steps follow the path to Sheol.
> She does not keep straight to the path of life;
> her ways wander, and she does not know it.
>
> (Proverbs 5:1–6)

The prudence and wisdom mentioned here inform, not only the *Tale of Melibee*, but also the *Merchant's Tale*'s exploration of the value of wifely counsel. May's lips drip the honey of placating words, the reward for Januarie's desire for 'mariage hony-sweete' (IV.1396). Her steps follow the path to Sheol, that is, to the garden presided over by Pluto. She is the two-edged sword, the 'knyf' that Januarie believes will never harm him: 'A man may do no synne with his wyf, / Ne hurte hymselven with his owene knyf' (IV.1839–40). The synonymity of wife and knife is stressed by the couplet's repetition (IV.2163–4) and by Damyan's introduction as 'a squyer, highte Damyan, / Which carf biforn the knyght ful many a day' (IV.1772–3).

The figurative meaning of the knife is not lost on the Host, who relates May to his own wife, just as later in the tales he compares her to Prudence: 'as trewe as any steel / I have a wyf' (IV.2426–7; cf. VII.3086, 3096–7). The Host, as is evident in the link following the *Tale of Melibee*, is a relentlessly literal reader; but even he can understand the allegory of the fabliau. Something of the old-style vertical allegory of the *Book of the Duchess* persists here in the tale's allegory of visual perception; but a new method of embedding figurative meaning also appears in the tale, one that is accessible to 'sondry folk' (I.25). The tale can be understood by a wide variety of interpreters on different levels and in different ways: people like the Host can interpret the metaphor of the knife in terms of misogyny, while people like the Clerk, for example, can interpret the knife in the more learned contexts of scripture and natural science.

The dominant framework for the figurative level of the *Merchant's Tale* is the most basic relationship in western epistemology: that of form and matter. The mirror of Januarie's imagination is like wax in that it receives the impression of the species from the common sense. Januarie believes that May is like that mirror, ready to receive his imprint both figuratively, in the form of his authority over her, and physically, in the imprint of his sperm during conception; his avowed purpose in marrying, of course, was to produce an heir (IV.1272, 1432–7). He marries May thinking that she will be as pliable as wax, for 'a yong thyng may men gye, / Right as men may warm wex with handes plye' (IV.1429–30). But when heated by Venus' 'fyrbrond' (IV.1727), or perhaps by Januarie's 'fyr of jalousie' (IV.2073),[55] the wax turns out to be receptive to the imprint of others as well. Damyan's desire for May causes her to 'take swich impression that day / Of pitee of this sike Damyan / That from hire herte she ne dryve kan / The remembrance' (IV.1978–81).

Within the terms of this analogy, May is the counterpart of the Virgin Mary as Dante describes her in his *Purgatorio*. After entering purgatory, Dante says, he saw a pure white wall on which was etched a scene of the Annunciation:

> [I]v'era imaginata quella
> ch'ad aprir l'alto amor volse la chiave;
> e avea in atto impressa esta favella
> 'Ecce ancilla Deï,' propriamente
> come figura in cera si suggella.

[T]here she was imaged who turned the key to open the supreme love, and these words were imprinted in her attitude: 'Ecce ancilla Deï,' as expressly as a figure is stamped on wax. (Pg.10.41–5)

Mary is like wax in that she passively receives God's imprint, conceiving Christ within the locked garden that, in exegetical interpretations of the Song of Songs, is her body.[56] The creation of Adam in the Garden of Eden is thus reenacted in this second creation. But there is an important difference: while Adam was created solely as a result of God's will, Christ is incarnated as a result in part of the will of a human being, Mary. Mary's body is like wax in that she receives the form that generates Christ's body; but her soul is like wax in that it conforms itself to the divine will. That act of conformity took place in the moment that she declared 'Ecce ancilla Deï,' a moment that Dante figuratively describes as the turning of a key in a lock: by surrendering her will to God, Mary opens the door to the garden. May, on the other hand, does not submit so willingly. When Januarie locks the garden gate, she finds a way to admit the person whose impression she chooses to take. In this sense, May is both a parody of Mary as Dante presents her in the *Purgatorio* and an illustration of the power of the individual will.

Conversely, Januarie's plight illustrates the weakness of the individual will. Because he allows his mind to be dominated by the passive faculty of imagination rather than the active faculty of reason, Januarie comes to be emblematic of passive matter rather than active form. Januarie's passivity is stressed, both in his telling May that 'Ye been so depe enprented in my thoght' (IV.2178) and in his crying out, when he regains his sight, 'As dooth the mooder whan the child shal dye' (IV.2365). The wax that figuratively represents May is literalized when she conspires with Damyan to meet secretly in the garden: 'This fresshe May ... / In warm wex hath emprented the clyket' (IV.2116–17). The slippery nature of warm wax allows the term itself to slip, from denoting May to denoting the agent she uses to take an impression of Januarie's key to the garden. May continues, like wax, to accept impressions, but she now accepts the impression of Damyan's key rather than Januarie's, conceiving a child whose father will never be known.

Januarie's double vision – his ability to conceive two visual images simultaneously, one realistic, one fantasical – is echoed in May's conception of a child with two fathers – one real and one apparent – for each case involves the same process of impressing form upon matter. Just as the sensible species is impressed upon the imagination, so the species of the sperm is impressed upon the inchoate matter supplied by the female. The last words spoken in the tale confirm the parallel: May hops down from the tree remarking 'Ful many a man weneth to seen a thyng, / And it is al another than it semeth. / He that misconceyveth, he mysdemeth' (IV.2408–10). Her explicit meaning is that misconceiving, that is, faulty exercise of the imagination, causes faulty judgment. Her implicit meaning is that misconceiving, that is, failing to conceive an heir, is the cause of Januarie's error.

The disordered state of Januarie's mental faculties, dominated by imagination instead of reason, is the result of a failure of Januarie's will. The Merchant forecasts this failure by declaring that a man 'may nat be deceyved ... / So that he werke after his wyves reed' (IV.1356–7). The irony is that a man is not deceived if he follows his wife's counsel only in the sense that she tells him that he is not deceived, as the Merchant suggests: 'as good is blynd deceyved be / As to be deceyved whan a man may se' (IV.2109–10). By inclining his own will to that of his wife, as Melibee and Pluto do, Januarie abnegates control of his rational faculty, allowing imagination to run free and memory to fail. The failure of his memory is evident when he fails to heed both Justinus's advice to 'have in youre memorie, / Paraunter she may be youre purgatorie' (IV.1669–70) and Pluto's praise of Solomon: 'Ful worthy been thy wordes to memorie / To every wight that wit and reson kan' (IV.2244–5). Because Januarie does not exercise 'wit and reson,' Solomon's warning regarding women is not retained in his memory.

Januarie's blindness, both the temporary eclipse of his physical vision and his subsequent inability to accept the reality he sees, is the result of a lapse in his rational faculty. The nature of this lapse is revealed by Chaucer's allusion to Argus, who 'hadde an hondred yen, / For al that evere he koude poure or pryen, / Yet was he blent' (IV.2111–13). Argus, however, was not physically blinded but rather put to sleep by the song and stories of Mercury.[57] This equation of blindness and sleep reappears in the tale's conclusion when May argues that, just as a man who has suddenly awakened may not see 'parfitly, / Til that he be adawed verraily,' so a man that has been blind may not see properly until he has 'a day or two yseyn' (IV.2397–2404).[58] Thus Januarie's blindness is the result of his dormant will.

A similar inability to see afflicts the narrator of the prologue to the *Legend of Good Women*. Although he sees Alceste with his bodily eye, he does not recognize her (F505–6). Dumbleton explains such failures as the result of inadequate prior knowledge: a person who hears or reads the term 'white' recalls a prior experience of whiteness, uniting sense perception and memory by an act of the intellect. A person who has had no such experience, however, cannot make that connection: that is why, when we see words to which there are no corresponding things in our memory, we call them 'unintelligible.'[59] The narrator does not recognize Alceste until the God of Love stimulates his memory by reminding him that he *has* seen her before: that is, he has seen her in books.

'Hastow nat in a book,' asks Love, 'The grete goodnesse of the quene Alceste / That turned was into a dayesye?' (F510–12). 'Yis,' the narrator responds, 'Now knowe I hire' (F517–18). For the narrator of Chaucer's dream visions, including the prologue to the *Legend of Good Women*, books serve as a kind of collective memory by means of which objects in the here and now are recognized – or not. Januarie has no such resource or, rather, chooses to have no such resource. In spite of the abundant information on wives and marriage recounted to him by the bookish Justinus (IV.1521ff.), Januarie refuses to 'wake up,' as it were, to what his eyes tell him is so. His dormant will is like that of the narrator of the *Legend of Good Women*, to whom the God of Love says reproachfully, 'Thy litel wit was thilke tyme aslepe' (F547). Sleep – whether the bodily sleep of Argus or the intellectual sleep of Januarie – allows the sleeper to be deceived and even, in the case of Argus, destroyed. The failure is not in the bodily eye, but in the eye of the mind: as the narrator of the *Merchant's Tale* puts it, 'as good is blynd deceyved be / As to be deceyved whan a man may se' (IV.2109–10). And what blinds a man is language, whether the songs of Mercury or the 'household rhetoric' of wicked wives.

It is, in the end, appropriate that the *Merchant's Tale* is a tale without a genre, of no fixed type. It can be read allegorically on a number of levels, but it is not an allegory. In general, allegory is a genre designed to be accessible only to the elect, following the model of Jesus' parables (cf. Matthew 13:10–17); conversely, the *Merchant's Tale* is all things to all people, able to be read with Argus's eyes, from as many different perspectives as there are viewers.[60] Its lack of genre – or plurality of genre – makes it accessible to each of the 'sondry folk' which make up the group of storytellers. Because each tale is filtered through the individual perspective of one of the pilgrims, the reader sees from a different point of view in each tale.[61] If allegory is vestigial here, it is thus fragmented so that each person may have a part.

By following the evolution of vision's role in Chaucer's works, we can better understand how he progressively alters and finally uses up the genre of allegory. Although the early dream visions are recounted by a narrator, the reader sees the images not from a unique point of view but generally, with an all-encompassing eye modelled on the sight of God, who sees all things at all times. By the *House of Fame*, however, it has become evident that reality does not look the same to every person; therefore, in the *Canterbury Tales* Chaucer finds a way to take into account the position, both physical and spiritual, of the viewer. He enables the reader to see from the perspective of the teller, and thus to receive a picture of the world that does not pretend to be true or sufficient, but merely an accurate rendering of what it looks like from one point of view. By means of the verbal exchange of storytelling, the hearer or reader accumulates many of these perspectives, finding in each of them much that is 'fals' and much that is 'soth.'

Chapter Nine

DIVISION AND DARKNESS

And God said, Let there be light: and there was light. And God saw the light, that it was good: and God divided the light from the darkness.

Genesis 1:3–4

Therefore is the name of it called Babel, because the Lord did there confound the language of all the earth: and from thence did the Lord scatter them abroad upon the face of the earth.

Genesis 11:9

Vision is a powerful metaphor for the act of knowing: the flash of insight, the thrill of illumination, the dawn of enlightenment. Yet this exuberantly positive view of the accessibility of knowledge, whether mediated through vision or through language, had already by the mid-thirteenth century come to be regarded with a more jaundiced eye. The fecundity of vision essential to the pseudo-Dionysian theory of the multiplication of light put forth by Grosseteste came, in the hands of Bacon, to be tainted with the suspicion that the species reproduced in the medium might be vulnerable to degradation. This concern, explicitly articulated by Olivi, reached its logical conclusion in Ockham's absolute rejection of the visible species. If the power of vision could not ensure the subject's accurate sense perception of the object seen, how much less certain must be other forms of knowledge, mediated not only through the senses but also through spoken language and the written word?

In the allegories discussed in the preceding chapters, the problem of certain knowledge is continually at the fore, whether it be knowledge of the self, carnal knowledge, knowledge of God, or knowledge of the world. For early allegorists, the link between vision and knowledge is strong; consequently, they use simple extramission models of vision to emphasize not only the stability of the link between seeing subject and visible object, but also, implicitly, the strength of the chain linking all created things to the divine. Allegory is the mediator that connects heaven and earth: it is the ladder depicted on Philosophy's robe, linking practical knowledge to theory. In the twelfth century, Alanus de Insulis showed how man could ascend through the multitude of changing images on Nature's *integumentum* or robe, where 'as the eye would image a picture seen in a dream was a packed convention of animals of the air.'[1] Nature's gown figuratively represents the *integumentum* of allegory, which the reader must unwrap in order to discover the meaning hidden within. In the fourteenth century, Chaucer rewrote that same scene, giving over the action to the oral exchange of the birds in an examination of the power of the individual will to choose not to choose. His birds are not images but real (and loud), while Nature presides over them only nominally. Alanus pointed upward, offering ascent through the mediation of the image; Chaucer instead directs attention to verbal exchange, to man's place here on earth and his ability to govern himself.

For Alanus, Nature's robe also offered the narrator an opportunity to gaze at the 'forma rosae, picta fideliter':[2] this faithfully depicted rose is *formosa* or beautiful because it is insubstantial, pure form. Guillaume de Lorris's rose is similarly insubstantial, not a real woman but rather an intangible reflected image of the self, eternally unattainable and unknowable. Conversely, Jean's rose is not form but matter, the body of a woman to be penetrated by the narrator and thus fully known by being known carnally. Dante's heavenly rose is insubstantial, because it is a reflected image of God: but it has within it the promise of reuniting body with soul, lover with beloved, God with man. Chaucer's daisy similarly offers a nexus where difference is resolved as the celestial 'eye of day' merges both with the mundane 'dayseye' that blooms in the grass and with the intent gaze of the narrator.

For each of these writers, vision is the primary metaphor for knowledge. But as their belief in the possibility of any certain knowledge waned, their presentation of vision became increasingly complicated. Guillaume de Lorris recognized that vision could be distorted, so much that a man might not even recognize himself; therefore he used William of Conches's terminology for the three kinds of vision to create a symmetrical, tripartite structure for his poem. There is no doubt, in Guillaume's poem, that direct vision is possible; only an internal defect, the sin of pride, makes the lover unable to render the mystery of the rose 'aperte.' Even more than Guillaume, Jean de Meun recognized the limitations of vision: he relentlessly emphasizes the gap separating the mind from certain knowledge and words from the things they represent. For that reason, Jean uses Roger Bacon's theory of the multiplication of species to show that every experience is mediated thousands of times over, and that there can be no certain knowledge. Like Jean, Dante uses sophisticated optical theories in the service of allegory; but, unlike Jean, he uses them to demonstrate that subject and object can indeed be united through the power of vision. By using what is demonstrably true – both historical and scientific fact – as the literal level of his allegory, Dante for the first time creates allegory that has no beautiful lie veiling the truth. Chaucer, conversely, suggests that truth can be found nowhere and everywhere: there is no one truth, just as there is no one cause of any action (as we learned in the *House of Fame*). Instead, something similar to truth – verisimilitude – can be had if we are willing to participate in the marketplace of storytelling, collecting the experiences of others in order to reconstruct the fragmented body of truth.

In the years after 1400, the use of vision as a metaphor for knowledge continued to decline in the realm of secular writing, though it continued to be used as part of the rhetoric of mystical experience and affective piety well into the fifteenth century. Changes in the genre of allegory, especially the increasingly varied use of personification, are accompanied by an altered depiction of light as a metaphor for intellectual or spiritual illumination, and by a new emphasis on division, degeneration, and decay, both social and moral. In John Lydgate's early fifteenth-century *Temple of Glass*, for example, light appears in the opening lines not as a source of illumination, but as a dazzling light which makes the narrator, for the moment, functionally blind.[3] The splendour that dominates the poem proves to be not the spiritual light of the divine Son, or even the intellectual light of Philosophy: instead, it is the seductive light of 'Esperus,' which appears both as the evening star of Venus (1341–61) and the bright 'stremes of [the] eyen clere' of his lady (582). Similarly, in the allegories of his contemporary, Christine de Pizan, light appears not as a ray linking the source of divine light with the human world, but rather as a diffuse 'clarté' which radiates dimly from objects such as precious stones and regal thrones in the *Chemin de long estude*.[4] The only entity in the poem who exudes light is Raison herself, who initially appears as 'tel lumiere / Qu'ou firmament pareille n'iere' ['a light which had no parallel in the firmament' (2495–6)]. Yet the illumination she offers proves to be less intellectual than affective, as the connection between Raison and the narrator is articulated not in terms of knowledge, but of desire and love: Raison addresses the narrator as 'Christine, chiere / Amie, qui science as chiere' ['Christine, dear friend, who hold knowledge dear' (6323–4)]. The *clarté* of Raison, like that of precious objects in the poem, draws attention to the virtue or potency of that which glows without explicitly illuminating anything. The light acts as a marker or signal rather than a medium of revelation.

The decreasing importance of light in fifteenth-century allegory is unsurprising when seen in the context of contemporary changes in the genre, for all of these changes reflect an increased focus on human actions in the present time rather than transcendent ideas which exist eternally. Changes in the genre include: minimization of the difference between human narrator and superhuman personification; more frequent coining of new personifications; increased emphasis on the personality (often gender) of the personification; and combination of personifications with classical gods and even with real, historical figures. Here, I can only give a few examples to illustrate the development of personification in the literature of the late fourteenth and early fifteenth centuries. They serve to point out, however, the extent to which allegory had become a rhetorical technique useful mainly to describe the ailments of contemporary society, and to prescribe appropriate remedies. It is true that (for example) both the twelfth-century *De planctu Naturae* and the late fourteenth-century *Somnium super materia scismatis* of Honoré Bouvet lament the degenerate state of society: yet while Alanus's remedy requires that each man recognize his role in the propagation of the species as decreed by Nature, Bouvet's recommendation is simply that the king of France ought to write some letters to the College of Cardinals in Rome.[5] The former text appeals to the transcendent realm of ideas as its source of authority and its organizing principle; the latter speaks only to human reason and the desire for a peaceful consensus.

The difference between human narrator and superhuman personification, so much in evidence in texts such as Boethius's *Consolation* and Alanus de Insulis's *De planctu Naturae*, is minimized in later allegory. In the early fifteenth-century allegories of Christine de Pizan, for example, the narrator emphasizes her own friendship and even kinship with the personifications with which she interacts. In the *Mutacion de Fortune*, she identifes her mother as Nature, who 'm'ama tant et tint si chiere / Que elle meismes m'alaicta' ['loved me so much and held me so dear that she herself nursed me'].[6] In the *Cité des dames*, Christine literally gets down into the trenches with Raison, Droiture, and Justice to begin the task of laying the foundations of the City of Ladies. This side-by-side labour, evident not only in the text but in the manuscript illustrations supervised by the author herself, not only increases the prominence of the autobiographical narrator central to Christine's works, but also brings the personification, as it were, down to a human level.[7]

During the same period, personifications came to be coined with increasing frequency. That is not to say that personified abstractions were not invented by earlier writers of allegory: in the *Book of the Duchess*, for example, Chaucer coins the figure of 'good faire White' in order to renew the debased genre of allegory, providing a fresh intermediary between the natural and the supernatural, an intermediary who can bridge the gap between the linear time of mourning and the eternal time of memory. In the late fourteenth and early fifteenth centuries, however, personifications come to be invented with abandon. The one-to-one correspondence of abstraction and personification comes to an end; instead, we find cases such as Anima in Langland's *Piers Plowman*, who declares that he is 'Anima' when he animates the body, 'Animus' when he exerts his will, 'Mens' when he actively knows, 'Memoria' when he appeals to God, 'Reson' when he judges, and so on.[8] Anima is a personification, an embodiment of something intangible; yet he has many functions, and just as many names. A similar fluidity of personification can be found in the *Somnium super materia scismatis* of Honoré Bouvet, where the guardians of a palace, formerly named Peace and Concord, are now named War and Opinion. The drawbridge they guard, formerly Compassion, is now called Without Humanity, while the surrounding land, formerly the Valley of Inquisition, is called Blind Ignorance. It is appropriate that the attendant of the palace, Sweet Talk, is identified as an 'ystrio' or actor: like the other personifications in the text, his identity is no more permanent than a costume.[9]

Already in Chaucer's *Tale of Melibee* we have seen the extent to which personification comes, in the course of the fourteenth century, to be contiguous with personality. Comparable emphasis on the personality of personifications, especially with regard to gender, can be seen in the *Tractié contre le 'Roman de la Rose*,' written by Jean Gerson in 1402 in connection with his participation, along with Christine de Pizan, in the infamous *Querelle de la Rose*. In it, Gerson uses numerous conventional personifications, including Raison, Prudence, and Sapience; but he also coins a character called 'Fol Amoureux,' who is prosecuted in court by a lawyer ['advocat' (24)] called 'Eloquance Theologienne.'[10] It is nothing new to personify a lover: in the very text being debated in the *Querelle*, the *Roman de la rose*, all events centre on the plight of Amant. What is new is to qualify his love as 'foolish.' Fol Amoureux may be a lover, but the merit of his love is called into question not by the events which unfold in the text (as in the *Rose*), but in the essential nature which is, by definition, manifested in the name of a personification. Gerson's figure of Eloquance Theologienne is even more novel. As a personified abstraction, Eloquance Theologienne is female, and the reader accordingly expects her to be referred to as *elle*. When she is in the act of prosecuting Fol Amoureuse, however, she carries out a task restricted (in Gerson's day) to men; therefore, when she prosecutes, Gerson calls her 'il.' The masculine pronoun reflects the lawyer's 'grande auctoritée et digne gravité' (179–80). It is instructive to compare Gerson's personification of Eloquance Theologienne with Guillaume de Lorris's personification of Bel Acueil: while Guillaume, writing in the early thirteenth century, seems to have taken pains to choose an abstraction that was grammatically masculine in order to represent the rose's 'fair welcome' as a young man, Gerson is unconstrained by the grammatical gender of his personification.

Finally, while personifications and classical gods had been combined in the past (think, for example, of Nature and Venus in the *De planctu Naturae*, or the marriage of Mercury and Philology celebrated by Martianus Capella), in later allegory the juxtaposition becomes even more evident. In John Gower's *Confessio Amantis*, Genius appears not as the priest of Nature (as he had been in the *De planctu*), but of Venus. In Christine de Pizan's allegorical works, personifications are combined freely with goddesses: for example, in the *Cité des dames*, Justice leads a procession of women which includes historical figures, classical goddesses, and virgin martyrs, ranging from Semiramis (1.15) to the Virgin Mary (3.1). Christine both coins the names of new goddesses, such as 'Othéa' in her *Epistre Othéa*, and emphasizes the kinship of narrator and goddess, as she had done with regard to personifications. In her *Livre des fais d'armes et de chevalerie*, for example, Christine appeals to Minerva not to be displeased by her effort because, 'like you I am an Italian woman.'[11] Perhaps the most peculiar use of personification by Christine de Pizan appears in that same work: she is guided in her task not by a personified abstraction such as Reason or Justice, but instead by a real person. Honoré Bouvet, in his capacity as author of the *Arbre des batailles*, appears to her in a dream in the guise of 'a solemn man in clerical garb' speaking 'as a right authorized judge' in order to instruct her regarding the various legal bases of war.[12]

In introducing the figure of Honoré Bouvet into her dream vision, Christine was simply enlarging upon a strategy Bouvet himself had used in his *Apparicion Maistre Jehan de Meun* (1398).[13] In that work, Bouvet, living by chance in the same house that Jean de Meun had inhabited more than a century earlier, has a dream vision in which the Master himself appears to lecture Bouvet on the divisive nature of contemporary French society. Jean de Meun is not the only 'real' character to appear in this dream vision: he is accompanied by a physician, a Jew, a Muslim, and a Jacobin, all of whom (for different reasons) were outcasts in the Parisian society of Bouvet's day. The *Apparicion* follows some of the conventions of the medieval dream vision, including the enclosed setting of the garden where the dream takes place; the passive ignorance of the narrator (who alone among the characters in the dream speaks in prose rather than verse); and the didactic position of the Master who addresses the narrator in language comparable to that of Philosophy in Boethius's *Consolation*. Yet it differs from other dream visions in its repeated recourse to the mundane world of daily events rather than the transcendent realm of ideas. In the dream, Jean de Meun does not tell the narrator to lift up his mind and contemplate higher things, as Philosophy does Boethius; instead, he laments the 'tres perilleux temps' (27) afflicting France in Bouvet's time.

Bouvet's central concern, throughout his works, is the divided state of both nation and Church. In this respect, he resembles many French and English writers of the period, tormented by their awareness of the papal schism, the ongoing Hundred Years War, internal civil conflict, and the stirrings of religious dissent. In the prologue to his handbook of battle strategies, the *Arbre des batailles*, Bouvet laments 'the great dissension which is today among Christian princes and kings' and 'the great grief and discord which exist among the communities.' Christendom is 'burdened by wars and hatreds, robberies and dissensions,' while France is consumed by 'the division of opinion between nobles and commons.'[14] Similarly, in Bouvet's *Apparicion Maistre Jehan de Meun*, the Jacobin regrets that, if no reconciliation is forthcoming, 'this schism will last a long time' (1325). This concern is most emphatically expressed, however, in Bouvet's *Somnium super materia scismatis*, which, as its title suggests, is not only a dream vision but a *somnium*, defined by Macrobius as 'an enigmatic dream [which] requires an interpretation for its understanding.'[15] Bouvet's *somnium*, however, seems to require little in the way of interpretation, for it is allegorical only on a very simple level. The majority of the text is devoted to a sequential encounter with the various kings of

Europe who have been 'friends' to the Church, ending (of course) with the king of France. The allegorical content is restricted to the outset of the work, when the Church appears as a personified woman, of beautiful form and elegant attire when seen from the front, but consumed with maggots and covered only with shredded rags (literally, 'materia schismatis' [104]) when seen from the rear.

Bouvet's preoccupation with division – that is, the breakdown and factionalization of political, moral, legal, and religious structures – is characteristic of his age. Similar concerns can be found in John Gower's *Vox Clamantis* and *Confessio Amantis*, where the statue seen in Nebuchadnezzar's dream, made up of various materials, is emblematic of the divisiveness and decay of contemporary society. 'The world empeireth every day,' says Gower:

> The cause hath ben divisioun,
> Which moder of confusioun
> Is wher sche cometh overal,
> Noght only of the temporal
> But of the spiral also.[16]

In the next generation, John Lydgate would continue in Gower's footsteps, addressing the breakdown in moral and social order in his *Serpent of Division*, while Christine de Pizan would describe the civil disruption within France in her *Lamentacions sur les maulx de la guerre* and *Lavision-Christine*. In the latter work, a 'Dame Couronné,' identified by Christine in her prologue as France herself, describes how her children have torn her body apart: she hopes that the people of France will 'espargner ses doulces mamelles quilz ne la succent jusques au sang' ['spare her sweet breasts, so that they do not suck until they draw blood'].[17] The pervasive 'chaos' found in Honoré Bouvet's *Somnium super materia scismatis*, generated by the 'rumor and tumult of the present schism' (105), is a fitting emblem of the ailment late medieval allegorists sought to diagnose and to remedy. Division, of course, is not invariably negative. In Genesis, division is creation: in the beginning, God 'divided the light from the darkness' (Gen. 1:3–4), the waters from the sky. Similarly, in both Robert Grosseteste's *Chastel de amur* and Guillaume de Lorris's *Roman de la rose*, the division of light into a spectrum of colours which results when light is refracted represents not decay, but regeneration: even, paradoxically, a kind of multiplication, whether of grace (in the *Chastel*) or of identities (in the *Rose*). By the later Middle Ages, however, division had come to represent not the creative capacity of Genesis, but the social and linguistic fragmentation of Babel, where language was confused and the people scattered (Gen. 11:9).

Having outlined the contours of the fifteenth-century afterlife of allegory and the role of vision in it, I will conclude by returning to the comparative medieval-modern framework discussed in the opening section of chapter 2. Some medievalists have recently suggested that knowledge of the past takes place not just on an intellectual plane, but on an affective one as well. Carolyn Dinshaw develops this argument at length, claiming that it is worthwhile to pursue the 'impulse' that allows the modern reader to 'mak[e] connections across time ... extend[ing] the resources for self- and community building into even the distant past.' Such connections are based on analogical relationships: of Margery Kempe, Dinshaw states, 'Margery and I are both queer – in different ways, in relation to our very different surroundings – and thus are queerly related to one another.'[18] While recognizing the alterity of the past, through analogy (Margery is to her environment as I am to mine) past and present can come together, if only in an elusive 'touch.'[19] Like Dinshaw, Nicholas Watson argues that an affective understanding of the past is both possible and desirable. One does not merely acquire knowledge about the past: one learns from the past itself how to approach it. Watson describes feelingly his own experience of 'a sort of hermeneutic osmosis' in the course of his work on (with?) Julian of Norwich, which brought him to the realization that 'bodies also think, minds also feel ... Julian and I had come to mirror one another across the hermeneutic gap.'[20] Both Watson and Dinshaw argue persuasively that a nexus between past and present is imaginable and can be realized through an empathetic effort. Watson makes explicit his own reliance upon Caroline Bynum's method of approaching the history of medieval women's spirituality,[21] and Dinshaw too posits a 'touch' of past and present based on analogical relationships, an approach which owes much to Bynum's recommendations that the relationship of past and present be understood as 'analogous and proportional.'[22] Bynum, however, points out that it is imperative to take into account the categories and the vocabulary in each culture, medieval and modern: a detailed understanding of these systems of language, these ways of ordering the world, is crucial to any reflective understanding of the past.

What I have endeavoured to do here is to fill in one aspect of the analogical comparison: that is, to anatomize the language of vision as it functioned in the allegory of the later Middle Ages. An affective knowledge of the past is most readily imaginable through the lens of a medi-eval writer who is herself suffused with desire and who cannot conceive of any knowledge divorced from that desire: it is no coincidence that Dinshaw and Watson's arguments are founded on readings of Margery Kempe and Julian of Norwich. Medieval allegory, conversely, represents the acquisition of knowledge in a much colder light. Even Dante's *Commedia*, where all knowledge and all movement is grounded in divine love, lacks the abnegation of artistic authority essential to devotional literature.[23] The role of desire in knowledge, so crucial to the epistemol-ogy of Augustine and pseudo-Dionysius, remained central in late medi-eval theology; Aquinas and, especially, Bonaventure continued to foreground the reciprocal desire of the soul and God. The persistence of vision as a metaphor for knowledge in devotional writing reflects an epistemology grounded in desire, where faith acts as the guarantee that the communion of subject and object is really possible. More secular texts, however, depict knowledge as contingent and, at best, fragmen-tary. The emphasis on division and splitting found in the late medieval allegories of Lydgate, Honoré Bouvet, and Christine de Pizan reflects the breakdown not only in the coherence of civil society, but in systems of knowledge. The language of multiplication and division appearing in the texts discussed in this chapter is related to the increasing assimilation of terms and concepts drawn from the exchange of currencies and commodities which took place with increasing frequency during the thirteenth and fourteenth centuries; to treat this topic properly will require a separate study. The discourse of economic exchange intersects, however, with the discourse of vision examined in the preceding chap-ters.[24] Their interaction illustrates how systems of knowledge fertilize one another, perpetually generating new systems of knowledge, new ways of ordering the world – however temporarily.

NOTES

1: Illumination and Language

1 'Visiones enim duae sunt; una sentientis; altera, cogitantis.' Augustine, *De trinitate* 11.9.16; ed. Mountain and Glorie 2: 353, lines 13–14; trans. McKenna 338.

2 '[O]mnis sensus in multis sit falsus, uisus tamen omnium est falsissimus.' William of Conches, *Dragmaticon*; 3.2.9, ed. Ronca 60; my translation.

3 Plato, *Timaeus* 106–7 (47 A, 47 C).

4 Aristotle, *De anima* 102–9 (2.7).

5 See the first page of Jay's Introduction to *Downcast Eyes*, in which he uses twenty-one visual metaphors to illustrate the ubiquity of the link between vision and understanding. On that link as it is manifested in various Indo-European languages, see Tyler, 'Vision Quest.'

6 See Harvey, *Inward Wits*, esp. 43–61; for an older but still useful survey, see Wolfson, 'Internal Senses.'

7 See Yates, *Art of Memory*, for a remarkable account of classical, medieval, and Renaissance systems of memorization based on visual images. On vision and memory, see Carruthers, *Book of Memory*, esp. 17–18, 32; Coleman, *Ancient and Medieval Memories*, esp. 39–59.

8 See Trottman, *Vision béatifique*; for a brief account of the history of the doctrine, see Redle, 'Beatific Vision.'

9 In making this observation, I do not mean to suggest that vision is actually primary, in any absolute or essential sense, but rather that ancient, medieval, and modern epistemologies generally begin from that assumption. For an unusual medieval exception to the rule, see Burnett, 'Superiority of Taste.'

10 Williams, *Christian Spirituality* 74. On the relationship of love and knowl-

edge in Augustine's *De doctrina christiana*, see Louth, *Discerning the Mystery* 78–80.

11 Louth, *Origins* 132–58; quotation from 144. Cf. Williams 86.

12 Louth, *Discerning* 81–2.

13 Milbank, 'Sacred Triads' 465.

14 On pseudo-Dionysius and Proclus, see Louth, *Origins* 159–78. For a more general study, see Louth, *Denys*.

15 Williams 118.

16 On apophasis and cataphasis in pseudo-Dionysius, see Louth, *Denys* 87–8; *Origins* 164–78. See also the useful summary of the theology of pseudo-Dionysius and its late medieval reception in Minnis and Scott, *Late Medieval Theory* 168–73.

17 Louth, *Origins* 167.

18 John of Damascus, *On the Divine Images* 25 (I.17).

19 Carruthers, *Book* 122–4, 150.

20 [Pseudo-]Cicero, *Rhetorica ad Herennium* 209 (III.xvii.30).

21 Augustine, *De Genesi ad litteram* 1:118 (9.12.20), trans. Taylor 1: 84; Dante, *De vulgari eloquentia* 10–12 (I.7).

22 On the iconography of Prudence and Laziness (Otia or Luxuria), see Katzenellenbogen, *Allegories* passim.

23 '[D]exterae partes videntur esse sinistrae, et sinistrae videntur esse dexterae.' Alanus de Insulis, 'Summa de arte praedicatoria' 118. Natura is called 'speculum caducis, / Lucifer orbis' ['mirror for mortals, light-bearer for the world'] in the *De planctu Naturae* 2: 458 (metrum 4); trans. Sheridan 128. The dichotomy of good and bad mirrors seems to have persisted at least into the Renaissance. The sixteenth-century writer Raphael Mirami states that, 'for some, mirrors constitute a hieroglyph of truth in that they uncover everything which is presented to them ... Others, on the contrary, hold mirrors for a symbol of falsity because they so often show things other than as they are.' Quoted in Baltrusaitis, *Essai* 6, translated in Nolan, *Through a Glass Darkly* 291.

24 The best known of these works is probably Fletcher's *Allegory: Theory of a Symbolic Mode*, while Quilligan's insightful study of allegorical language makes its purpose explicit in its subtitle, 'Defining the Genre.'

25 On Benjamin, see Buck-Morss, *Dialectics of Seeing*; also Teskey, *Allegory and Violence* 12–14. On the importance of spectacle and the redemptive potential of allegory in Benjamin, see Kelley, *Reinventing Allegory* 256–7 and, on de Man's rhetorical notion of allegory, ibid. 10–11.

26 Fletcher, *Allegory* 28.

27 Tuve, *Allegorical Imagery* 145.

28 'The genesis of this theory was quite simple. I noticed that William Langland punned a great deal ... I knew that Edmund Spenser had done the same. So I began reading allegory, counting puns. ... From this shared fact – the generation of narrative structure out of wordplay – the members of the genre grouped themselves' (Quilligan, *Language* 21–2). Note the passive construction in the last phrase.

29 Honig, *Dark Conceit* 183. Whitman similarly concludes by characterizing allegory as an ultimately undefinable mode that is always '[a]t once slipping away from its object and toward it' (*Allegory* 262).

30 Copeland and Melville, 'Allegory' 179–80.

31 Martianus Capella, *De nuptiis* 28 (IV.327–8); trans. Stahl and Johnson 106–7. On the mnemonic function of allegory, see Carruthers, *Book* 142.

32 *Rhetoric* 3.11.1–2; 1411b. Aristotle uses the phrase 'πρὸ ὀμμάτων' repeatedly in his discussion of metaphor; the translator most often renders the phrase as 'vivid.' See *Rhetoric* 3.10.7; 1411a–1411b.

33 *Rhetoric* 3.11.5; 1412a.

34 *Poetics* 22.17; 1459a. The word Aristotle uses for the act of discerning similarity in apparent dissimilarity, θεωρεῖν, literally means 'to see.'

35 *Rhetoric* 3.11.6; 1412a.

36 The search for a language that will make meaning visible goes on: Ernest Fenellosa writes that the 'etymology [of the Chinese written language] is constantly visible ... After thousands of years the lines of metaphoric advance are still shown, and in many cases actually retained in the meaning.' In the Chinese written language, we find 'the visibility of the metaphor' (*Chinese* 25).

37 Augustine, *De Trinitate* 15.9.15, 15.9.16; trans. McKenna 471, 472.

38 Alanus de Insulis, *De planctu Naturae*, trans. Sheridan 221.

39 Bloomfield notes that, for Quintilian, personification is a kind of 'vivid illumination,' a trope which provides 'animation' and 'a sense of activity' (171). On the history of the term 'personification,' see Whitman, *Allegory* 269–72; for a survey of theories of personification from antiquity to the twentieth century, see Paxson, *Poetics* 8–34.

40 On the distinction between pseudo-Dionysian symbol and Augustinian sign in the twelfth century, see Chenu, *Nature* 124–7. On the pervasiveness of pseudo-Dionysian notions of symbolism, see Dronke, *Fabula* 44–5, 67. It is important to note that there is a dramatic distinction between medieval notions of symbolism and the familiar Romantic dichotomy of allegory and symbol. (See, for example, Dronke, *Fabula* 9n2 and Whitman, *Allegory* 267–8.)

For the Romantics, transcendent meaning is located in the symbol, if anywhere; allegory, by contrast, represents nothing more than a literary trope which obscures meaning rather than revealing it.

41 See Hagen, *Allegorical Remembrance*; Boyde, *Perception and Passion*. Nolan's *Through a Glass Darkly* surveys a wide range of classical and medieval texts with respect to the metaphor of the mirror, but does not discuss the development of optics, saying merely that there was a 'general muddle regarding sight, light, and the phenomena of refraction' (287). In these studies, Chaucer has solicited a disproportionate amount of attention: see Brown, 'Chaucer's Visual World'; Holley, *Chaucer's Measuring Eye*; Torti, *Glass of Form*; Klassen, *Chaucer on Love*.

42 Delany argues that the decline of allegory is the result of growing disbelief that knowledge could be acquired through analogical thought: 'the importance of the logic, science and political theory of the [fourteenth century] was its variegated attack on analogical thought. In this tendency I believe Chaucer participated, turning intuitively from allegory to other modes in order to express his vision of a complex and contingent world' ('Attack' 58). On the role of nominalism, see ibid. 56–7; Bloch *Etymologies* 149–58; Sturges *Medieval Interpretation* 29.

43 Copeland, *Rhetoric* 63–5, 81.

44 Minnis, *Authorship* 33–9.

45 '[B]iblical exegetes borrowed and built on the techniques of arts commentators, and ... the techniques of scholastic biblical exposition ... were widely applied to the exposition of pagan *auctores*.' Copeland, *Rhetoric* 112; cf. Minnis, *Authorship* 22, 55–7.

46 On the *Ovide moralisé*, see Copeland, *Rhetoric* 107–26.

47 See the perceptive account of Dante's allegory in Minnis and Scott 382–7.

48 Quilligan, *Language* 28; 236.

49 Copeland and Melville 162.

50 Minnis and Scott 3; cf. Minnis, *Authorship* 73–159, esp. 112.

51 Copeland, *Rhetoric* 93.

52 Kelly, *Medieval Imagination* 178.

53 Bloch 28, 153.

54 Sturges 2, 33.

55 'Celestial Hierarchy' 154 (3.2).

56 Quilligan, *Language* 236, 21–2.

57 Copeland and Melville 162, 184.

58 Copeland, *Rhetoric* 93–5; she states that the secondary translation 'occurs in later stages of the development of vernacular textual traditions,' (95). On the relationship of allegory and translation, see Copeland and Melville 179.

59 'Metaphora est verbi alicuius usurpata translatio.' Isidore, *Etym.* I.xxxvii.2. On the use of the terms *translatio* and *translative* in the twelfth century, see Dronke, *Fabula* 21.

60 On etymology and origins, see Bloch 30–63 and also Jacquart and Thomasset, *Sexuality and Medicine*, especially their first chapter, 'Anatomy, or the Quest for Words.' On the development of etymology from *origo* to *expositio*, see Klinck, *Die lateinische Etymologie* (Thanks to Anders Winroth for this reference.)

61 'Est vero "etimologia" nomen compositum ab "ethimos" quod interpretatur "verum" et "logos" quod interpretatur "sermo" ut dicatur "ethimologia" quasi "veriloquium," quoniam qui ethimologizat veram, id est primam, vocabuli originem assignat.' Pierre Elie, quoted by Klinck 15.

62 Dominicus Gundissalinus, *De divisione philosophiae*, ed. Ludwig Baur (Munster, 1903) 140; cited in Minnis, *Authorship* 20. Klinck uses the same metaphor to describe the relationship between allegoresis and etymology: 'Etymology ... cracks open ... the chaff, removes the *integumentum* and displays the seed of the spiritual meaning, so that it can be effortlessly taken up by the reader.' Klinck 164, my translation; for etymology as a key, see Klinck 160.

63 'Then boys in whom the seed is for the first time working its way into the rough seas of their youth ... are encountered by images from outside, images from any and every body, images heralding a gorgeous face and a beautiful complexion which awakes and stimulates the parts, swollen with excess seed, so that just as if the whole job had been done they often spurt out the flood of a massive river.' *De rerum natura* 66–7 (IV.1030–6). Godwin, the translator, stresses the 'internal coherence' of the fourth book, which is unified by its focus on the difference between voluntary and involuntary actions (7). The flow of forms from objects and the emission of seed from the body are both involuntary actions.

64 Elford, 'William of Conches' 314n29.

65 Bernardus remarks that the surface of the eye must be smooth because 'the images of things adhere most clearly to a smooth surface.' *Cosmographia* 2.14; trans. Wetherbee 123, 126.

66 On cerebral spermatogenesis, see Jacquart and Thomasset 53–6.

67 *Cosmographia* 2.14, trans. in Stock, *Myth and Science* 218; *Cosmographia* 2.13, trans. in Stock 212. On the relationship of brain physiology and sexual reproduction in Bernardus, see ibid. 207–19.

68 Roger Bacon, *Opus maius* 4.2.1; trans. Lindberg in Grant, *Source Book* 393.

69 'Hoc speculum mediator adest, ne copia lucis / Empiree, radians uisum, depauperet usum'; 'speculum ... dissona rerum / Paret in hiis facies. Hic

res, hic umbra uidetur, / Hic ens, hic species, hic lux, ibi lucis imago'
(*Anticlaudianus* 6.127–8, 6.119–23; trans. Sheridan 160).

70 '[U]surpo michi noua uerba prophete. / ... uerbisque poli parencia cedent
/ Verba soli. ... / Carminis huius ero calamus, non scriba uel actor.' Alanus
de Insulis, *Anticlaudianus* 5.269–73; trans. Sheridan 146.

2: The Multiplication of Forms

1 Jay, *Downcast Eyes*, 590.
2 Spearing, for example, declines to choose a specific psychoanalytic theory
in order 'to retain freedom of manoeuvre in deploying the large categories
[of] looking, listening, secrecy, desire, gender, and power' (*Poet as Voyeur*
2). The medieval optical context he posits is not much more specific (10).
See also Nolan, *Through a Glass*, which includes a brief summary of David
Lindberg's survey of medieval optics among a melange of paragraphs on
Lacan, Eco, and 'Cultural History from Colie to Kristeva.'
3 Stanbury, 'Lover's Gaze' 230–1.
4 Bynum, 'Material Continuity' 244–52.
5 Bynum, 'Why' 29.
6 Vescovini, *Studi* and 'Vision et Réalité'; Tachau, *Vision and Certitude* and
'Maxime visus.'
7 Smith, 'Big Picture' 569.
8 In using the term 'verisimilitude,' I am following Tachau's distinction
between the terms 'veridical,' 'truth,' and 'verisimilitude': 'That a concept
is a "veridical" image entails that it depicts its object as that object "truly"
(*vere*) is. ... To term a concept a "verisimilitude" is weaker, for all that is
claimed is that it is *somehow* "like the truth" (*veri similitude*) ... [N]either
label should be taken as a synonym for "true"' (Tachau, *Vision and Certitude*
17n48).
9 See Edgerton, *Renaissance Rediscovery*, esp. Chapter 5, 'The Fathers of
Optics' (64–78), and also idem, *Heritage*; Wack, *Lovesickness* 56–9, 90–3,
101–2, and 132–5, and also 'Mental Faculties.' For an occasionally unreliable history of optics, see Ronchi; for a not very technical yet wide-ranging
account, see Park, *Fire Within*.
10 '*Species*, from *spec* ... translates the Greek *eidos*, derived from *eido* "to see,"
perfect form *oida*, "to know" ("idea" is from the same root). Hence both
terms literally denote "what a thing looks like," yet both underlie terms
denoting intellectual acts. In view of its equivocal nature, it is no wonder
that *species*, even when intended as purely intelligible, was taken as a sort of
pictorial representation in the mind' (Smith, 'Big Picture' 574n24). On

medieval usage of the term *species*, see Michaud-Quantin, 'champs semantiques' 113–50, esp. 143–4; on the use of the term in the theological context, see Phillips, 'John Wyclif.'

11 See Colish, *Stoic Tradition* 1:51–2; Verbeke, *L'évolution.*

12 *Timaeus* 45 C-D, trans. in Cornford 152–3. Cf. Cornford 151–6 on Plato's
theory of vision.

13 *De sensu* 2.438a26; see also Lindberg, *Theories of Vision* 217n39.

14 On the influence of pseudo-Dionysius, see Lindberg, *Bacon's Philosophy of
Nature* xxxiii–xliv, and Tachau, 'Maxime visus' 204–6.

15 See Thonnard 125–75; O'Daly, *Augustine* 80–105.

16 On sight and self-knowledge in the *De Trinitate*, see Stock, *Augustine* 259–73.

17 Colish, *Stoic Tradition* 2:173–7.

18 For a brief summary of Macrobius's influence, see Stahl's introduction to
his translation of the *Commentary* 39–55; for a more detailed assessment, see
Duhem, *Système du monde* 3:44–162.

19 This formulation appears in Diogenes Laertios, Galen, and Gellius, as
noted in Sambursky, *Physics* 23; cf. Colish, *Stoic Tradition* 1:51.

20 Boethius, second commentary on Aristotle's *Peri hermeneias*, 1.1, in *Anicii
Manlii Severini Boetii: Commentarii in librum Aristotelis* Peri hermeneias, ed.
Carolus Meiser (2 vols. Leipzig, 1877–80) 2:34; on Boethius's use of Stoic
accounts of perception in his commentary on the *Peri hermeneias*, see
Magee, *Boethius* 101–2.

21 See Colish, *Stoic Tradition* 2:289–90.

22 See Hoenen and Nauta, especially the fine essays by Nauta on the commentaries of William of Conches (pp. 3–39) and Nicholas Trevet (pp. 41–67).
See also Gibson, *Boethius*; Minnis, *Boethius*; Minnis, *Chaucer's* Boece.

23 For a brief survey of the transmission of galenic theories, see Siraisi, *Medieval* 4–16.

24 Galen derived the theory of humours from Hippocrates, that of the three
souls from Plato. See De Lacy, 'Third Part.' Siegel argues that Galen actually refers to three aspects of a single soul, but was frequently misunder-
stood as referring to three souls or a tripartite soul (*Psychology* 114–30).

25 On Galen's theory of vision, see Siegel, *Sense Perception* 40–126.

26 On Constantinus Africanus's translation of the *Isagoge*, see Jacquart, 'A
l'aube'; Newton, 'Constantine' in Burnett and Jacquart. (This volume also
includes several articles concerning the translation of the *Pantegni*.)

27 On ocular anatomy as found in the *Pantegni*, see Gül Russell, 'Anatomy of
the Eye'; on Hunain's use of Galen, see Eastwood, *Elements*.

28 Haefili-Till 101–3.

29 Gül Russell 263.

30 On the translations of the 'school of Salerno,' see Jacquart and Micheau 96–129.

31 On Benvenutus Grassus's position in ophthalmic tradition, see Eldredge, 'Anatomy.'

32 *The Wonderful Art of the Eye* 49, lines 15–16 (and cf. 52, lines 87–90). For the Latin text with English translation, see *De oculis eorumque egritudinibus et curis*, ed. and trans. Wood.

33 On the date and distribution of Benvenutus's work, see Eldredge, 'Textual Tradition.'

34 Jacquart, *Médecine médiévale* 413.

35 D'Alverny, 'Translations and Translators.'

36 Elford, 'William of Conches' 323n80; also Elford, 'Developments' 233–7. Lindberg (*Theories of Vision* 248n40) and Ricklin ('Vue et vision' 39) agree that, if there was influence, Adelard's work was probably prior.

37 Ronca, 'Influence' 285.

38 See Gregory, 'Platonic Inheritance' and *Anima mundi*.

39 Southern has argued that William taught at Paris; this has been disputed by Dronke, while Häring, Elford, and Ronca support Dronke's position, to varying extents. The relevant articles are cited in Elford, 'William of Conches' 309n7, and a summary of the debate can be found in Luscombe 23–5.

40 The *Dragmaticon*, formerly available only in the edition of Guilielmi Gratarolus, *Dialogus de substantiis physicis ante annos ducentos confectus, à Vuilhelmo Aneponymo philosopho* (Strasbourg, 1567; rpt. Frankfurt: Minerva, 1967), has been recently edited by Italo Ronca. An English translation has been provided by Ronca and Curr.

41 *Saturnalia* 7.9.16–25 (pp. 431–3); cf. *De placitis* 7.8 (De Lacy 2:376–7), and cf. 1.7 (1:84–5) and 6.3 (2:380–1).

42 The opening of *Saturnalia* 7.14 concerns the magnification of objects seen through a spherical glass vase filled with water, as does the opening of the fifth book of Ptolemy's *Optics* (5.5, prop. 79). Ptolemy uses as his example the magnification of a coin, while Macrobius describes how an egg, an onion, and the fibres seen in a liver appear enlarged. See Ptolemy, ed. Lejeune 225; see also Smith, 'Psychology.'

43 Flamant, *Macrobe* 494–524; Gersh, *Middle Platonism* 2: 493–596.

44 On the dates of William's works, see Ronca's edition of the *Dragmaticon* xix–xxii. On William's glosses on Macrobius, see Jeauneau, *'Lectio Philosophorum'* 267–308.

45 For editions of William's commentaries, see Ronca's edition of the *Dragmaticon* lxxix–lxxx.

46 'Idem aiunt videre nos vel tuitione, quam phasin vocant, vel intuitione, quam emphasin vocant, vel detuitione, quam paraphasin nominant' (ccxxxix; ed. Waszink 251).

47 *Glosae super Platonem*, ed. Jeauneau 243–5.

48 See Lindberg, *Theories of Vision* 247n16.

49 Ibid. 91, and Elford's corrective in 'William of Conches' (310) and 'Developments' passim. See also Ricklin 39–40.

50 On William's use of Constantinus Africanus, see Ronca, 'Influence'; O'Neill, 'Cerebral Membranes' and 'Descriptions.' O'Neill has even suggested that William's use of Constantinus Africanus's *Pantegni* and *Isagoge*, which is well documented, is supplemented by Stephen of Antioch's later, improved translation which corresponds to the *Pantegni* (the *Liber regalis*, ca. 1128) in his *Dragmaticon* ('Descriptions' 214).

51 For a persuasive account of how the role of the *obstaculum* in vision is related to William's account of solar eclipses, see Ricklin 34–7.

52 Lindberg, 'Alkindi's Critique' 488–9; cf. Lindberg, *Theories of Vision* 31–2, 227n66.

53 On the lens and retina in Islamic optical theory, see Lindberg, *Theories of Vision* 54–6.

54 See Dewan, 'St. Albert'; Steneck, 'Albert.'

55 See Lindberg, 'Alhazen'; also *The Optics of Ibn al-Haytham, Books I–III*, trans. with commentary by A.I. Sabra. On dissemination and influence, see Lindberg's introduction to the reprint of Risner's 1572 edition, v–xxxiv.

56 Lindberg declares that he 'prefer[s] to substitute the expression "philosophy of light," since much of it has nothing to do with metaphysics.' *Theories of Vision* 95. On the influence of Grosseteste, see A.C. Crombie, *Origins* and 'Grosseteste's Position.' Lindberg considers Crombie's evaluation of Grosseteste's influence to be somewhat overstated (*Theories of Vision* 102).

57 *De luce* 51–9, trans. Riedl 10–17; quotations from ed. 51 and 56, trans. 10 and 15. See also the paraphrase of the *De luce*, with an exposition of the mystical significance of light, in Grosseteste's *Hexaëmeron* 2.10, trans. 97–100.

58 See McEvoy, *Philosophy* 69–123 and 351–68.

59 On the Neoplatonic context of Grosseteste's optical theory, see Eastwood, 'Medieval Empiricism.'

60 See Richard Lemay, 'Roger Bacon.' On the chronology of new translations, see 'Appendix: The Translation of Optical Works from Greek and Arabic into Latin,' 209–13 in Lindberg, *Theories of Vision*.

61 Lindberg, 'Alhazen's Theory' 337.

62 See Tachau for a lucid yet detailed account of the profound influence of

Bacon's theory of the multiplication of species on medieval philosophy. Tachau also notes that the debate generated by Bacon's theory continued to influence both 'subsequent generations of intellectuals and ... culture more generally' (*Vision and Certitude* xvii). See also Tachau, 'Maxime visus' 201–24.

63 Both edited by Lindberg; on dating of the works, see Lindberg's introduction to Bacon, *Perspectiva* xxiii. Lindberg's introduction to *De multiplicatione specierum* includes a substantial biography of Bacon; his introduction to *Perspectiva* includes a detailed account of Bacon's theories and their place in the history of optics. For a briefer survey of Bacon's notion of species, see Lindberg, 'Roger Bacon.'

64 On the wide dissemination of Pecham's *Perspectiva communis*, see Lindberg's introduction to the edition 29–32; cf. Lindberg, *Theories of Vision* 120–1.

65 On the relationship of Bacon, Pecham, and Witelo, see Lindberg, 'Bacon, Witelo' 103–7; see also Lindberg, 'Alhazen' 336–41; Lindberg, *Theories of Vision* 116–21.

66 See Smith's introduction to his edition of Witelo's *Perspectiva*, pp. 13–72, esp. 58–66. Books I, II, and III of Witelo's *Perspectiva* have been edited by Unguru.

67 On the study of perspectiva in medieval universities, see Vescovini, 'Perspectiva'; see also Lindberg's introduction to Bacon, *Perspectiva* xciv–c.

68 See Tachau, *Vision and Certitude* 130–5.

69 Tachau summarizes Olivi's rather complex view as follows: 'if species are representations, and if a chain of species is generated between the sensible object and the sense, then the final species in this chain when received in the sense would first, most strongly, and most properly represent to the eye the species from which it had been multiplied, rather than the object. In other words, although the Baconian account purports to provide real and direct contact with extramental objects, in fact the direct contact is *only* with the final mediator in a chain of mediators' (Tachau, *Vision and Certitude* 44).

70 Ockham, *Opera Theologica* V, quoted in Tachau, *Vision and Certitude* 131.

71 On sense perception, see Wood, 'Adam Wodeham,' and Wood's introduction to the edition of Wodeham's *Lectura secunda* 5–49, esp. 20–30. More generally, see Courtenay, *Adam Wodeham*; Tachau, *Vision and Certitude* 275–310.

72 On Dumbleton, see the pioneering yet largely unpublished work of James A. Weisheipl, including his thesis, 'Early Fourteenth-Century Physics'; his edition of Dumbleton's *Summa logicae et philosophia naturalis*; 'The Place of John Dumbleton'; and 'Ockham and Some Mertonians.' See also Sylla,

Oxford Calculators 130–44 and passim; also her 'Oxford Calculators and Mathematical Physics.' A cursory account of Dumbleton's optics also appears in Molland.

73 A useful survey of the use of these texts appears in Twomey. On the dissemination of Isidore's encyclopedia, see Bischoff; Reydellet; Vollmann. On that of Bartholomaeus Anglicus, see Seymour, 'English Owners'; Seymour, 'French Readers'; Lidaka, 'John Trevisa'; Lidaka, 'Bartholomaeus Anglicus.' On that of Vincent of Beauvais, see Paulmier-Foucart, 'L'Atelier'; Voorbij, '*Speculum Historiale*'; and several of the essays in Paulmier-Foucart, Lusignan, and Nadeau.

74 Isidore of Seville, *Etymologiae* 11.1.20, trans. Sharpe 39.

75 *De proprietatibus rerum* 62–6 (3.17), quotation from 62. A fourteenth-century Middle English translation can be found in *On the Properties of Things: John Trevisa's Translation of Bartholomaeus Anglicus*, De proprietatibus rerum. Seymour refers to a planned edition of the Latin text in his edition of the Middle English translation (3:41); until then, a better (though partial) Latin text can be found in *Bartholomaeus Anglicus: On the Properties of the Soul and Body*, ed. R. James Long.

76 On Vincent's use of Albertus Magnus's commentary on the *De anima*, see Lieser. Vincent cites the *De anima* commentary explicitly in books 24–7 of the *Speculum naturale*, but Lieser argues that it is used, unattributed, earlier in the text as well (Lieser, *Vinzenz von Beauvais*, 68).

77 See Jonsson, 'Le sens du titre' 11–32.

78 For a fine survey of the philosophical and medical ways of schematizing the mental faculties, see Harvey, *Inward Wits*; see also Wolfson, 'Internal Senses'; Steneck, 'Problem of the Internal Senses.'

79 See Sudhoff, 'Lehre' 149–205.

80 For the galenic division of the faculties, see, for example, Hunain ibn Ishaq, in *Le Livre des questions sur l'oeil de Honaïn ibn Ishaq* 94–5 (quest. 53–6).

81 On imagination, see Bundy, *Theory of Imagination*; on reason, see Klubertanz, *Discursive Power*; on memory, see Carruthers, *Book*.

82 For a more detailed account, see Harvey 43–5; Steneck 9–17; and Tachau, 'Senses and Intellect' 656–8.

83 See Bundy 199–224.

84 See Augustine's description of how one can imagine impossibilities such as a black swan or a four-footed bird by conflating images one has seen (*De Trin.* 11.10.17); cf. Bartholomaeus Anglicus's similar account of how one can imagine a mountain made of gold (*De prop. rer.* 3.11; ed. 53, trans. 1:99).

85 Boethius distinguishes between *ratio* and *intellegentia*, which is a higher faculty. See Magee 128n140 and 142–4.

86 See the *Rhetorica ad Herennium* 3.16.29–3.22.37; cf. the artificial memory system of Thomas Bradwardine which appears in Carruthers, *Book* 281–8 (Appendix C). On memory techniques, see Carruthers, 'Poet as Master Builder' and *Craft* 10–35; Yates, *Art of Memory*; Hajdu, *Mnemotechnische*.

87 Carruthers, *Book* 18; 47–60. Modern neuropsychology also attests to the strong visual substrate of memory: both expert mnemonists and otherwise handicapped idiot-savants, when asked how they can recite long strings of numbers or lists of unrelated items after studying them only briefly, respond 'I see it.' For a survey of such 'remembering as seeing,' see Bellezza.

88 Noted and translated in Rigg, *Anglo-Latin Literature* 225.

89 *Château d'Amour* lines 557–628; quotation from 567–8. See also the Middle English translations of this popular work edited by Sajavaara.

90 The phrase famously coined by Kuhn in his *Structure of Scientific Revolutions*.

3: Guillaume de Lorris's *Roman de la rose*

1 The term 'courtly love' was coined by Gaston Paris in the nineteenth century, and so some writers claim that the use of the term in relation to medieval literature is anachronistic. See Hult, 'Gaston Paris'; Newman, *The Meaning of Courtly Love*; Boase, *The Origin and Meaning of Courtly Love*; Ferrante, 'Cortes Amor.'

2 References to the *Roman de la rose* are to Lecoy's edition; translations are my own.

3 Walters, 'Author Portraits' 362. On medieval reception of the *Rose*, see Huot, *The* Romance of the Rose *and Its Medieval Readers*.

4 See Gunn, *Mirror of Love* 95, 120. Dahlberg himself summarizes the debate regarding the poem's unity in the 'Preface to the 1995 Edition,' pp. xvii–xxvi in his translation, *The Romance of the Rose*.

5 For a survey of modern perspectives on the completion of the first *Rose*, see Hult, *Self-Fulfilling Prophecies* 5. Although Hult mentions Lejeune and Poirion, he takes no note of Strohm ('Guillaume as Narrator and Lover in the *Roman de la Rose*'), whose conjecture precedes those of Lejeune and Poirion by several years.

6 'The overall image of Jean's public that emerges from his apologia ... involves an expansion of Guillaume de Lorris's courtly public into a potentially universal audience. ... Guillaume thus serves as a point of departure. ... What emerges is thus a new kind of vernacular poetic discourse that is potentially universal' (Brownlee, 'Reflections in the *Miroër aus Amoreus*' 70).

See also Brownlee, 'Jean de Meun and the Limits of Romance' and 'The Problem of Faux Semblant.' Uitti ('From *Clerc* to *Poète*') has also shown how the divergent aims of the two parts of the poem are manifested in the two different narrative personae.

7 'Ce roman de Guillaume est un cadre qui contient un cadre qui, à son tour, contient un troisième cadre qui encadre, et ainsi de suite, *ad infinitum*.' Verhuyck, 'Guillaume de Lorris' 286.

8 Barney, *Allegories of History* 184.

9 Verhuyck, 'Guillaume' 288.

10 Kelly, 'Translatio Studii' 300.

11 See Vitz's perceptive analysis of how Guillaume uses 'dichotomies between inside and outside, opened and closed ... to express structures of the self' (64) in 'Inside/Outside,' esp. 77–9. See also Blumenfeld-Kosinski's account of the use of the terms 'overte' and 'coverte' in both parts of the poem ('*Overt* and *Covert*').

12 Kruger has shown in detail that, even as medieval theories of dreaming divide all dreams into categories of true and false, they simultaneously recognize that it may be difficult to determine the degree of verity of a particular dream. See *Dreaming in the Middle Ages* 17–34.

13 Blumenfeld[-Kosinski], 'Remarques' 390. The usage of the rhyme pair *songe/mensonge* in Chrétien de Troyes's *Yvain* is particularly interesting, for it appears at the end of an account of how sense perception takes place (specifically, the mechanism of hearing): 'Et qui or me voldra entandre, / cuer et oroilles me doit randre, / car ne vuel pas parler de songe, / ne de fable, ne de manconge' ['Now, whoever will hear my words, must surrender to me his heart and ears, for I am not going to speak of a dream, an idle tale, or lie']. *Le Chevalier au lion (Yvain)*, ed. Roques, lines 169–72; trans. Comfort, *Chrétien de Troyes: Arthurian Romances*.

14 'Cum aspicit tamen non libero et directo lumine videt sed interiecto velamine quod nexus naturae caligantis obducit.' Macrobius, *Commentarii in somnium Scipionis* I.3.18; ed. Willis 12, trans. Stahl 92.

15 'Hujus igitur imaginariae visionis subtracto speculo, me ab extasi excitatum in somno prior mysticae apparitionis dereliquit aspectus.' Alanus de Insulis, *De planctu Naturae* prosa 9; ed. Wright 2: 522, trans. Sheridan 221.

16 For translation of *ainz* and *ançois* as 'instead' or 'rather,' see Hult, *Self-Fulfilling Prophecies* 130.

17 'Cura interpretationis indigna sunt, quia nihil divinationis adportant.' Macrobius I.3.3; ed. Willis 8–9, trans. Stahl 88.

18 The dream is identified as a *somnium* by Langlois, *Origines et sources* 58; Dahlberg, 'Macrobius' 575; and Pickens, 'Somnium' 176.

19 'Somnium proprie vocatur quod tegit figuris et velat ambagibus non nisi interpretatione intellegendam significationem rei quae demonstratur.' Macrobius I.3.10; ed. Willis 10, trans. Stahl 90.

20 Pickens, for example, has suggested that 'the narrator reveals himself to be an inadequate interpreter of his own *somnium.* ... Nowhere is [this] more evident ... than in the prehistory of Narcissus' (183).

21 Richards ably refutes earlier interpretations of Oiseuse as a figure of Luxuria and suggests that her name, at least in Guillaume's portion of the poem, is better interpreted as 'verbal frivolity' ('Reflections' 310).

22 'Dexterae partes videntur esse sinistrae, et sinistrae videntur esse dexterae.' Alanus de Insulis, 'Summa de arte praedicatoria,' PL 210: 118.

23 Robertson associates the tower in Chrétien de Troyes's *Cligés* with the Tower of Babel ('Doctrine of Charity' 40n68).

24 Hult, 'Closed Quotations' 267; see also Pickens, 'Somnium' 182.

25 The parallel of dream and tower is more subtly reinforced by the presence of another tower which is briefly mentioned in the *Rose* when Reson descends briefly from her ivory tower to counsel Amant to abandon his desire: 'Lors est de sa tor devalee, / si est tot droit a moi venuc' ['Then she came down from her tower, and came right to me' (2960–1)]. She comes 'tot droit' both in the sense of moving directly and of herself being 'tot droit,' 'entirely right.' She descends from a tower, implying a transition from what is literally another plane. Just as a dream takes place on another level of consciousness, so the tower is a physical representation of movement to another plane – in the case of Bel Acueil's prison, an inaccessible one.

26 '*Speculantes* dixit, *per speculum* videntes, non de specula prospicientes. Quod in graeca lingua non est ambiguum unde in latinam translatae sunt apostolicae litterae. Ibi quippe speculum ubi apparent imagines rerum ab specula de cuius altitudine longius aliquid intuemur etiam sono verbi distat omnino. Satisque apparet apostolum ab speculo, non ab specula dixisse *gloriam domini speculantes.*' *De Trinitate* 15.8.14; ed. W.J. Mountain and Fr. Glorie 479, trans. McKenna 469–70.

27 Hillman appears to be the only reader to claim that 'the Fountain episode, including the signification of the crystals, has at times been somewhat over-interpreted' (229n2). Hillman asserts, rather absurdly, that the episode at the fountain of Narcissus is too brief to be of significance: 'Guillaume is not known as a poet of obscure or hidden meaning, and it is the nature of allegory to identify the critical elements which compose the drama. If the poet had ascribed a greater significance to the crystals, he would undoubtedly have made his intentions crystal clear' ('Another Look' 238).

28 On the crystals as the lady's eyes, see Lewis, *Allegory of Love* 125; Frappier, 'Variations' 151; Köhler, 'Narcisse' 160. On the crystals as the lover's own eyes, see Robertson, *Preface to Chaucer* 95; Fleming, *Roman de la Rose* 93.

29 Hult, *Self-Fulfilling Prophecies* 279; 66.

30 Knoespel, *Narcissus* 80; 84–5.

31 Ibid., 87.

32 In this respect, Hunain departs from Galen, who placed the lens not at the centre of the eye, but nearer the cornea. See Lindberg, *Theories of Vision* 34; also Eastwood, 'Elements of Vision' 6–7.

33 Jeauneau, 'L'Usage' 35–100.

34 Though William of Conches's glosses on Macrobius remain unpublished, excerpts appear in Dronke (*Fabula* 69–78) and in the unpublished dissertation of Helen Rodnite [Lemay], 'The Doctrine of the Trinity in Guillaume de Conches' *Glosses on Macrobius*' (Columbia University, 1972). On William's theory of allegory, see Dronke, *Fabula* 13–55, esp. 25; cf. Copeland and Melville, 'Allegory and Allegoresis' 159–87.

35 William of Conches, *Glosae super Platonem*, ed. Jeauneau 89–90; cf. Jeauneau, 'Integumentum' 63. Subsequent references are in the text by page number; translations are my own.

36 *Commentum super sex libros Eneïdos Virgilii*, ed. Jones and Jones, trans. Schreiber and Maresca; see also *The Commentary on Martianus Capella's* De nuptiis Philologiae et Mercurii *Attributed to Bernardus Silvestris*, ed. Westra, especially 23–33 on the use of the *integumentum* in the latter commentary. William of Conches is known to have written a commentary on Martianus as well, which is apparently lost. It is tempting to speculate what differences in the notion of *integumentum* a comparison of the two might have revealed.

37 Gregory, *Anima mundi* 123–74, esp. 133–48; cf. Jeauneau, 'Integumentum' 66ff.

38 Jeauneau, 'Integumentum' 64, 86.

39 Ibid. 39–40; cf. Klinck 156–60.

40 '*Unus, duo, tres* ... Queritur cur Plato, quem constat nichil sine causa fecisse, librum suum a numeris incepit. ... Plato igitur, ut pitagoricus, sciens maximam perfectionem in numeris esse, quippe cum nulla creatura sine numero possit existere, numerus tamen sine qualibet potest existere, ut perfectionem sui operis ostenderet, a perfectis scilicet numeris incepit.' William of Conches, *Glosae* 71.

41 For the text, see William of Conches, *Glosae* 153–7; cf. Jeauneau, 'Integumentum' 69, 75.

42 'Conus geometrica forma est oblonga que ascendando protenditur de lato in acutum ... [F]orma divina teres, id est perfecta, dicitur quia principio

caret et fine. Ab hac ergo forma anima descendens in conum protenditur quia conus est in initio indivisibilis, sed in imo dividi potest. Similiter anima ... [quia] per potentias dividitur quas diversas in diversis corporibus exercet.' Text in Rodnite [Lemay] 202; cf. Jeauneau, 'Macrobe' (rpt. in *Lectio Philosophorum*) 295.

43 'Sed quia melius retinentur que oculis percipiuntur, predictam dispositionem aquarum in figura oculis subiciamus.' William of Conches, *Glosae* 89.

44 Ricklin, 'Vue et vision' 34–5.

45 Rodnite [Lemay] 128.

46 William of Conches, *Dragmaticon* 5.4; ed. Ronca 145–51, trans. Ronca and Curr 98–100.

47 'Quamuis omnis sensus in multis sit falsus, uisus tamen omnium est falsissimus. Baculus, cum sit integer, fractus uidetur in aqua; duae turres, si a longe prospiciantur, quamuis loco distent, uidentur coniunctae.' William of Conches, *Dragmaticon* 3.2.9; ed. Ronca 60, trans. Ronca and Curr 40.

48 On manuscripts of the *Dragmaticon*, see Elford, 'William of Conches' 308. On the importance of the Timaean glosses, note Klibansky's statement that 'the *Timaeus* was long considered as a textbook for scientific cosmology. ... the influence of the school of Chartres and in particular of William of Conches reigned supreme until it was supplanted by Aristotle in Paris in 1255' (*Platonic Tradition* 162n1).

49 William of Conches, *Glosae* 242–3.

50 *Intuitio*: 'quod in eius superficie aliquod apparet simulacrum ut in speculo'; *detuitio*: 'quando non in superficie sed in profondo apparet simulacrum ut in liquoribus.' *Glosae* 244.

51 'Tuitione, quam phasin uocant, uel intuitione, quam emphasin appellant, uel detuitione, quam paraphasin nominant.' Calcidius, *Timaeus a Calcidio translatus commentarioque instructus* cap. 239, ed. Waszink 251.

52 'Non in cute speculi, sed introrsum.' Calcidius, cap. 242, ed. Waszink 254.

53 William of Conches, *Glosae* 244.

54 Jacquart and Thomasset have shown how medieval writers used etymology to illustrate the meaning of a word rather than to reveal its philological source, focusing particularly on the anatomical vocabulary in Isidore's *Etymologiae* to demonstrate that 'philology in the service of theology played a very important part in medieval thought. Etymology was merely the consequence of a certain conception of language' (*Sexuality and Medicine* 16). On the link of etymology and memory, see Yates, who notes that in the *Dialexeis*, the 'earliest *Ars memorativa* treatise, the images for words are

formed from primitive etymological dissection of the word' (*Art of Memory* 30).

55 *Altfranzösisches Wörterbuch*, ed. Adolf Tobler and Erhard Lommatzsch (11 vols. Berlin, 1925–.), s.v. 'deduire.'

56 Dahlberg, trans., *The Romance of the Rose* 361, note to line 590.

57 *Deduit* as a noun first appears in the *Eneas* (ca. 1160), though the verb *deduire* appears earlier, in the eleventh-century *Vie de saint Alexis*. Passages associating *deduit* with homoerotic desire include 'il voldroit deduit de garçon' (9132) and 'a garçon moine son deduit' (9137). *Deduit* is applied to the physical pleasure enjoyed with women only in a disparaging sense, when it is rejected: 'de feme n'a soing; il n'a de tel deduit besoig' (9146–7). On homoerotic desire in the *Eneas*, see Gaunt, 'Epic to Romance' 18ff.; Burgwinkle, 'Knighting the Classical Hero' 1–43; Baswell, 'Men in the *Roman d'Eneas*' 162–3.

58 See Harley, 'Narcissus' 334; Uitti, 'Cele [qui] doit estre Rose clamee' 40.

59 On Orpheus, see Jeauneau, 'Integumentum' 43–55.

60 Carruthers notes how, in one image-based memory system, 'One remembers abstract concepts by a concrete image: "sweetness" by an image of someone happily eating sugar or honey, "bitterness" by an image of someone foully vomiting. Wholly abstract ideas like God, angels, or the Trinity, can be attached to "an image as painters make it" or ... "as it is usually painted in churches." This is more direct evidence that every sort of image, whatever its source or placement, was considered to have memorial utility' (*Book of Memory* 131–45; quotation from 134–5). See also Yates, *Art of Memory* 50–81, for an account of the use of allegorical paintings to aid memory of abstract ideas. On the *Ad Herennium* system and the *Rose*, see van den Boogaard, '*Roman de la rose*' 85–90.

61 Hult, *Self-Fulfilling Prophecies* 226–7.

62 Tobler and Lommatzsch, *Altfranzösisches Wörterbuch*, s.v. 'paire.'

63 Ibid. s.v. 'lëece.'

64 'Qui uero dicunt irim non esse substantiam, dicunt illam esse imaginem solis. Sed quia omnis imago similis est illi cuius est imago, sol autem rotundus est, in ea rotunda apparet figura. Et quemadmodum iris non est substantia, sed imago substantiae, sic nec sunt in ea colores, sed coloris imagines ... Fuerunt qui dicerent irim non esse aliud quam nubem nec nimis obscuram nec nimis lucidam, habentem ex quatuor elementis quatuor principales colores: ab igne rubeum, ab aere purpureum, ab aqua glaucum, a terra propter herbas et arbores uiridem.' William of Conches, *Dragmaticon* 5.4.3–4; ed. Ronca 145–6, trans. Ronca and Curr 99.

65 Aristotle's *Meteorologica* was available in the medieval West, having been translated directly from Greek during the mid-twelfth century. See Devons, 'Optics' 224n17. William of Conches draws upon the *Meteorologica* in his *Dragmaticon.*

66 'Per Irim diversicolorem et soli oppositam figuratur sensus qui quidem diversis speciebus et potentiis est distinctus et rationi contrapositus ... Itaque sensum multiformem et rationi contrapositum intelligimus per Irim multicolorem et soli oppositam.' *Commentum super sex libros Eneïdos Virgilii,* ed. Jones and Jones 27; trans. Schreiber and Maresca 29.

67 'Dextra manus triplicis speculi flamata nitore / Splendet et in triplici speculo triplicata resultat.' Alanus de Insulis, *Anticlaudianus* I.450–1; ed. Bossuat 70, trans. Sheridan 63. Reason's threefold mirror illustrates the three aspects of the relationship of matter to form.

68 'Aliquando deus denuntiat vel per simile vel per contrarium. Per simile ut si somniarem me dentes emittere vel sanguine diminui amicos mori significaret. Per contrarium ut si somniarem me ridere significaret me flere vel econverso. Et hoc vocatur somnium.' [Rodnite] Lemay 276.

69 See Harley 327.

70 This view of the symbolism of Echo is shared by Kenneth Knoespel (136n26).

71 Goldin has argued that the narrator escapes the fate of Narcissus due to the benefit of experience: the narrator 'has learned to see not his own image but other things in the fountain of Narcissus.' While it seems obvious that the narrator sees himself reflected in the fountain, saying 'de fort eure m'i mire' ['in a painful hour I saw myself there' (1605)], Goldin claims that, although the lover sees himself, his reflection appears not on the surface of a pool but in the eyes of the lady, so that he experiences not the self-directed love of Narcissus, but the productive love of another. But whether the mirror is the surface of the pool or the eye of the lady matters little: clearly the lover gazes at himself, as does Narcissus, and the mirror's harmful effect is apparent in the lover's statement that 'Cil miroërs m'a deceü' ['this mirror deceived me' (1607)].

72 'An nescis Echo nympham hoc agere?'; 'Non sum Narcissus, quem ipsa insequatur: physicam rei rationem, non fabulam quero.' William of Conches, *Dragmaticon* 6.21.6; ed. Ronca 255–6, trans. Ronca and Curr 164.

73 Fleming, *Roman de la Rose* 197.

74 Ferrante, *Woman as Image* 109.

75 Ibid. 110; 110n15.

76 Dahlberg, 'First Person and Personification' 46.

77 Dronke has recently suggested that this work circulated in the clerical

milieu of the early thirteenth century rather than the 'courtly' world of the late twelfth century ('Andreas Capellanus' 107).

78 Jung notes that 'Jalousie fait enfermer Bel Accueil et les roses dans un château fort. Guillaume de Lorris reprent ici, en le modifiant, le château décrit dans le *De Amore* [de] André le Chapelain' (*Etudes* 309). See also Demats, 'D'*Amoenitas*' 217–33.

79 Andreas Capellanus, *The Art of Courtly Love*, trans. Parry 73. See also the edition with translation by P.G. Walsh, *Andreas Capellanus: On Love*.

80 'Dextre partes videntur sinistre et e converso, in speculo plano, quia si movet homo dextram manum, simulacrum sinistram et e converso.' William of Conches, *Glosae* 245.

81 See, for example, Harley 333–4.

82 In her analysis of an important early manuscript of the *Rose*, Walters notes that the illuminator consistently depicts Bel Acueil as a woman in order to stress the character's feminine nature (being a personified aspect of the lady), even though, 'because of the masculine gender of the name, Bel Acueil is represented as a man in the majority of manuscripts' ('Illuminating the *Rose*' 179). Here, the illustrator's desire to emphasize that the character is an aspect of the lady conflicts with the author's choice of a grammatically masculine personification.

83 'Prius ancillam captandae nosse puellae / Cura sit: accessus moliet illa tuos, ... / Hanc tu pollicitis, hanc tu corrumpe rogando: / Quod petis, ex facili, si volet illa, feres.' Ovid, *The Art of Love, and Other Poems*, trans. Mozley, lines 351–6.

84 Perhaps the most famous Old French translation of the *Ars amatoria* is that of Chrétien de Troyes, now lost, mentioned in the opening lines of *Cligés*. Another, by 'Maistre Elie,' dates from the beginning of the thirteenth century. There is also an anonymous prose translation from the first third of the thirteenth century; see the Introduction to *L'Art d'Amours (The Art of Love)*, trans. Blonquist.

85 Harley 333–4.

86 Poirion, 'Narcisse et Pygmalion' 161.

87 'Iste ego sum: sensi, nec me mea fallit imago; / uror amore mei: flammas moveoque feroque.' Ovid, *Metamorphoses* 3.463–4.

88 'Ista repercussae, quam cernis, imaginis umbra est: / nil habet ista sui; tecum venitque manetque; / tecum discedet, si tu discedere possis!' Ovid, *Metamorphoses* 3:434–6.

89 Fleming, *Roman de la Rose* 6.

90 Luria, *Reader's Guide* 42.

91 Barney 190.

92 Harley 334; cf. Uitti 40.

93 Poirion, 'Narcisse et Pygmalion' 158.

94 Vitz, 'Inside/Outside' 88.

95 Quilligan, 'Words and Sex' 207.

96 Alanus de Insulis, *De planctu Naturae*, prosa 1; ed. Wright 439, trans. Sheridan 93.

97 'Forma rosae, picta fideliter, / A vera facie devia paululum.' Alanus de Insulis, *De planctu Naturae*, metrum 2; ed. Wright 444, trans. Sheridan 106.

98 Dahlberg stresses the importance of the rose as a symbol of sensuous love, and notes the medieval association of the five petals of the rose with the five senses ('Love' 577–8).

99 Vitz notes that 'the lover's relationship to any ostensible "object" becomes increasingly problematic as the story progresses. For it is not just the rose – that one erotic thing – that the lover wants, but also Fair Welcome. ... Increasingly, no one object or act would be adequate to satisfy the lover'; Guillaume's poem 'is about *desire*, about being *outside*' ('Inside/Outside' 87–8).

100 'Narcissus etiam, sui umbra alterum mentita Narcissum, umbratiliter occupatus, seipsum credens esse alterum se, de se sibi amoris incurrit periculum.' Alanus de Insulis, *De planctu Naturae*, prosa 4; ed. Wright 463, trans. Sheridan 136.

101 Harley 331.

102 Ferrante, *Woman as Image* 108.

103 Hult discusses this question in terms of authorial authority: 'Guillaume's own title, *Le Roman de la Rose*, has replaced (or subsumed) [Ovid's] *Art of Love*; implicitly, romance has replaced art (doctrine), and the Rose – the ultimate symbol – has replaced Love' (*Self-Fulfilling Prophecies* 135).

104 Hult points out that Narcissus is described in the *Roman* as being 'morz toz envers' (1572), which he interprets as not only 'envers' but 'en vers'; that is, not merely 'dead, flat on his back,' but 'dead totally in poetic verse' (*Self-Fulfilling Prophecies* 297). Similarly, Guillaume's narrator is 'versez' both in the sense of being turned and in the sense of being put into verse.

4: Jean de Meun's *Roman de la rose*

1 These include Robertson, *Preface to Chaucer*; Fleming, *Roman de la Rose*; Dahlberg, in his translation, *Romance of the Rose*, as well as in his numerous articles on the poem; and Gunn, *Mirror of Love*. Gunn assumes that Guillaume and Jean's intentions were congruent, and thus refers to 'the completed allegory' (95); at the same time, he acknowledges that Jean produced a 'vastly extended amplification of Guillaume's theme' (120).

2 Jackson, 'Allegory and Allegorization' 167. Köhler also stresses the difference in Guillaume and Jean's approaches: 'A l'hédonisme courtois, si
raffiné, si spiritualisé de Guillaume de Lorris, Jean de Meung oppose un
hédonisme naturaliste, fondé sur les idées averroïstes du XIIIe siècle'
('Narcisse' 156). As early as 1936, C.S. Lewis had emphasized the disparity
of the two parts of the *Rose*: 'Jean de Meun did not write a mere conclusion.
Rather he made the original *Romance* the pretext for a new poem which
completely dwarfed it' (*Allegory of Love* 137).

3 On sexuality and language in Jean's continuation, see Quilligan, 'Words
and Sex' 197–200.

4 In his translation, Dahlberg comments on the last line of Jean's continuation: 'This line suggests that the Lover has experienced a rather erotic
dream. On this level, the dream is an *insomnium*, according to Macrobius's
classification, and therefore unworthy of notice' (*Romance of the Rose* 424–
5).

5 Wetherbee, 'Literal and Allegorical' 270.

6 Lecoy, ed., *Roman de la Rose* 280, note to lines 4249–328. All quotations
from the *Rose* are from Lecoy's edition and are noted in the text by line
number; translations are my own.

7 Brook, 'Continuator's Monologue' 16.

8 Brownlee, 'Problem of Faux Semblant' 266.

9 Hult, *Self-Fulfilling Prophecies* 40.

10 Huot, 'Authors, Scribes, Remanieurs' 204. See also Huot's fuller account of
the contemporary reception of the *Rose* in *The* Romance of the Rose *and Its
Medieval Readers*.

11 Kelly, 'Li chastiaus' 75.

12 Brownlee, 'Orpheus' Song Re-sung' 203.

13 Frappier claims that this 'graine' is not the seed of love, or semen, but
rather coloured dye: 'Jeu de mot sur *graine* qui signifie aussi "écarlate"'
('Variations' 150n45). In view of the importance of sterility and fertility in
the first *Rose*, as seen in Guillaume's parable of the peasant near the poem's
end, it is difficult to discount the meaning 'seed' in the word 'graine.'

14 The cognate of Guillaume's term 'espi,' Latin *spica*, appears in the passage
quoted from Mark.

15 In the line notes to his translation of the *Roman de la rose*, Charles Dahlberg
follows Langlois in claiming that the comparison of the two statues 'as a
mouse is to a lion' is actually inverted in lines 20785–6, the statue of
Galatea likened to a mouse and the figure in the tower to a lion. This is
based on the assumption that, in the comparison which concludes the
passage on Pygmalion, 'cele que tant ci pris' ['that which I here esteem so
greatly' (21197)] refers to the image in the tower. But the statue Jean has

just described for the last 410 lines is that of Galatea; it seems implausible that Jean de Meun would have intentionally produced such a misleading grammatical construction.

16 'Jean de Meun forgets for thousands of lines together that Bialacoil is a "young bachelor." He identifies Bialacoil with the heroine and describes female conduct accordingly' (Lewis, *Allegory of Love* 140).

17 Gunn 287.

18 Camille, *Gothic Idol* 330.

19 Hill has pointed out the role of Venus in helping both Pygmalion and the lover to reach their amorous goals ('Narcissus, Pygmalion' 414).

20 Hill 422.

21 This is described in lines 21697–700. Also see Hill 414–15, and Wetherbee, 'Literal and Allegorical' 268, who agree that the rose is inseminated.

22 Camille 327.

23 Alanus de Insulis, *De planctu Naturae*, ed. Wright 430 (metrum 1), trans. Sheridan 70; ed. Wright 463 (prosa 4), trans. Sheridan 135–6.

24 Camille 330.

25 Gunn 267.

26 On the meaning behind Jean de Meun's alternative title, see Eberle, 'Lovers' Glass' 244–6.

27 Knoespel, *Narcissus* 96–7.

28 *Le Roman de la Rose*, ed. Lecoy xix.

29 Wetherbee 265; 274.

30 Eberle 243; 253; 250.

31 Ibid. 250; for Lecoy's suggestion, see his edition of the *Rose*, 3: 172.

32 Eberle 249.

33 Eberle employs the phrase coined by Crombie in her sketch of Grosseteste: 'Grosseteste's major optical treatise, his work *De iride et speculo*, represented a critical synthesis of all the information known to him, which he tested by his own experiments, and set in the context of his own metaphysics of light derived from the neo-Platonic, Christian tradition' (Eberle 248).

34 Lindberg, *Theories of Vision* 94–5.

35 Eberle 245.

36 In his introduction to the reprint edition of Alhazen, Lindberg notes that 'the *Perspectiva* was not sufficiently well known in the first half of the century to have had any influence on the optical works of Robert Grosseteste,' and states that 'it is abundantly clear from the content of Grosseteste's optical works that they were written without knowledge of Alhazen's *Perspectiva*' (vii, vii n.14).

37 'Et ut plane appareat, non accidere hoc ex comprehensione formae:

obturetur medietas forminum instrumenti, & in aliquo obturatorum sit scriptura aliqua, si inspiciatur speculum regulae per foramen scripturam respiciens: comprehendetur in speculo scriptura: per quodcunq, aliud minime: quod si scripturae forma speculo esset impressa, per quodcunq; foramen instrumenti posset percipi.' Alhazen 113 (IV.4.20). Books 1–3 only, concerning direct vision, have been edited by A.I. Sabra, *Optics of Ibn Al-Haytham.*

38 Lindberg, Introduction to Alhazen, *Opticae Thesaurus* xxi.

39 Ibid.

40 In the Introduction to his edition with translation of Bacon's *De speculis comburentibus*, Lindberg states that 'Bacon conceived the central purpose of the treatise to be the solution of a problem about burning mirrors' (*Roger Bacon's Philosophy of Nature* xxix).

41 Text and translation of book 5 of the *Opus maius* appear in Lindberg, *Roger Bacon and the Origins of* Perspectiva 332–5.

42 *Rose* 18151–66. Compare Alhazen 214–18 (lib. 6, cap. 7, 39–44), and also Bacon, *Opus maius* lib. 5, pars 3, dist. 1, cap. 4 (ed. and trans. Lindberg 270–1) and lib. 5, pars 3, dist. 3, cap. 4 (Lindberg 332–3). Alhazen, like Bacon, states that the phenomenon is caused by concave mirrors; Bacon adds that 'Among all mirrors, concave spherical ones give rise to the greatest deception' (Lindberg 269), a comment echoed in Jean's statement that mirrors producing phantasms 'les regardeürs deceit' ['deceive the viewers' (18166)].

43 *Rose* 18179–90. Cf. Alhazen 265 (lib. 7, cap. 5, 33), and also Bacon, *Opus maius* lib. 5, pars 3, dist. 3, cap. 3 (Lindberg 330–1).

44 'De visione fracta maiora sunt, nam de facili patet per canones supradictos quod maxima possunt apparere minima et econtrario, et longe distantia videbuntur propinquissime et econtrario ... Et sic ex incredibili distantia legeremus litteras minutissimas, et pulveres ac harenas numeraremus propter magnitudinem anguli sub quo videremus. Et maxima corpora de prope vix videremus propter parvitatem anguli sub quo videremus ... Et sic posset puer apparere gigas et unus homo videri mons, et in quacunque quantitate, secundum quod possemus hominem videre sub angulo tanto sicut montem, et prope ut volumus.' Bacon, *Opus maius* lib. 5, pars 3, dist. 3, cap. 3 (Lindberg 332–5).

45 *Le Roman de la Rose*, ed. Lecoy viii.

46 Lindberg, *Roger Bacon's Philosophy of Nature* xxv.

47 Ibid. xxiii.

48 Ibid. xvii, xix.

49 Lindberg, *Theories of Vision* 255n59.

50 For a concise account of Bacon's theory of the multiplication of species, see Lindberg, *Theories of Vision* 107–16, esp. 113–14. See also Lindberg's Introduction to *Roger Bacon's Philosophy of Nature* and, of course, Bacon's own *De multiplicatione specierum*, published in the same volume.

51 Bacon, *Opus maius* lib. 4, pars 1, dist. 2, cap. 1; trans. Lindberg in Grant, ed., *A Source Book in Medieval Science* 393. See also Bacon, *De multiplicatione specierum* 6.2, in Lindberg, *Roger Bacon's Philosophy of Nature* 258–9.

52 Eberle suggests that the tale appears here 'as a parody of the moral applications found in Seneca. ... Venus and Mars use mirrors, not to know themselves nor to correct their senses, but to enable them to indulge themselves in the pleasures of the senses' (258).

53 'Oculus igitur habet tres tunicas seu panniculos et tres humores et unam telam ad modum tele aranee. Et prima tunica eius ... ramificatur ad modum retis concavi in prima parte sui. Que ideo vocatur rete vel retina.' Bacon, *Opus maius* lib. 5, pars 1, dist. 2, cap. 2 (Lindberg 26–7).

54 *Opus maius* lib. 5, pars 3, dist. 3, cap. 1 (Lindberg 324–5).

55 See Macrobius, *Commentarii in Somnium Scipionis* I.3, ed. Willis 8–10.

56 Economou, *Goddess Natura* 114.

57 Knoespel 99.

58 Albertus Magnus, *De mineralibus*, excerpt trans. Wyckoff in Grant, ed., *Source Book* 629.

59 Albertus Magnus, *De mineralibus*, trans. Wyckoff 629. Wyckoff notes that Albert's immediate source, 'Constantine ... was quoting the *Lapidary* of Aristotle, which distinguishes three kinds of *hyacinthus* (corundum gems), red, yellow, and blue, of which the red (*granatus*, "like pomegranite seeds"), that is, rubies, are said to be the best. But Albert takes *granatus* to be the best of the *red* stones ...' (629n29).

60 Verhuyck concludes his excellent article on the structure of the first *Rose* by stating that Jean de Meun 'démystifie le jeu des miroirs. ... *Jean de Meun ou la destruction des cadres* serait un tout autre article' ('Guillaume' 289).

61 'Although the Lover is seeking an art of love that will reveal the "shortest path" ("le plus brief chemin," 10,032) to the Rose, the discourses in the second part of the *Romance* are "digressions" in the etymological sense of the word: they divert the lover from the direct pursuit of his chosen path and they use love as a point of departure for a series of excurses on a bewildering variety of other topics' (Eberle 241).

62 Wetherbee 281.

63 Hult, *Self-Fulfilling Prophecies* 40.

64 Quoted in ibid. 48.

65 Ibid. 48. For a useful account of Gui de Mori's adaptations of the *Rose*, see Walters, 'Illuminating the *Rose*' 170–5.

66 Regalado, 'Contraires Choses' 72.

67 Ibid. 75.

68 Ibid. 77.

69 Saturn is a particularly good choice to represent the masculine power of creation, for the writer of the commentary on Martianus Capella's *De nuptiis* attributed to Bernardus Silvestris states that Saturn's copulation with Rhea allegorically signifies God's infusion of *silva* with *forma* to produce the elements. See Westra's introduction to his edition of Bernardus Silvestris, *Commentary on Martianus Capella* 32.

70 Hult, 'Language and Dismemberment' 114–15.

71 Poirion, 'Mots et choses' 9; see also his 'De la signification selon Jean de Meun.'

72 For an analysis of the presentation of Heloïse in the *Rose*, see Baumgartner, 'De Lucrèce à Heloïse.'

73 Jean's translation now survives in a single manuscript, indicating that it was probably infrequently copied; and, as Brook notes, it had no imitators: 'Pour la traduction des *Lettres d'Abélard et d'Héloïse*, aucun imitateur; silence total' ('Comment évaluer?' 63–4).

74 Jean de Meun, *La Vie et les epistres Pierres Abaelart et Heloys sa fame*, ed. Hicks 1: 18. In her edition of the text, Schultz compares selections from the *La Vie et les epistres* to corresponding passages in the *Roman de la rose*, including the episode concerning Venus and Mars found in Nature's discourse.

75 Hult, 'Language and Dismemberment' 109.

76 Huot, 'Medieval Readers' 405. Huot does, however, mention one manuscript which has several markings, apparently by different readers, at this passage. Significantly, this manuscript also contains letters from the Querelle de la *Rose*, indicating that 'possibly its readers took a special interest in the linguistic issues raised by Raison' (405n5). Elsewhere, Huot mentions another manuscript in which 'toutes les additions appartiennent au discours de Raison et d'Ami; on peut imaginer que notre copiste appréciait surtout ces deux parties du poème' ('Notice' 121).

77 Huot, 'Medieval Readers' 413.

78 Hult, 'Language and Dismemberment' 118.

79 Quilligan, 'Words and Sex' 199–200.

80 Hult, 'Language and Dismemberment' 120.

81 Maloney, 'Roger Bacon' 210. See also Bacon's *De signis*, which was written in 1267, and thus represents an earlier version of Bacon's theory of significa-

tion than that appearing in the *Compendium studii theologiae* (written in 1292); the text appears in Fredborg et al., 'An Unedited Part of Roger Bacon's *Opus maius: De signis.*'

82 Hult, 'Language and Dismemberment' 122.

83 Bernardus Silvestris, *Commentary on Martianus Capella*, ed. Westra 32.

84 Whitman, *Allegory* 272; see also Whitman's extremely useful Appendix 'On the History of the Term "Personification"' (269–72).

85 Zumthor, 'Narrative and Anti-Narrative' 193.

86 Muscatine's interpretation of the *Rose* is based on the assumption that the poem 'is largely a representation of a single female psyche, whose elements engage and react to each other according to psychological laws' ('Emergence of Psychological Allegory' 1162).

87 The term 'symbol' is at best problematic, not least because it does not appear in medieval definitions of figurative discourse. Jauss rightly chastises C.S. Lewis for his attempt to distinguish between allegory and symbol, instead proposing a distinction based on biblical typological allegory and Prudentian personification allegory ('Transformation' 113).

88 Knoespel 80.

89 Hult notes that such is also the case with Franchise; however, here there is a possibility of corruption in the text (*Self-Fulfilling Prophecies* 230–1).

90 Whitman 270.

91 Alanus de Insulis, *De planctu Naturae*, ed. Wright 444 (metrum 2), trans. Sheridan 106.

92 Fleming, *Roman de la Rose* 193–4.

93 Wetherbee 283; 283n73. Note that Lewis considers the Genius figure to be descended from the Oyarses only (*Allegory of Love* 361–3).

94 Wetherbee 281; 285n76.

95 'La response maistre Pierre Col, chanoine de Paris aux deux traitiés precedens' (*Débat*, ed. Hicks 100); Christine de Pizan, 'A maistre Pierre Col, secretaire du roy nostre sire' (*Débat*, ed. Hicks 132).

96 Lewis, *Allegory of Love* 140.

97 Richards, 'Reflections on Oiseuse's Mirror' 306.

98 Ibid. 305.

99 Wetherbee 286; 290.

100 Vegetius's *De re militari*, translated as *Li Abregemens noble homme Vegesce Flave René des Establissemenz apartenanz a chevalerie*, ed. Löfstedt; see also *Végèce, l'art de chevalerie*, ed. Ulysse Robert.

101 This is the *Marvels of Ireland* by Giraud de Barri. Jean's translation of this work and of Ailred of Rievaulx are lost. It is interesting to note that Ailred

also wrote a work titled the *Speculum caritatis*, or *Mirror of Love*. The title is startlingly similar to the alternative title Jean proposes for the *Rose*: the *Mirror of Lovers*.

102 *Li Livres de Confort de Philosophie*, ed. L.V. Dedeck-Héry in 'Boethius' *De Consolatione* by Jean de Meun'; quotation from 168. Due to a typographical error at the head of his article, the late Professor Dedeck-Héry continues to be referenced as '*V.L.* Dedeck-Héry' in spite of the fact that his Christian names were 'Louis Venceslas.'

103 Jean de Meun, *Confort* 168.

104 Dembowski, 'Learned Latin Treatises' 259–60.

105 Eric Hicks, Introduction to *La vie et les epistres Pierres Abaelart et Heloys sa fame* xxvi.

106 See Copeland's distinction between 'primary' and 'secondary' translation: the former adheres closely to the original text, while the latter is more loosely based on the original and instead stresses the creative process. See also her discussion of Jean de Meun's translation of Boethius in the context of other medieval translations of the same work in *Rhetoric, Hermeneutics, and Translation* 133–44.

107 Jean Gerson, 'Le traictié d'une vision faite contre *Le Ronmant de la Rose* par le Chancelier de Paris [1402]' (*Débat*, ed. Hicks 76).

108 Kelly, 'Translatio Studii' 293.

109 Ibid. 291.

110 See Dronke, *Fabula*.

111 Maloney 206.

112 The *Testament* is infrequently discussed by critics, with the exception of Gallarati's book-length study. Dembowski doubts its attribution to Jean de Meun; he does not make clear, however, on what basis he questions the attribution ('Learned Latin Treatises' 259). In his Introduction to Jean's translation of the letters of Abelard and Heloïse, Hicks accepts the accuracy of the traditional attribution (xxvii). Gallarati offers the most detailed argument for Jean de Meun's authorship, based on internal textual evidence ('Mots sous les mots'; *Testament*).

113 Quotations from the *Testament* are from Gallarati's edition, and are cited within the text by line number; translations are my own. On the emendation of 'aimme' to 'n'aimme,' see Gallarati 123n63. A non-critical edition of the *Testament* can be found in *Le Roman de la Rose par Guillaume de Lorris et Jehan de Meung*, ed. M. Méon (4 vols. Paris: 1814) 4: 1–116.

114 This stanza, missing in the manuscript edited by Gallarati, is quoted from Méon's edition, p. 95.

5: Dante's *Vita nuova* and *Convivio*

1 Charles S. Singleton's interpretations of the allegory of the *Commedia* have exerted a great influence on subsequent readings of Dante. The running gloss on Dante's text provided in Singleton's edition with translation is probably the most effective vehicle for his approach; but see also his *Journey to Beatrice* and 'Commedia.' For a reading of the *Commedia* as a non-fictional account of a real experience, see Nardi, *Dante* 265–326; as Teodolinda Barolini notes, historians of religion include the *Commedia* in their accounts of visionary literature. More recently, Barolini has offered a much-needed corrective to the tendency to read the truth-claims of the *Commedia* uncritically in a close examination of the work's 'textual metaphysics' (*Undivine Comedy* 144, 20). On the political allegory of the *Commedia*, see Ferrante, *Political Vision.* Two particularly stimulating interpretations of Dante's writings in the context of medieval philosophy and science are Mazzotta, *Dante's Vision*, which argues that the *Commedia* can be seen as an encyclopedia of late medieval thought; and Durling and Martinez, *Time and the Crystal.* Though the latter centres on a small group of Dante's lyrics, reading them in the context of medieval science leads to an illuminating interpretation of the *Commedia.*

2 Pertile, 'Desire for Paradise' 160; cf. Pertile, 'Paradiso'.

3 Newman, 'St. Augustine's Three Visions'.

4 Barolini, *Undivine Comedy* 149.

5 See O'Rourke, *Pseudo-Dionysius* 234–59.

6 On the role of the medium in Albertus's philosophy and theology, especially with regard to the refution of Averroës in Albertus's commentary on the *De anima*, see de Libera, *Albert le Grand* 237–42.

7 Singleton's stress on the importance of light in the *Commedia* has been influential; note particularly his *Elements of Structure*, especially 'The Substance of Things Seen' (61–83). Like Busnelli, Singleton identified Aquinas as Dante's source with regard to the use of light in a theological context; Busnelli proposes that the different levels of the narrator's vision in the last cantos of *Paradiso* can be understood in the context of Aquinas's commentary on I Corinthians (*Il concetto* 212–13). See also Mazzeo, *Medieval Cultural Tradition* 56–90, *Structure and Thought*, and 'Light, Love and Beauty'; Brandeis, *Ladder of Vision*; Nardi, *Saggi di filosofia dantesca* 81–109 and 167–214. A few writers have considered Dante's work in the context of medieval optics. Cantarino has suggested that Dante's 'light metaphysics' can be understood best in the context of the writings of ibn Gabirol (Avicebron) ('Dante and Islam'). In their work on the *rime petrose*, Durling and Martinez

note a 'possible relationship' between Dante's description of the act of vision in the *Convivio* and Alhazen's theory of the visual ray, but do not pursue the connection any further (*Time* 412n99). Parronchi has shown that Dante frequently alludes to the principles of perspectivist optics, but accounts for Dante's apparently inconsistent use of perspectivist theory in the *Commedia* by suggesting that the author uses it only here and there ('Perspettiva dantesca' 102). Despite the promise implicit in his title, Patrick Boyde cites, rather vaguely, Aristotle and 'his commentators and followers' (*Perception and Passion* 47) as the source of Dante's knowledge, without taking account of what Parronchi had shown more than three decades earlier. More recently, however, Rutledge has built on Parronchi's work, showing how Dante uses optical terminology in order to articulate the relationship of the material and spiritual realms ('Dante, the Body, and Light'). The historian of science Vasco Ronchi uses Dante as something of a straw man in his *Storia della luce* in an effort to illustrate the weak understanding of perspectivist optics in medieval and Renaissance Europe. Ronchi claims that Dante was aware of the innovations of perspectivist optics, but understood them poorly; on the contrary, as is shown below, the apparent inconsistencies in Dante's use of optical terminology are deliberate variations. Lindberg has pointed out several serious flaws in Ronchi's account of the dissemination of perspectivist optics ('Alhazen's Theory of Vision' 333).

8 On Albertus as a source, see Toynbee, 'Unacknowledged Obligations' 399, and Nardi, 'Raffronti fra alcuni luoghi di Alberto Magno e di Dante,' in *Saggi* 63–72. Durling and Martinez establish Dante's use of Albertus's *De mineralibus* (*Time* 37), as does Cioffari ('Dante's Use of Lapidaries').

9 I will refer throughout this chapter to both narrator and poet as 'Dante,' not pausing to take note of the fictional nature of the events related both here in the *Vita nuova* and in the *Commedia*. As Musa remarks in the Preface to his translation of the *Vita nuova*, 'it goes without saying that ... we must suspend our skepticism and accept as "true" the events of the narrative. For only by doing so can we perceive the significance that the author attributed to his poems by placing them where he did' (xi). Quotations from the *Vita Nuova* are from de Robertis's edition and Musa's translation, and are cited parenthetically in the text by section number of the edition and page number of the translation.

10 'Est igitur illa passio innata ex visione et cognitatione' ['So this emotion of love is inborn, arising from seeing and thinking']. Andreas Capellanus, ed. and trans. Walsh 34–5 (1.1.13).

11 Even as he uses the geometrical ray as a model for the visual ray, Albertus

notes that there is a fundamental difference, for a ray of light can be bent while the geometrical ray is always straight: 'In hoc enim differt radius a lumine, quod radius non procedit immutando nisi recta linea, lumen autem etiam non ad rectam immutat' (*De homine*, quaest. 21.1, p. 184). For other instances of the term, see also *De homine* (in *Summae de creaturis*), quaest. 21.3; pp. 199–200; also Albertus Magnus, *De anima* 2.3.15; p. 122, line 37.

12 'Quando longe est, terminatur ejus visus ad superficiem aquae, quae cum est plana, diffunditur super ipsam multum lumen, et ideo apparet albedinis vehementis. Cum autem fit prope, non sistit visus in superficie, in qua multum diffunditur lumen, sed penetrat in profundum, in quo minus est de lumine: et ideo apparet magis obscurum' (*De homine*, quaest. 21.3; p. 189).

13 'Quod *corpus terminatum* dicitur corpus densum et solidum: dividitur enim corpus terminatum contra corpus pervium, et tunc dicitur corpus terminatum, quia non suscipit lumen nisi in suis terminis exterioribus: pervium autem quod suscipit lumen in suis terminis et in suo profundo' (*De homine*, quaest. 21.3; p. 190).

14 Musa is typical in interpreting the geometrical terminology in this passage as indicative of Dante's 'interest in sheer geometrical form.' He suggests that the image of the line conveys the lover's 'feeling of fatality' (97–8). Similarly, in his edition of the *Vita nuova*, De Robertis remarks that 'la geometria dei rapporti costituisce una specie di trama fatale' (46n).

15 For example, in his *De anima*, Albertus Magnus uses the term *medium* as a synonym for the diaphanous in considering whether or not colour can be seen without the diaphanous being illuminated by light: 'Sed nos prius quaeremus, cum color non immutet visum nisi diaphano illuminato, utrum lumen illuminans exigatur ad visum propter colorem vel propter medium, hoc est quaerere, utrum color non possit generare intentionem in medio, nisi sit ipse color actu illuminatus, vel ipse color sine lumine sufficiat generare intentionem suam, sed medium non sit natum ipsam intentionem suscipere, nisi sit actu illuminatum' (*De anima* 2.3.7; p. 108, lines 39–47).

16 Albertus Magnus uses the term *simulacrum* most often to refer not to the visible species but to a reflected form, such as that seen in a mirror: 'Saepe enim contingit, quod in una visione et in uno tempore videmus rem visam in seipsa, et etiam simulacrum ipsius ejusdem in speculo illi opposito' (*De sensu et sensato* 9; p. 22). He does so, I think, in order to protect the integrity of the visible form transmitted from the object to the subject: since the term *simulacrum* denotes something which is similar but not the same, it cannot serve to denote the visible form unless we wish to imply that the

visible form, even under the best of circumstances, is never transmitted perfectly. Albertus's awareness of this difficulty is evident in his use of the term *species* for the visible form, for he defines it differently in his *De homine, De sensu et sensato,* and *De anima.*

17 As Musa notes, many readers interpret Amore's words as a command to abandon the use of other women as a 'schermo' or screen. Musa rightly focuses on the meaning of the term *simulacrum,* 'an imitation as opposed to the original ... an appearance as opposed to what is real'; but instead of considering these *simulacra* to be images (which they always are, even when used figuratively), Musa goes on to suppose that they are 'the attitudes or actions of the young lover which were only false imitations of what true love for Beatrice should be. ... The greeting of Beatrice had seemed to the young lover to represent the ultimate in bliss ... but it was only a seeming, a *simulacrum.* Thus, Love's first words would seek to teach the lover, mourning the destruction of his happiness, the vanity of that happiness itself' (Musa 112, 113).

18 'Donna me prega,' ed Nelson 38–41; lines 1–3. For earlier considerations of whether love is a substance or an accident in a substance, see Nardi, 'La Filosofia dell'amore,' in *Dante e la cultura medievale* 14ff.

19 On the relation of 'Donna me prega' and the *Vita nuova* and their diverging views of the relation of love and light, see Barolini, *Dante's Poets* 143; see also Mazzotta, *Dante's Vision* 60–2.

20 Singleton, who calls Guido Cavalcanti 'the theorist of a love which is dark and tragic,' contrasts the two poets: 'Dante did not terminate the upward way of love at a point where the lover succumbed to the power of love. ... [He] found a way out of the tragic love of his first friend' (*Essay on the* Vita Nuova 73–4; 98–100). See also Mazzotta, who remarks: 'Dante installs his poetry at the point where Cavalcanti's poetry ... stops: between the dead body and the soul's existence' (*Dante's Vision* 62). On the *Vita nuova* as a response to Cavalcanti, see Harrison, *Body of Beatrice* 69–90.; Corti, *Felicità mentale*; and Malato, *Dante e Guido Cavalcanti.*

21 It is, at least, if you believe Averroës. For studies linking Guido Cavalcanti's philosophy to Averroism, see Lowry's introduction to his edition and translation, in which he disputes Nardi's claim that Guido was an Averroist (xliv–l). See also Nardi, 'Noterella polemica.' For an illuminating account of Aquinas's thoughts on Averroism, see McInerny, *Aquinas Against the Averroists.*

22 Musa is thus in error when he asserts that the scenario described by Dante in the fourth stanza of 'Donne ch'avete' 'differs from the traditional one in that there are involved not only the eyes of the one who beholds the beauti-

ful lady, but also the eyes of the lady herself, from which issue the flaming spirits that strike the eyes of him who looks upon her ...' (195n8).

23 'Era in penser d'amor,' lines 23–6 and 39–42 in *Poetry of Guido Cavalcanti*, ed. and trans. Nelson (44–7). I have translated 'nel cor' (25) as 'in your heart,' rather than 'over your heart,' as Nelson does. In another of his pastoral poems, Guido notes how from the eyes of the beloved there comes 'un gentiletto spirito d'amore' ['a noble little spirit of love']. 'Gli occhi di quella gentil foresetta,' line 7; pp. 48–9.

24 For a different reading of the visual experience of the 'gentile donna,' see Harrison, *Body* 115–16.

25 On the open-endedness of the *Vita nuova*, see Harrison, *Body* 129–43.

26 Musa's translation is misleading here: the lady *receives* light ('luce') and reflects it (her 'splendore'). In *Convivio* 3.9, Dante specifies that 'splendore' always denotes reflected light, not the light emanating from a source.

27 'Veggio negli occhi de la donna mia,' *Poetry of Guido Cavalcanti*, ed. Nelson (36–7).

28 I have emended Musa's translation here to render the text more literally.

29 The dates of composition are not certainly known, but the order has been (as Nelson puts it) 'established plausibly' by Favati in his edition of *Guido Cavalcanti: Le Rime*. See also Nelson's Introduction (xxxvi).

30 The editor of the text follows Mario Casella in adding 'de la memoria' before 'dinanzi,' thus implying that the images of both ladies are in the memory. (The translator rejects the emended reading.) In theories of faculty psychology, the memory is *always* behind or 'di dietro.' What is in front or 'dinanzi' is either, as the translator has it, direct 'sight of that lady' or the imaginative faculty, which receives the visible species from the common sense. Dante makes it clear that he knows where the imagination is conventionally located when he later refers to 'la parte del cerebro dinanzi, dov'è la sensibile virtude' (*Convivio* 3.9.9). On medieval faculty psychology, see Harvey, *Inward Wits*. Quotations from the *Convivio* are taken from Simonelli's edition and Ryan's translation, and are cited in the text by section number.

31 On the similarities of Beatrice (both as she appears in the *Vita nuova* and in the *Commedia*) to Filosofia, see Pelikan, *Eternal Feminines*.

32 It has five stanzas of eighteen lines each: each stanza ends with a line number that, like all multiples of nine, equals nine when its integers are added together (i.e., for line eighteen, 1+8=9, for line twenty-seven, 2+7=9, and so on). The *canzone* has a total of ninety lines, the greatest number of lines it could have without it being necessary to add the integers together twice in order to get nine.

33 There is, of course, one overwhelmingly important reason to offer a full
literal interpretation and only a sketchy allegorical one: if the matter
allegorically expressed in the *canzone* is so far removed from human understanding that it cannot be conveyed in non-figurative language, then it is
impossible for an allegorical interpretation (i.e., a translation, as it were,
from the language of allegory to regular language) to adequately convey
that meaning. The meaning can be reconstituted only in the mind of the fit
reader, one who can rightly interpret the allegorical code.

34 Albertus Magnus devotes much of the first book of *De sensu et sensato* to a
refutation of the 'falsa opinione Antiquorum qui visum igni adapterave-
runt,' who believed that we see by means of 'extra mittentes radios ... quasi
radios igneos.' See especially *De sensu et sensato* 1.2–7; quotations from 1.3,
p. 5 and 1.6, p. 11; he also devotes a large appendix to question 22 of *De
homine* to a refutation of extramission theories (215–28), including what
seems to be a version of Grosseteste's theory (223ff.).

35 Cf. Albertus Magnus, *De anima*: 'color est motivum visus secundum actum
lucidi' (2.3.7; p. 108, line 30). Albertus stresses the primacy of light over
colour: that is, light can be visible without the presence of colour, but
colour is never visible without the presence of light.

36 *De sensu et sensato* 26.

37 'Non proprie imago vel forma, sed species imaginis vel formae'; 'in anima
sicut habitus, in speculo vero sicut dispositio, et etiam in aere' (*De homine*
201).

38 'Species visibiles sunt accidentia: ergo non sunt in oculo et in medio'; 'in
oculo sicut in termino, sed in medio sicut in via: et propter hoc etiam in
oculo est aliquo modo ut in actu, sed in medio sicut in potentia' (*De homine*
210).

39 'Oculum ... bene recipiat species visibilium et etiam retineat.' *De anima*
2.3.14; p. 120, lines 65–7.

40 *De anima* 2.3.15, pp. 121–2. Lindberg has remarked that there seems to be a
gradual evolution in Albertus Magnus's treatment of vision, but does not
elaborate on his conjecture (*Theories of Vision* 105).

41 'In oculo dominatur aqua quantum ad illam partem in qua fit impressio'
(*De homine* 171).

42 'Pupilla igitur et totus oculus praecipue inter se componentia sunt aqua'
(*De sensu et sensato* 32). In his *De sensu et sensato*, Aristotle refutes Democritus's assertion that eyes are like mirrors and that sight is merely a process
of mirroring external images. Interestingly, although Avicenna largely relies
upon Aristotle's theory of vision, he too likens the eye to a mirror: 'the eye
is like a mirror, and the visible object is like the thing reflected in the
mirror by the mediation of air or another transparent body; and when light

falls on the visible object, it projects the image of the object onto the eye ...
If a mirror should possess a soul, it would see the image that is formed on
it.' Avicenna, *Danishnama*, trans. Achena and Massé, *Livre de science* 2: 60;
English translation in Lindberg, *Theories of Vision* 49. There is probably no
connection between Avicenna's statement and the intromission theory
used by Dante, since the *Danishnama* was evidently unknown in medieval
Europe (d'Alverny, 'Traductions d'Avicenne'; 'Notes sur les traductions
médiévales d'Avicenne'). Lindberg remarks that 'Avicenna's theory of
vision has been almost entirely neglected by historians of optics' (*Theories of
Vision* 234n70); for an exception, see Vescovini, *Studi sulla prospettiva
medievale*, chapter 5.

43 'Quod enim ibi repraesentatur forma quasi in speculo, accidit oculo, non
secundum quod est instrumentum visus, sed secundum quod lenis est et
terminatus' (*De sensu et sensato* 31).

44 *De multiplicatione specierum* 3.1; trans. Lindberg in Grant, ed., *Source Book*
419. Lindberg notes that, here, 'Bacon reasserts the point he has made
elsewhere that light is not corpuscular and that reflection is not an instance
of mechanical rebound; reflection is rather to be explained by the self-
diffusive properties of species' (419n133). Like Bacon and his source
Alhazen, the perspectivists Pecham and Witelo similarly believed that light
was not corpuscular, but rather a power or form (418n128).

45 Lindberg clearly demonstrates that, according to the atomists, the visible
form is corpuscular; as he puts it, 'without a material efflux, it isn't atomism' (*Theories of Vision* 3).

46 On Galen's development of the theory of visual spirit (also called optic
pneuma), see Lindberg, *Theories of Vision* 10. Alhazen, whose intromission
theory was the origin of western perspectivist optics, says that visual spirit
does exist within the eye though not, as the proponents of extramission
would have it, outside the eye. Bacon, Pecham, and Witelo avoid using the
term visual spirit, perhaps to avoid confusion with the visual spirit of
extramission theories; yet their explanation of the transmission of form
along the optic nerve seems to be a restatement of Alhazen's explanation
of the operation of visual spirit within the eye. See Lindberg, *Theories of
Vision* 84; also 110 and n32.

47 *De homine*, quaest. 20, p. 171. See also *De sensu et sensato* 1.5, p. 10; 1.10,
p. 25; 1.14, p. 35.

48 Albertus implies the correspondence when he cites Alfarabi: 'Instrumentum virtutis visibilis est oculus, et in isto instrumento dominatur
aqua, quae est substantia diaphana' ['the eye is the instrument of the
power of vision, and this instrument is dominated by water, which is a
diaphanous substance'], thus implying a similarity between the diaphanous

medium outside the eye and the 'substantia diaphana' within (*De homine* 171).

49 *De homine* 172.

50 *De sensu et sensato* 35.

51 Here I borrow I.A. Richards's terminology for the metaphysical conceit, since it is clearer to refer to tenor and vehicle rather than to the first and second terms of metaphor. The metaphysical conceit operates much the way allegory does, and particularly resembles what I have called 'vertical allegory'; I hope to pursue this idea in the future.

52 Ryan's translation reinforces this impression by characterizing the lady's 'attitude' as disdainful; the word Dante actually uses is 'sembiante' [3.10.3], a term that most immediately refers to a thing seen and only figuratively refers to a behaviour.

53 'Visus obscuratur in iratis: quia ira accendit sanguinem circa cor, et elevat vapores ad caput' (*De homine* 172).

54 'Quia vero illi spiritus clari et puri sunt, qui deferuntur ad oculos, super omnes spiritus sensibiles, ideo turbatio quaecumque contingens in cerebro, magis apparet in ipsis: et ideo contingit in ebriis, in quibus multus vapor fertur ad cerebrum, quod obscuratur visus' ['Because this spirit that is sent to the eyes is clear and pure, above all other sensitive spirits, therefore whenever turbulence occurs in the brain, it appears even more greatly there (i.e., in the eyes). This is what happens in drunken people, in whom a great deal of vapour is sent to the brain, which obscures the vision' (*De hominis* 171–2)].

55 'Cum enim premitur intus et cedit ad interius oculi, et tunc coangustatur foramen, et videtur res minor quam sit in veritate, aliquando cedit ad exterius: et tunc foramen dilatatur, et videbitur res major quam sit' (*De sensu et sensato* 1.11; p. 29).

56 For an account of the commentaries generated by this poem, see Courcelle, *Consolatione de Philosophie* 161ff. See Durling and Martinez on the importance of 'O qui perpetua' to Dante's cosmology (11–18).

57 On the importance of vision as a metaphor for understanding in Boethius's *De consolatione Philosophiae*, see O'Daly, *Poetry of Boethius* 120ff.; see also 157 and 176.

58 'Acies lacrimis mersa caligaret nec dinoscere possem, quaenam haec esset mulier' (1.pr.1.44–6).

59 'Statura discretionis ambiguae. Nam nunc quidem ad communem sese hominum mensuram cohibebat, nunc vero pulsare caelum summi verticis cacumine videbatur; quae cum altius caput extulisset, ipsum etiam caelum penetrabat respicientiumque hominum frustrabatur intuitum' (1.pr.1.8–13).

6: Dante's *Commedia*

1 Throughout this chapter, I will point out evidence of the strong relationship between the *Roman de la rose* and Dante's *Commedia*. In a number of articles and books, Contini has argued that Dante is profoundly interested in the *Rose*, and attributes to Dante an Italian adaptation of the *Rose*, called the *Fiore*. For a summary of the debate regarding the authorship of the *Fiore*, see Barber, 'Statistical Analysis.' On Dante and the *Rose*, see Baranski and Boyde, eds., *The* Fiore *in Context*.

2 All quotations from the *Commedia* are taken from the Petrocchi edition as presented by Singleton together with his translation and commentary, *The Divine Comedy*, and are cited in the text by canto and line number.

3 In his analysis of this episode from the *Commedia*, Miller gives a more concise explanation of what is actually seen in the three mirrors; it would seem, however, that he concurs with me in view of the fact that, in the caption to an illustration of the experiment in one medieval manuscript, Miller notes that the image of the flame appearing in the more distant mirror should rightly be depicted as smaller than the other two images of the flame ('Three Mirrors' 265, caption to illustration 1). For other treatments of Beatrice's experiment, see Nardi, 'La dottrina delle macchie lunari nel secondo canto del "Paradiso,"' in *Saggi* 3–39; Durling and Martinez, *Time* 224–32.

4 'The subject of the whole work, then, taken literally, is the state of souls after death, understood in a simple sense; for the movement of the whole work turns upon this and about this. If on the other hand the work is taken allegorically, the subject is man, in the exercise of his free will, earning or becoming liable to the rewards or punishments of justice.' *Epistolam X ad Canem Grandem della Scala* 8, ed. Pistelli, trans. Haller, *Literary Criticism* 99. The attribution of this letter to Dante has been questioned. For a brief history of the debate, see Barolini, *Undivine Comedy* 10 and n20. In *Dante's Epistle to Cangrande*, Hollander reviews the debate and argues that the epistle was indeed written by Dante.

5 Leyerle notes the resemblance between Grosseteste's concept of light and Dante's treatment of light in the *Commedia* ('Rose-Wheel Design' 301). On ibn Gabirol, see Cantarino, though he suggests that Dante's knowledge of ibn Gabirol's theory of light was not the result of direct acquaintance with his texts, but rather was filtered through a 'stream of thinking' ('Dante and Islam' 34).

6 Trans. in Lindberg, *Theories of Vision* 118–19. A critical text of the original of the prefatory letter is found in Baeumker, *Witelo, ein Philosoph* 128, lines

6–17. Lindberg notes that Witelo enjoyed the 'friendship of William of Moerbeke, who was ... the one who encouraged Witelo to take up optical studies' (255n61). For another, less technically accurate translation of the first part of the same quotation, see Durling and Martinez, who also remark upon the similarity of Witelo's description of the emanation of light to that of Grosseteste (*Time* 36).

7 'Alia vero naturalis actio fit per reflexionem a corporibus aliis, ut radii solis a corpore lunae reflectuntur. Quamvis enim propter raritatem lunaris corp-oris quiddam solaris transeat virtutis, plurimi tamen radiorum reflectuntur inferius, ut a speculo sphaerico convexo.' Witelo, Preface to *Perspectiva*, in Baeumker 127–31; quotation from 131, lines 21–4.

8 Love can alternatively be identified with Lucy, depending on whether Beatrice's statement means that love was the proximate cause or the first cause. In other words, Lucy moved Beatrice directly; Mary was the cause of that movement. Note that Bernard of Clairvaux later uses the same verb ['mosse' (Pa.32.117)] to describe how Lucy moved Beatrice to action, reinforcing the identification of Lucy with love. (For a fuller discussion of the identification of Lucy with love and with the Holy Spirit, see Ferrante, *Woman as Image* 140–1.) I would argue, however, that even if Lucy does represent love, she does so only insofar as she reflects that aspect of divinity. Like Beatrice, and like Mary herself, Lucy is an empty vessel to be filled by God. She represents love; but other souls also represent love, and Lucy represents, not only love, but also other things. Similarly, Lucy represents light; but so do other souls in paradise. For a cogent refutation of recent efforts to deny Lucy's association with light and vision, see Cassell, 'Santa Lucia.'

9 On the significance of embraces in the *Commedia*, see Economou, 'Saying Spirit in Terms of Matter.'

10 Dante also uses the term 'spezie' in Bonaventure's discourse on the meta-physics of light (Pa.13.71).

11 To describe the figure of Beatrice beyond her place in the optical allegory of the *Commedia* and her role as a quasi-personification would be to wander too far from my topic. For more on Beatrice as mediator, see Ferrante, *Woman as Image* 129–52; Harrison, *Body of Beatrice*; Kirkham, 'Canon of Women'; Singleton, *Journey to Beatrice*. It seems to me that Beatrice as she is portrayed in the *Vita nuova* splits into two figures in Dante's subsequent writings, corresponding to the two aspects of the sun: her capacity to illuminate appears in Filosofia of the *Convivio*, while her capacity to gener-ate heat appears in the lady of the 'rime petrose.' Filosofia is eternally insubstantial, separate from man. This is illustrated in the structure of the

second *canzone* of the *Convivio*, where three stanzas in turn describe her unified body and soul, her soul, and her body. Even in the stanza concerning her body, Dante focuses on the least corporeal aspect of the lady: her gaze. On the other hand, the lady of the 'rime petrose' is manifestly tangible: hard as stone, she is finally softened by the 'vendetta' waged by the frustrated lover in 'Così nel mio parlar voglio esser aspro.' In the *Convivio*, the lady offers intellectual satisfaction; in the 'rime petrose,' she offers physical satisfaction. On the eroticism of the 'rime petrose,' see Dronke, *Medieval Lyric* 164–6. See also Durling and Martinez, who use the poems as a kind of key to the *Commedia*; and Sturm-Maddox, 'Rime Petrose.'

12 'There is a kind of apostolic succession from Lady Philosophy, Philosophia as Woman, to Lady Beatrice, Theologia as Woman' (Pelikan, *Eternal Feminines* 52).

13 *De sensu et sensato* 29–33 (chs. 12–13).

14 Albertus begins 'De homine' with an account of the platonic extramission theory, which states that 'in oculo dominetur ignis vel lumen' (quaest. 20, p. 168); he then goes on ('Sed contra ...') to give an account of the aristotelian intromission theory, which states that 'in oculo dominatur aqua' (p. 169).

15 Dante emphasizes the relative brightness of Limbo by mentioning the 'lumera' (4.103) he approaches in the company of the philosophers, and their open meadow 'luminoso e alto' (4.116). Further underlining the acute intellectual vision of those in Limbo, Homer is identified as an eagle ['aquila' (4.96)], an animal renowned for the acuity of its material vision; Caesar has 'li occhi grifagni' ['the eyes of a falcon' (4.123)].

16 According to Michael Scot's *Liber physiognomiae*, crossed eyes are indicative of a deceptive nature; noted in Kay, 'Spare Ribs of Dante's Michael Scot.'

17 Durling and Martinez point out the aggressive nature of the lady's 'viso' not only in the *rime petrose*, but also in Dante's 'E' m'increscé di me sì duramente' (*Time* 401n30).

18 '... li occhi miei in uno / furo scontrati ... ïo a figurarlo i piedi affisi ... E quel frustrato celar si credette / bassando 'l viso; ma poco li valse, / ch'io dissi: "O tu che l'occhio a terra gette"' ['... my eyes were met by one of them ... I paused to make him out. ... And that scourged soul thought to hide himself, lowering his face, but it availed him little, for I said, "You there, casting your eye upon the ground"' (Inf.18.40–8)].

19 On the corporeality of hell, see Durling, 'Deceit and Digestion.' Caroline Bynum suggests that, although mutilation and fragmentation of the body worsens in the deeper circles of hell, it is not because body is in any way sinful or evil. On the contrary, paradise is peopled with souls whose aerial

forms resemble real bodies which serve as outward manifestations of their individual identities ('Faith Imagining the Self').

20 Ahern, 'Troping the Fig.'

21 Though Dante uses the form 'fiche' in canto 25, the author of the *Fiore* uses 'fica': 'e facciagli sott'al mantel la fica.' *Il Fiore*, ed. Contini 176, line 14; noted in Singleton's note to Inf.25.2. Interestingly, the word used to describe the penetration of the rose at the climax of the *Fiore* is *ficcare* (230, line 1), the same verb Dante uses to describe the motion of the aggressive visual gaze. If Dante was the author of the *Fiore*, as Contini argues, or even if he merely knew the text well, he would surely have made the connection between visual and sexual penetration.

22 Brunetto wrote his voluminous *Tresor* in French prose so that, according to Brunetto, it would reach a larger number of readers. He later completed a summary of the *Tresor* in Italian verse, called the *Tesoretto* or 'little treasure.' It is plausible that Dante sees Brunetto's linguistic transgression as 'unnatural' behaviour, especially since it is the French work that Brunetto commends to Dante in the *Inferno* (15.119).

23 In the *Purgatorio*, Statius praises Virgil, whose words in the *Aeneid* 'were the seeds ('seme') of my poetic fire' (Pg.21.94–7). His comment on Virgil's fertility retrospectively highlights the sterility of Brunetto's poetic achievement.

24 It has been suggested that Dante's knowledge of French literature and of the *Rose* in particular came through his association with Brunetto. See Contini, 'Un nodo della cultura medievale'; also Richards, 'Dante's *Commedia*.'

25 Quotations from the *Roman de la rose* are from Lecoy's edition and are cited in the text by line number.

26 For the account of the sin of Onan, who 'spilled his semen on the ground,' see Genesis 38:8–10.

27 'Nullam materiam matricis signat idea, / Sed magis et sterili litore vomer arat.' *De planctu Naturae*, metrum 1, ed. Wright 430, trans. Sheridan 69.

28 Dante reinforces the association of grains of sand with fertile seed later in the *Inferno* when he plays on 'rena' ('sands') and 'ren' ('loins'): Libya's sands give birth to foul creatures like the snakes in hell that 'ficcavan' ('thrust') through the loins of the damned souls (Inf.24.85–96). This image may have been inspired by the mid-twelfth-century *Vision of Tondal*, in which fornicators are punished by being filled with vipers that emerge from their bodies and then double back to penetrate them; the genitals, too, change into vipers and thrust themselves through the sinners. See *Visio Tnugdali: Lateinisch und Altdeutsch*, ed. Albrecht Wagner (Erlangen:

Deichert, 1882) 27–9; noted in Bynum, 'Faith Imagining the Self' 92 and n29.

29 See Harrison, 'Bare Essential' 297–9.

30 'Qui donne benefice pour espargnier sa bource, / Je di que ceste paie est diverse et rebourse, / Et em pert Dieu et s'ame qui tel avoir embourse, / Car le drap et la penne de discretïon bourse.' Jean de Meun, 'Testament' 581–4, ed. Gallarati 144.

31 'In *Timaeo* Platonis scribitur: tunc accideret videre exeunte ab oculo lumine ignis per tunicas oculi pervias, sicut lumen egreditur per vitrum vel per pellem perviam ex lucerna.' *De sensu et sensato*, cap. 12, p. 29.

32 Albertus explains that the apparent ability of certain animals to see at night is a process of reflection and multiplication of rays: 'in corpore reflectente lumen et radios, colliguntur radii in punctum pyramidis, et ampliantur plus et plus secundum quod radii distant magis a corpore reflectente lumen et radios. Et cum tale corpus sit oculus, per hunc modum fit reflexio lucis ab ipso, et non sic quod radii procedant ex ipso sicut ex primo illuminante.' *De homine*, quaest. 20, p. 172; cf. 185–6. Things that shine in darkness are 'noctiluca, sicut oculi quorumdam animalium et capita piscium et corpora putrefacta.' *De homine* 172.

33 See chapter 6, p. 147 and n13 above.

34 See chapter 6, pp. 144–6. Within the terms of this analogy, Dante is both the *individuum* of the atomists (since according to Democritus and his follower, Lucretius, the species is the indivisible particle in vision) and the individual in our modern sense of the term.

35 The original statement reads 'God is a circle whose centre is everywhere and whose circumference is nowhere'; in the twelfth century, Alanus de Insulis made the conceit three-dimensional, as it were, in his 'Sermo de sphera intelligibili.' For the origins of the original phrase and information on Alanus's sermon, see Sheridan's Introduction to his translation of the *De planctu Naturae* 25.

36 See Singleton's note to Pg.4.21.

37 For 'panno' as the membrane of the eye, see Albertus Magnus, *De sensu et sensato* 1.14, p. 34: 'videmus quod visus post excellentem claritatem aliquam luminis vel albedinis in quam diu adspexit, si convertat se postea ad colores minus claros, videtur ei primum quod tecti sunt panno subtili albo qui paulatim deficit.' Albertus uses the word 'tunicae' to refer to the membranes of the eye (e.g., 31); but Dante himself uses the term *gonna* elsewhere in the *Commedia* in the technical sense, referring to the tunics of the eye (Pa.26.72).

38 'Forament tunicae quae dicitur *uvea*: et per ipsum foramen uveae deveniunt ad cellam quae vocatur *aranea*.' Albertus Magnus, *De sensu et sensato* 1.13, p. 31.

39 John Pecham, *Perspectiva communis*, 1, proposition 31: 'Within the cornea is the tunic called the uvea. The uvea is black like a grape to darken the humor in which [the power of] sight resides, for unless that humor is darkened, the visible species will not appear in it. This is a strong tunic to prevent the enclosed humor from exuding, and so that species may pass through, it has a circular aperture in front.' Trans. Lindberg in Grant, ed., *Source Book* 398.

40 Aristotle repeatedly uses the analogy of a seal pressing upon wax to describe impression in sense perception; e.g., *De anima* 3.12.434b–435a, ed. Hett 198–9; cf. Conv.2.9.4. This passage in the *Purgatorio* (10.28–46) is Chaucer's source for the allegorical play on form and matter in the *Merchant's Tale*; see chapter 8 below.

41 'Nervi abscissi fuerunt qui dicuntur *optici*, per quorum poros virtus visiva decurrit ad oculos.' Albertus Magnus, *De sensu et sensato* 1.14, p. 34.

42 Note, too, that the souls of contemplatives in heaven are described as 'holy flowers': 'Questi altri fuochi tutti contemplanti / uomini fuoro, accesi di quel caldo / che fa nascere i fiori e 'frutti santi' (Pa.22.46–8).

43 It is difficult to know whether Dante was aware of the play on Deduit and *deduitio* in the first *Roman de la rose*, since even without identifying Guillaume de Lorris's play on words, Dante would surely have noted that the narcissism of the lover is expressed through reflection and refraction. But Dante almost certainly was familiar with the text in which William of Conches uses the term *detuitio*, the *Glosae super Platonem*, as Margherita de Bonfils Tempier has shown in a series of scholarly articles ('Due ineffabilitadi del *Convivio*'; 'Genesi di un'allegoria'; 'Il dantesco "amoroso uso di Sapienza"'; 'Ragion e intelletto').

44 It is revealing that the sixteenth-century commentator on Dante, Benedetto Varchi, gave a series of talks on the *Paradiso* in which he stressed the importance of optics to an understanding of the poem, and stated that, in certain passages in the *Paradiso*, Dante writes as a 'prospettivo.' See Benedetto Varchi, *Opere* (2 vols. Trieste, 1858–9) 2: 372–8; noted by Frangenberg, 'Perspectivist Aristotelianism' 138. See also Vescovini's argument that the study of perspectiva was increasingly central to medieval curricula ('"Perspectiva" nell'enciclopedia').

45 Rutledge rather misses the point when she states that 'Dante, knowing well the Platonic theory of extramitted vision, quite deliberately reserved that

power for Beatrice, the radiant centre of the *Comedy*' (164). Surely making Beatrice the source of light would be, for Dante, a form of idolatry; her goodness lies in her mediation, that is, her ability to reflect God.

46 On the significance of this passage, see Barolini, 'Arachne, Argus, and St. John.'

47 Dante attributes this view to Avicenna; it also appears in Albertus Magnus's *De homine* 184 (where Albertus attributes it to Avicenna) and his *De anima* 110 (where he does not).

48 For example, the only late medieval optical source to state that light travels instantaneously is Witelo: Dante seems to follow his lead when he states that 'in vetro, in ambra o in cristallo / raggio resplende sì, che dal venire / a l'esser tutto non è intervallo' ['in glass, in amber, or in crystal, a ray shines so that there is no interval between its coming and its pervading all' (Pa.29.25–7)]. Witelo says that 'Unimpeded light is necessarily moved in an instant through the whole of a medium proportioned to it.' Witelo, *Perspectiva* 2, theorem 2; trans. Lindberg in Grant, ed., *Source Book* 395. Lindberg notes that 'Witelo's view that light is propagated instantaneously is opposed to the opinions of Alhazen and Roger Bacon' (395n17).

49 O'Rourke, *Pseudo-Dionysius* 257, citing *Summa Theologica* I, 104, 1; note O'Rourke's correction to the Blackfriars' translation (257n174).

50 On Dante's use of language in the *Paradiso*, see Ferrante, 'Words and Images.'

51 'Haec et huiusmodi inducere consueverunt quidam lucem corpus esse dicentes. Adhuc autem, dicunt lucem multiplicatam divaricare et disgregare aërem et calefacere et ignire, sicut in speculis comburentibus apparet, quae omnia non nisi corpus efficere potest.' Albertus Magnus, *De anima* 2.3.10, p. 113, lines 71–5. Durling and Martinez rightly identify the burning mirror as an example of Dante's use of optical phenomena as metaphors, and suggest that the *rime petrose* are meant to act as burning mirrors to ignite the lady's passion; the only source they cite, however, is Albertus Magnus's *De mineralibus* 2.2.3 (*Time* 388n30).

52 Witelo, *Perspectiva* 9, theorem 43; trans. Lindberg in Grant, ed., *Source Book* 417–18. See also Witelo, *Perspectiva* 5, prop. 65, ed. Smith.

53 It is similar with the other souls in paradise: their light seems brighter as their joy increases: 'Per letiziar là sù fulgor s'acquista, / sì come riso qui' ['Through rejoicing, effulgence is gained there on high, even as a smile here' (Pa.9.70–1)]. Since joy is a manifestation of love, presumably their increased 'fulgor' is manifested as both heat and light.

54 Albertus Magnus, *De homine* quaest. 21, p. 201. It also recalls Alanus de Insulis's description of the threefold mirror held in the hand of Reason, in

which she sees how 'in mundo fantasma resultat ydee' ['the image of the idea is reflected in the universe']. *Anticlaudianus* 1.500; ed. Bossuat 71, trans. Sheridan 65.

55 For a learned and sensitive treatment of this episode, see Rossi, 'Miro gurge.'

56 Dante uses the word 'parlar' here, probably to emphasize Virgil's role as a poet in the vernacular, that is, the spoken language of his day.

57 It is interesting that this *terzina* is the same one that contains the phrase 'vagliami 'l lungo studio' ['may the long study ... avail me' (Inf.1.83)] that Christine de Pizan is so struck by that she uses it for the title of her *Livre du chemin de long estude*. This may suggest that she not only knew the importance of reading Virgil but also noted the significance of that *terzina* within the overall structure of Dante's work.

58 On the doctrine of the Beatific Vision as it is presented by Dante, see Mazzotta, *Dante's Vision* 170ff. and 269n32–4. On the importance of this doctrine, see Douie, 'John XXII'; Sandler, 'Face to Face with God.' For a brief account of the history of the doctrine, see Redle, 'Beatific Vision.'

59 Durling and Martinez rightly point out the connection of the crystals at the base of the fountain of Narcissus in the *Roman de la rose* and the lady's eyes in the *rime petrose*: 'In this, as in many other respects, the *rime petrose* must be seen against the background of this important thirteenth-century masterpiece' (*Time* 389n35). They add that, in the *Paradiso*, Beatrice's eyes take the place of those of the lady in the *rime petrose*. 'Beatrice's eyes thus correspond ... to the nearly all-seeing *deus pierres de cristal* in Narcissus's fountain in Guillaume de Lorris's part of the *Roman de la rose*' (*Time* 426–7n99).

60 On the figure of Narcissus in the *Commedia*, see Shoaf, *Dante, Chaucer, and the Currency of the Word* 21–100; Nolan, *Now Through a Glass Darkly* 105–9.

61 Baldassano points out that, by kicking the sinner who, by giving him information, is his benefactor, Dante mirrors the sin being punished in this circle ('Dante's Hardened Heart').

62 Shoaf argues that Narcissus appears at the summit of paradise, not in God's gazing upon Himself, but in Dante's vision of God: 'When Dante-Narcissus looks into the Trinity, he does not see himself as God, but he does see God as himself ... This is the genius of Dante: in assimilating himself to Narcissus, he is able to insist that he sees himself in God but does not see himself as God' (*Dante, Chaucer* 99). Shoaf is right to point out that Dante sees 'nostra effige' ['our image' (Pa.33.131)] in God, and thus sees the humanity in the divine, the result of Christ's love for man. But Dante takes pains to emphasize that he sees *our* image, not his own; thus he is, if anything, an anti-Narcissus, as he was earlier in the *Paradiso* when he made 'the contrary

error' to that made by Narcissus. God alone gazes upon Himself, for He is the proper object of all vision; Dante gazes upon God. If anyone here is Narcissus (in the bad sense), it is the reader who takes 'nostra effige' to mean that humanity is divine. Dante has, I think, built in the capacity for the reader to read wrongly, and thus to sin.

63 On the rich symbolism of the rose, see Dronke, 'Symbolism and Structure' 29–48, especially his discussion of Peter of Capua's *De rosa* (41–2). For a broader examination of the significance of emblems in medieval literature, see Hanning, 'Poetic Emblems.'

64 Barolini notes Eugene M. Longen's comment that the 'punto solo' is both 'the moment of vision' and 'the substance of the vision as well,' and relates his observation to Francesco Tateo's discussion of the 'punto solo' as the Augustinian instant. See Longen, 'The Grammar of Apotheosis: *Paradiso* XXXIII, 94–9,' *Dante Studies* 93 (1975): 209–14; Tateo, 'Il "punto" della visione e una reminiscenza da Boezio,' in *Questioni di poetica dantesca* 203–16; both cited in Barolini, *Undivine Comedy* 347n56.

65 The seminal work on this subject is Yates's *Art of Memory*; see also Coleman, *Ancient and Medieval Memories* 39–59 and Carruthers, *The Book of Memory* 130–44.

66 Freccero, 'Introduction to the *Paradiso*' 212–13; see also Chiarenza, 'Imageless Vision.'

67 Singleton stresses the role of will; more recently, readers have stressed the role of memory. See Carruthers, *The Book of Memory* 57–8; also Paul W. Spillenger, '*Dante's Mnêmnosunê.*' The identification of every instance of *mente* as memory seems to me limiting. *Mente* refers to memory only when the mind is in the act of remembering, that is, when an image is called up from the storehouse and flashed upon the mirror of imagination. For a representative passage in which Dante clearly distinguishes between the two, see Pg.9.13–19. Here, the *mente* is dominated by the faculty of imagination, receiving the impress of a vision from without. Regarding the meaning of *mente* in Dante's writings, see Mazzotta who, citing the definition Dante offers in the *Convivio*, declares that '"Mente" is defined as that power of the soul capable of "virtù ragionativa, o vero consigliativa"' (*Dante's Vision* 261n2).

68 For example, fear affects the will: a man loses his will to carry out an endeavour because of fear, just as a beast shies due to 'falso veder' (Inf.2.48). All the faculties, whether imagination, will, or memory, are moved by impressions ['impresa' (2.47)].

69 On the relation of Virgil's description of the Harpies in the *Aeneid* to Dante's vision of the 'femmina balba,' see Hollander, 'Purgatorio XIX.'

70 Mazzeo suggests that sirens found elsewhere in *Purgatorio* (31.45) represent 'the temptations of the mind as well as the temptations of the flesh' (*Medieval Cultural Tradition* 209). More recently, Mazzotta has argued that the dangerous power of imagination rears its ugly head at various points in the *Commedia*, notably in the dream of the 'femmina balba': 'the dream of the Siren discloses Dante's fascination with death, his temptation to yield to the insubstantial world of phantasms or, to say it differently, to be swallowed in the eddies of the imaginary' (*Dante's Vision* 150). But Mazzotta claims that the power of imagination is dangerous primarily because it distorts one's perception of reality, not because it is also a productive faculty that challenges the creative authority of God. For a thorough treatment of contemporary scientific and philosophical views of the role of the imagination, see Mazzotta 116–34.

71 'Maiorem nunc tendo liram totumque poetam / Deponens, usurpo michi noua uerba prophete. / ... / Carminis huius ero calamus, non scriba uel actor' ['I now pluck a mightier chord and laying aside entirely the role of poet, I appropriate a new speaking part, that of the prophet. ... I will be the pen in this poem, not the scribe or author']. Alanus de Insulis, *Anticlaudianus* 5.268–9, 273; ed. Bossuat 131, trans. Sheridan 146. Dante's knowledge of the *Anticlaudianus* has been demonstrated by Dronke in 'Boethius, Alanus and Dante'; cf. Rossi, who posits a relation between Dante's river of light and the *fons* in the *Anticlaudianus* ('Miro gurge' 81; see also the sources cited at 96n10).

72 Dante's careful ambiguity in this regard is, of course, the starting point of Barolini's *Undivine Comedy*.

7: Chaucer's Dream Visions

1 '[T]he relationship between Chaucer's self-reflexivity, his concern with the grounds of discourse, and certain shifts in fourteenth-century philosophy and theology (*not* to be confused with epistemological "skepticism") is greatly in need of close investigation.' Aers, 'Parliament of Fowls' 17n36.

2 Referring to Aers's article, Lynch writes 'I see my own work as in part responding to that call' ('Late Medieval Voluntarism' 15n17).

3 See Grennen, 'Calculating Reeve'; Holley, 'Medieval Optics'; Roney, *Chaucer's Knight's Tale*; Collette, 'Seeing and Believing'; Yager, 'A Whit Thyng'; Stanbury, 'Lover's Gaze.' Also note Stanbury's brief but provocative treatment of perception in Chaucer in the conclusion of *Seeing the* Gawain-*Poet*.

4 It is disappointing that even studies surveying several of Chaucer's works

(including Holley, *Chaucer's Measuring Eye*, Klassen, *Chaucer on Love, Knowledge, and Sight*, Collette, *Species, Phantasms, and Images*) do not take note of the gradual evolution of his use of vision, both as it appears explicitly as a metaphor for understanding and its implicit presence in what Holley calls 'Chaucer's verbal perspective' (1).

5 Burnley has previously observed apparent discrepancies in Chaucer's use of faculty psychology, concluding that the poet did not consistently use any one theory (*Chaucer's Language* 103). Burnley provides an extremely useful analysis of Chaucer's philosophical vocabulary; see especially 99–115 on the vocabulary of faculty psychology.

6 Windeatt translates some of the French *marguerite* poetry which seems to have influenced Chaucer's prologue to the *Legend of Good Women*, including Guillaume de Machaut's 'Dit de la Marguerite' and 'Dit de la Fleur de Lis et de la Marguerite,' and Jean Froissart's 'Dit de la Marguerite' and 'Lay de Franchise' (*Chaucer's Dream Poetry* 145–55). See also Lowes, 'Prologue.'

7 All references to Chaucer's works are in *The Riverside Chaucer*, ed. Larry D. Benson (3rd ed. Boston: Houghton Mifflin, 1987), and are cited in the text by line number.

8 Chaucer reproduces one widely known version of this theory in his translation of Boethius' *De consolatione philosophiae*: 'The lookinge, bi castinge of his beemes, waiteth and seth from afer al the body togidere, withowte moevinge of itself ... Algates the passioun (that is to seyn, the suffraunce or the wit) in the qwyke body goth byforn, excitinge and moevinge the strengthes of the thoght. Ryht so as whan that clerness smyteth the eyen and moeveth hem to sen, or ryht so as voys or sown hurteleth to the eeres and commoeveth hem to herkne, than is the strength of the thoght imoeved and excited, and clepeth forth to semblable moevinges the speces that it halt withinne itself; and addeth tho speces to the notes and to the thinges withowteforth, and medleth the ymages of thinges withowteforth to tho formes ihidde withinne hymself' (*Boece* 5.pr.4.148–51; 5.m.4.46–59).

9 Ovid, *Metamorphoses* 4.228. The trope also appears in Dante (*Purgatorio* 20.132).

10 Bernard of Clairvaux, 'De consideratione' 3.5, trans. Evans 151.

11 Later in the prologue, Love rebukes the narrator for omitting mention of Alceste 'that ylke tyme thou made / "Hyd, Absolon, thy tresses," in balade' (F538–9), implying that the song does not venerate Alceste. Yet the narrator clearly indicates that the song (or at least the version included in the prologue) is 'in preysyng of this lady fre' (F248), that is, Alceste, who is described in the preceding lines (F241–6).

12 Travis draws attention to this confusing 'infinitude of analogues' in his

perceptive analysis of Chaucer's use of metaphor: 'But surely there is a major philosophical problem here ... : how can Alceste BE the sun, let alone BE the daisy and the sun at once? And if Alceste IS the sun, why should anything else that is also sunlike, such as the daisy or the God of Love, NOT BE the sun?' Travis, 'Chaucer's Heliotropes' 414–15.

13 Bernard of Clairvaux, *De consideratione* 3.5, trans. Evans 151.

14 Ibid.

15 Cherniss comes to a similar conclusion on the basis of Chaucer's representation of the narrator and his dream in the prologue of the *Legend of Good Women* ('Chaucer's Last Dream Vision' 198).

16 It seems likely that the daisy of the Prologue to the *Legend of Good Women* was also chronologically the last allegorical theme explored by Chaucer, though as with Chaucer's other works the dating of this poem (probably 1386–7) is far from certain. Although two of the *Canterbury Tales*, the *Second Nun's Tale* and *Tale of Melibee*, are allegorical, both appear to be early works later modified for inclusion in the *Canterbury Tales*.

17 Such a distinction corresponds to Copeland's description of 'primary' and 'secondary' translations: the former follow the original text closely, while the latter paraphrase loosely, using the original text as the basis of a text that will convey quite a different meaning (*Rhetoric, Hermeneutics, and Translation* 94–5).

18 I have repunctuated the line, following Fisher, *Complete Poetry and Prose*.

19 On Jean de Meun and Chaucer's common interest in translation, see Fleming, 'Smoky Reyn' 3–4.

20 Deschamps's poem, addressed to Chaucer and accompanying a gift of his own writings, appears in Eustache Deschamps, *Oeuvres*, ed. le marquis de Queux de Saint-Hilaire and G. Raynaud (11 vols. Paris: SATF, 1878–1903) 2: 138–9; also in J.M. Manly, ed. *Canterbury Tales* (1928) 23–5. The latter is reproduced with emendations in Fisher's edition of Chaucer, pp. 952–3.

21 Windeatt, *Dream Poetry* xvi.

22 Edwards, *Dream* 32.

23 Copeland, *Rhetoric* 94–5.

24 Ibid. 193.

25 Although the *Book of the Duchess* has been convincingly dated to the mid-1370s, dating of the later two allegories is less conclusive and even their order is uncertain. It seems unlikely that Chaucer would move from the four-stress line of the *Book of the Duchess* to the five-stress line of the *Parlement of Fowls* and then return to the four-stress line in the *House of Fame*, unless, of course, the *House of Fame* was already in progress and Chaucer completed it using the old metrical pattern rather than rewrite

the portions written earlier. It seems to me that the concerns raised in the *Parlement of Fowls* are more deeply explored in the *House of Fame*, and so I would suggest that the *House of Fame* was at least finished later, though it may well have been started before the *Parlement of Fowls*. My argument is not dependent upon this order of composition; but I will discuss the works in this sequence in order to highlight the development of Chaucer's presentation of how the act of perception takes place, and how perception leads to knowledge.

26 Elaine Tuttle Hansen stresses the significance of the lady's absence in the *Book of the Duchess*, suggesting that her death actually frees the Man in Black to form other bonds, both social and personal (*Chaucer and the Fictions of Gender* 58–86, esp. 72–3).

27 Hardman suggests that the poem may have been written to commemorate the construction or dedication of a memorial tomb for John of Gaunt and his wife Blanche ('*Book of the Duchess* as a Memorial Monument').

28 Fisher, ed., *Chaucer* 543; but see Hanning's convincing argument that the story of Ceyx and Alcione is central to the poem ('Chaucer's First Ovid' 126–41).

29 *Metamorphoses* 11.731–48. Machaut does not omit the transformation in his *Fonteinne amoureuse*.

30 The question of the *Book of the Duchess*'s success as a consolation has been much debated: on the pro-consolation side, see Fyler, 'Irony and the Age of Gold'; for an 'anti-consolatory reading,' see Ellis, 'Death.' A survey of the terms of the debate can be found in Walker, 'Narrative Inconclusiveness.'

31 Noted in Edwards, *Dream* 77; see also his discussion of Somnis and Morpheus in Ovid, Machaut, and the *Book of the Duchess* (74–82).

32 William of Conches uses this terminology in his glosses on the *Timaeus*: 'Somniorum igitur quedam cause sunt interiores, quedam exteriores. Interiorum alia ex anima, alia ex corpore. Ex anima sunt reliquie cogitationum; ut corpore ut qualitas complexionis, sacietas, fames, cibus.' *Glosae super Platonem* CXLI, ed. Jeauneau 242. Similarly, Macrobius states that one of the two types of non-prophetic dreams, the *insomnium*, 'arise[s] from some condition or circumstance that irritates a man during the day and consequently disturbs him when he falls asleep' (I.3.5, trans. Stahl 89). Peden notes the importance of William of Conches's glosses on Macrobius in promoting interest in Macrobius and dream theory in general in the twelfth and early thirteenth centuries ('Macrobius' 64–6).

33 Russell, *English Dream Vision* 74.

34 Kruger, *Dreaming in the Middle Ages* 83–99; quotation from 96.

35 Russell, *English Dream Vision* 81.

36 Quilligan, *Language of Allegory* 33.

37 Prior argues that three senses of *hart* ('hart/heart/hurt' [3]) are present in the poem, and that the operation of the pun is integral to 'the movement from the literal hunt to the metaphorical one ... [T]he quest becomes one for the *herte*/human heart, and especially for the *herte* (wounded) heart' ('*Routhe* and *Hert-Huntyng*' 11).

38 'Hart' as the deer appears at 540; as the lover's corporeal organ, 'membre principal / Of the body' at 488–9; as the seat of emotion of the Man in Black at 768, 772, 776, 1211, 1222, 1224, 1226, 1275; of the Duchess at 884, of others at 575, 594, 713; as the Duchess herself at 1233; and even as a verb, to 'hert' (hurt), at 883. This is not a complete list, and of course each example only refers to the primary meaning of the word in the passage: often 'hart' refers to at least two of these meanings at once.

39 The extra *canto* in the *Inferno*, making thirty-four, is conventionally excluded as introductory; alternatively, it can be taken as a manifestation of the imperfection of hell.

40 Hart, 'Medieval Structuralism'; on the principles of order and harmony which inform the poetic composition and structure of *Troilus and Criseyde*, see Vance, *Mervelous Signals* 274–7.

41 Peck, 'Theme and Number'; see also North, *Chaucer's Universe* 343–8.

42 Walker 13. Walker also rightly stresses Chaucer's characterization of the lady as an ideal, rather than an individual: 'While it is clear that Chaucer intends a reference to Blanche of Lancaster, the real historical Blanche surrenders to the conventional and symbolic White' (12).

43 Peck asserts that the Man in Black's description of his mind having been 'a white walle or a table' ('Nominalist Questions' 780) illustrates Chaucer's acceptance of a faculty psychology traceable to Ockham: 'The intellective process Chaucer describes is in keeping with Ockham's notion of intuitive and abstractive cognition' (751). Peck, however, neglects the importance of the mechanism of vision in the *Book of the Duchess*: it seems unlikely that Chaucer would couple the Platonic extramission theory of vision with a nominalist view of the mental faculties, particularly since the optical model proposed by Ockham omits the role of species entirely. Burnley more plausibly links the same passage with sources in Plato and Calcidius ('Some Terminology' 17–18).

44 I have slightly repunctuated Fisher's text here: since line 1011 begins with 'And,' it seems unlikely that a full stop should precede it. Burnley also repunctuates the passage as I do ('Some Terminology' 21–2).

45 Although more detailed medieval accounts of the structure of the mind often describe five faculties, distributed among the three chambers of the

brain, most writers simplify these to the three faculties of imagination, judgment or wit, and memory (Harvey, *Inward Wits* 43–61). For example, in his encyclopedia, Bartholomaeus Anglicus states that the mind is divided into three parts, 'ymaginativa,' 'logica,' and 'memorativa,' 'for in the brayn beth thre smale celles' (*De proprietatibus rerum* 3.10, trans. Trevisa, ed. Seymour 98).

46 I believe Chaucer intends a pun on 'white' and 'wit' at line 1010; a similar pun appears in the *Legend of Good Women* ('Hire white corowne berith of hyt witnesse' [527; see also 83–8]).

47 The narrator shares the Man in Black's melancholy state: in the poem's opening, he tells how 'sorwful ymagynacioun / Ys alway hooly in my mynde' (14–15); 'Suche fantasies ben in myn hede / So I not what is best to doo' (28–9). Neaman also interprets the poem as an enactment of the ordering of the mind's three faculties, both that of the Man in Black and that of the narrator: 'Even before he can begin to understand grief, the Narrator must experience the emotion of "routhe" that others know in the middle chamber of reason. He must recall his feelings by exercising his memory. ... He must next comprehend the relations between the past and the sensations experienced in the present (an activity which is also the province of reason). At last, he will realize that the past, like art, is memory fixed in image' ('Brain Physiology' 108–9).

48 Yates cites Quintilian's description of the art of memory, based on the structure of a building: 'In order to form a series of places in memory, he says, a building is to be remembered. ... The images by which the speech is to be remembered ... are then placed in imagination on the places which have been memorised in the building. This done, as soon as the memory of the facts requires to be revived, all these places are visited in turn' (*Art of Memory* 3). Carruthers rightly notes that 'The architectural mnemonic was certainly not the only nor even the most popular system known in the Middle Ages for training the memory' (*Book of Memory* 80); see her summary of Quintilian's system of artificial memory (71–5). See also Coleman, who discusses not only the antique system of architectural mnemonics (*Ancient and Medieval Memories* 39–59), but also later medieval reception of such theories (417–18).

49 Noted in Rendall, '*Gawain* and the Game of Chess' 194. On the evolution of the pieces of the game, see Murray, *History of Chess*.

50 Rowland gives an alternate interpretation of the chess analogy, suggesting that the 'ferses twelve' alluded to by the narrator is a veiled reference to the sign of the zodiac, thus attesting to the extent of Chaucer's interest in the influence of the stars ('Chess Problem').

51 Chaucer's intentional ambiguity has occasioned some debate regarding the extent of his reliance on Macrobian categories of dreams, and even regarding Macrobius's general popularity in the later Middle Ages. For example, Bennett stresses the importance of Macrobius in the *Parlement of Fowls*, stating that 'Macrobius ran second only to Boethius ... It is noteworthy that these are the only two philosophical texts that Chaucer chose to summarize or translate' ('Some Second Thoughts' 136). Conversely, Peden asserts that 'the primary inspiration of the poem was Cicero's *Somnium Scipionis*, not Macrobius' ('Macrobius' 69). It is not necessary to concur in this assessment to agree with Paden that attempts to classify Chaucer's dream visions according to Macrobian types 'are bedevilled by Chaucer's slippery terminology' (68). Chaucer's intentional ambiguity makes it ultimately impossible to limit one of his narrator's dreams to a single category, thus demonstrating that such categories are irrelevant, for it is impossible to ever know if a dream is prophetic or not, externally or internally caused, an *oraculum* or an *insomnium*.

52 Aers 2.

53 Macrobius, *Commentary on the Dream of Scipio* I.3.8, trans. Stahl 90.

54 As Collette has shown, in the *Franklin's Tale*, Dorigen's ability to see the black rocks is affected by her will: 'Dorigen's *derke fantasye* is an excellent example of sense experience producing, through *species*, *phantasms* which, uncontrolled by will operating through the rational faculties, multiply in strength and number' ('Seeing and Believing' 404). See also Stanbury's discussion of the relation of will and vision in *Troilus and Criseyde* ('Lover's Gaze' 228–9).

55 Scipio only speaks to the narrator briefly, urging him on physically rather than verbally. See 120–1, 169–70.

56 Latin *errare*, to wander; Chaucer stresses the narrator's state by using the word twice (146, 156).

57 Ferster, 'Reading Nature' 193, 195–6.

58 Strangely, Lynch introduces her essay by stating that the poem's 'most central, motivating, or pivotal problem has not yet been identified as one of will rather than of understanding ...' ('Voluntarism' 1). She adds that 'virtually nothing in modern criticism of the poem's philosophy attempts to uncover the theory of the will that underlies the significant moments of choice ...' (5). Her omission is all the more striking for the fact that she cites Ferster's work in her conclusion (88n53). Though Lynch's essay is not the first to treat the question of will in the *Parlement of Fowls*, it is nonetheless the fullest treatment to date.

59 Lynch, 'Voluntarism' 6.

60 Ibid. 6. That is not to say the role of the will is not stressed earlier, perhaps most notably by Augustine. But fourteenth-century voluntarism, as Lynch presents it, accords the will a new prominence.

61 See Lynch, 'Voluntarism' 7–11, on medieval philosophers' use of the example of the animal who cannot choose (Buridan's ass).

62 Ibid. 10–11. She also rightly notes that the passage may be related to the 'adamaunt [which] / Can drawen to hym sotylly / The iren that is leid therby' in the *Roman de la rose* (*Romaunt of the Rose* 1182–4). In the *Rose*, however, there is no suggestion of a divergent pull from two magnets; rather, the magnet is an illustration of the power of gold and silver to attract people's affection.

63 Sylla, *Oxford Calculators* 308–427.

64 'Quilibet terminus cuiuscumque rei est significativus ad placitum' (Dumbleton, *Summa* 1: 5.27–8; and cf. book 1, cap. 1, passim [1: 1–9]). Quotations from Dumbleton are taken from Weisheipl's unpublished edition, described in chapter 2, and are cited by volume, page, and line number; translations are my own.

65 'De universalibus quae ideae apud Platonem dicuntur' (2: 5.1–2).

66 Boucher, 'Nominalism' 214. Boucher opposes Chaucer's and Boccaccio's 'nominalism' to the 'Platonic epistemology' of Dante and the anonymous author of the *Queste del Sainte Graal*. Though this dichotomy is unnecessarily reductive, Boucher is right to distinguish between two kinds of writing divided by nominalism's dissolution of 'the firm bonds between signifier and signified' (215). For other assessments of Chaucer's relation to nominalism, see Steinmetz, 'Late Medieval Nominalism'; Delasanta, 'Nominalism'; and Russell, who asserts that nominalism profoundly influenced the development of the dream vision genre in England (*English Dream Vision* 109–14).

67 Utz and Watts have issued an important corrective to such loose characterizations of Chaucer's 'nominalism' ('Nominalist Perspectives'); see also Minnis, 'Looking for a Sign,' esp. 144; Penn, 'Literary Nominalism.'

68 Sylla, *Calculators* 12.

69 Fletcher, 'Developments in the Faculty of Arts' 340–1.

70 For the relevant passages in Dumbleton, see his *Summa* book 1, ch. 6–11 (1: 17–27). On Strode, see Ashworth and Spade, 'Logic in Late Medieval Oxford' 57. For more on Strode's sources in his *Insolubilia*, see Spade, *Medieval Liar* 87–91; Spade, 'Robert Fland' 60. A complete edition of Strode's *Logica*, which includes as one of its six parts the treatise on insolubles, is promised by Alfonso Maierù ('Le ms. Oxford canonici misc. 219 et la *Logica* de Strode').

71 Delasanta, 'Chaucer and Strode' 211, 205.

72 Courtenay, 'Dialectic of Divine Omnipotence' 111–21.

73 Ashworth and Spade 37; see also Wilson, *William Heytesbury*.

74 Spade, *Medieval Liar* 87–91.

75 For a review of the evidence connecting Strode in Oxford with Strode in London, see Delasanta, 'Chaucer and Strode' 206–10.

76 Books 4–9 of Dumbleton's *Summa* remain unedited, but see Sylla's detailed outline of the later books in her *Calculators* 565–625; magnetism is at 616 (book 6, ch. 11).

77 I plan to treat this topic in greater detail elsewhere; see 'Intention in *Troilus and Criseyde*,' forthcoming. On the 'process by which Troilus and Criseyde fall in love with each other' and the exercise of '"fre chois"' in book 2, see Minnis, *Chaucer and Pagan Antiquity* 71–4. On the 'coeval philosophical superstructure' of *Troilus*, see Utz, 'As Writ Myn Auctor' 134.

78 The *Oxford English Dictionary* notes that 'As the female ... was greatly superior for purposes of sport, the sense of *formel* in this application may be "regular," "proper"' (s.v. 'formel, formal, sb.').

79 Entzminger 6, 11.

80 On Chaucer's use of Dante, see Schless, *Chaucer and Dante*; Taylor, *Chaucer Reads the* Divine Comedy. For a useful survey of the debate regarding Chaucer's attitude toward Dante, see Ellis, 'Chaucer, Dante, and Damnation.'

81 Steadman also emphasizes the use of the eagle to represent contemplation ('Chaucer's Eagle'). Rowland claims that the eagle represents 'eloquent speech' ('Bishop Bradwardine' 49), which overlooks the significance Chaucer himself ascribes to the bird in the *Parlement of Fowls* (331).

82 See Travis, 'Chaucer's Trivial Fox Chase' 197–9.

83 '[D]a quella imagine divine, / per farmi chiara la mia corta vista, / data mi fu soave medicina' (Pa.20:139–41).

84 See 66 ('at my gynnynge'), 109 ('Now herkeneth'), 151 ('First sawgh I'), 509 ('Now herkeneth'), 512 ('For now at erste'), 525 ('now shal men se'), 1109 ('Now entre in my brest').

85 Grennen, 'Chaucer and Chalcidius' 262.

86 '[L]icet omnes pyramides in uno lumine contente sint essentialiter lux una, differunt tamen virtualiter, id est efficaciter, sicut cum lapis proicitur in aquam generatur circuli diversi, qui tamen aquam non dividunt.' John Pecham, *Perspectiva communis* 1.6, ed. and trans. Lindberg 64–5. Lindberg describes the wide dissemination of Pecham's work, stating that 'No other medieval optical work is extant in nearly as many copies, and we are forced to conclude that the *Perspectiva communis* became the standard elementary

optical textbook of the late Middle Ages' (29).

87 Quoted in Irvine, 'Grammatical Theory' 865.

88 This connection has been noted previously; see Leyerle, 'Chaucer's Windy Eagle.'

89 '[L]atitudo hesitationis,' book 1, ch. 21 passim. (1: 61–9); on truth and falsehood in propositions, see book 1, cap. 21 (1: 62).

90 On the memorial function of these images, see Kolve, *Chaucer and the Imagery of Narrative* 41–2; Rowland, 'Bishop Bradwardine'; and Carruthers, 'Italy, *Ars Memorativa*, and Fame.'

91 Delany gives an account of contemporary usage of the word 'fantome,' stating that 'As a strictly technical term, then, *phantom* denotes a mental process, or the product of a mental process, which is deceptive in that it does not accurately mirror the phenomenal world' ('Phantom' 70).

92 Beryl Rowland, 'Bishop Bradwardine' 49, 48.

93 See Knapp, 'Relyk of a Saint.'

8: Chaucer's Personification and Vestigial Allegory in the *Canterbury Tales*

1 Quilligan, *Language of Allegory* 42. On the interrelation of personification and allegory, see Paxson, *Poetics of Personification* 38–41, 70–4.

2 Whitman, *Allegory* 270.

3 Delany, 'Undoing Substantial Connection'; Paxson 138. See also the critique of Delany in Penn, 'Literary Nominalism.'

4 Penn 186. See also Minnis, 'Authorial Intention' 17.

5 The story of Ceyx and Alcione, recounted by the narrator, includes gods from classical literature (Juno, Morpheus) who can, in a sense, be considered as personifications; but they are contained in a story within the story and are not part of the interaction of the narrator and Man in Black.

6 Lynch, 'Book of the Duchess' 293, 297.

7 Dumbleton, *Summa* book 1, ch. 5 (1: 17.5–7).

8 'Cum homo intelligens istum terminum "album" viderit in scripto vel audierit in voce, statim reducitur ad memoriam illius hominis vera intentio albi. ... [T]erminus intentio et intentio rei idem nomen habent propter hoc quod per terminum aliqua intentio ad actum nostrum reducitur.' Dumbleton, *Summa*, book 1, ch. 2 (1: 3.9–11, 26–7).

9 Although Priapus appears 'in sovereyn place' in Venus's temple (254), he is clearly secondary to the goddess herself, who is tucked away in the innermost recess of the temple (260). The phallic god is (to put it allegorically) the point of entrance into Venus's sacred precincts.

10 Ferrante notes that 'Venus is degraded progressively in the allegories – from her role in Martianus as essential to marriage and life, to the corrupted aide of Nature in the *De planctu* and the cohort of evil in the *Anticlaudian*. In the *Roman de la Rose*, she becomes a powerful figure of lust, who uses Nature and Genius to do *her* will' (*Woman as Image* 61n38). See also Economou, 'Two Venuses' and *Goddess Natura* 136–9; Tinkle, *Medieval Venuses*.

11 Dronke notes Chaucer's alteration of Alanus's description of Nature's palace in the *Anticlaudianus*: 'Chaucer has a subtle detail that goes beyond Alan, or indeed deliberately contradicts him: where in Alan's garden Natura has a house made of gold, silver and gems, in Chaucer's, "Of braunches were here halles and here boures"' (Dronke and Mann, 'Medieval Latin Poets' 166). Hewitt describes how Chaucer magnifies the gap between the abstraction Nature and her earthly domain: 'What Chaucer's dreamer has done to the *Plaint of Nature* is to turn the allegorical birds on Nature's gown into natural fowls without separating them absolutely from their prior frame of reference. The *Parliament of Fowls*' feathered assembly, that is, transfers the two-dimensional fowls from the fabric of Nature's magnificent robe in Alain's poem onto the earth, lake, and tree-branches of a three-dimensional world' (Hewitt, 'Ther It Was First' 26).

12 These include Cupid, Wille, Plesaunce, Aray, Lust, Curteysie, Craft, Delyt, Gentilesse, Beute, Youthe, and at least fifteen others (211ff.)

13 Ferster, 'Reading Nature' 198.

14 Ibid. 194.

15 For Chaucer's allusions to Ovid's *Metamorphoses* in his description of Fame, see Fyler, *Chaucer and Ovid* 23–64.

16 'If a film were made of the *Anticlaudianus* she would be played by Susannah York.' Trout, *Voyage of Prudence* 41.

17 The earliest example I have seen cited is a manuscript of ca. 1470 noted by Tuve in *Allegorical Imagery* (74 and fig. 17).

18 Isidore of Seville, *Libri differentiarum* 2.39.154 (*PL* 83: 94); noted in Katzenellenbogen, *Allegories of the Virtues and Vices* 55. (Note that the illustrations are larger and therefore more legible in the 1939 edition than in the 1989 reprint.)

19 Katzenellenbogen, *Allegories* 56 and 56n4, 76n1.

20 Wallace, *Chaucerian Polity* 225.

21 On the relationship between Boethius's Philosophy and Chaucer's Prudence, see Johnson 141.

22 Owen notes that Prudence speaks as 'a woman rather than an abstract quality' (270). Wallace more matter-of-factly calls Prudence 'a stay-at-home

wife' (*Chaucerian Polity* 224). Johnson points out that Chaucer's changes to the original treatise emphasize Prudence's wifely status: where Reynaud refers to her as having produced 'sage compagnie,' Chaucer calls her 'a wyf of so greet discrecioun' (144). Collette suggests that Chaucer's Prudence represents a specifically feminine form of 'aristocratic virtue' ('Heeding the Counsel' 419).

23 This was first argued in 1957 by Hartung, 'Textual Affiliations.' Building on Hartung's work, Matthews suggests a chronology which 'track[s] *Melibee* from its early composition near 1373 to its experimental flirtation with the *Canterbury Tales* in 1388 as the lead piece to be narrated by the Man of Law, and thence to Fragment B2 in the mid-1390's where it finally came to rest' ('Date' 231). Patterson questions Matthews's early dating and suggests that a date shortly after 1377 might be more likely ('What Man Artow?' 140n74).

24 Severs, 'Tale of Melibeus' 568n14. The original Latin text by Albertano of Brescia reads 'oculis.' Severs notes two passages near the end of the *Tale of Melibee* that do not appear in the French version but which do resemble the corresponding passage in the Latin (614, n. to 1170–1, 1174–5), suggesting that, though Chaucer was translating the French version, he remembered the general content of the Latin.

25 Strohm seems aware of the problem here, but dismisses it by suggesting that 'Chaucer's understanding of his obligations as a translator' outweighed his desire to maintain consistency within the allegory ('Allegory' 40). In light of the many minor alterations to Reynaud's text noted by Severs, Palomo, and Bornstein, such extreme fidelity to the original seems unlikely. Waterhouse and Griffiths also suggest that Chaucer's retention of the reading 'piez' is intentional, but this leads them to suggest that the first of Sophie's two wounds, to the hands and feet, allude to the crucifixion, identifying her as a figure of Christ ('Sweete Wordes' 345). Such an interpretation is suggestive: if Sophie is identified with Christ, and Christ as the second person of the Trinity is the model for one of the triune faculties of the mind, then injury to Sophie indicates injury to one of the three mental faculties.

26 Wallace, *Chaucerian Polity* 231.

27 See Burnley, *Chaucer's Language* 49–63, on the distinction between prudence and wisdom, prudence being a specifically human, rational power.

28 Wallace suggests that the incoherence of 'Prudence's allegorization' is not a significant shortcoming: 'it is only of local and limited effect [and] serves to help Melibee see who he is and where he is at this stage of the dialogue' (*Chaucerian Polity* 240).

29 Patterson, 'What Man Artow?' 158.

30 On the relationship of divine and human will in Ockham, see Adams, 'Ockham on Will'; for a brief but cogent summary, see Colish, *Medieval Foundations* 313–15.

31 See Sylla, *Calculators* 137. On the significance of Dumbleton's use of latitudes to quantify intellective acts, see North, 'Natural Philosophy' 88.

32 Courtenay, *Schools and Scholars* 365–8.

33 For more on the role of intention in the movement of Melibee's will ('entente,' VII.1871), see my 'Intention in *Troilus and Criseyde*,' forthcoming.

34 Wallace, *Chaucerian Polity* 228.

35 Patterson, 'What Man Artow?' 158.

36 Matthews comments that 'the particular indebtedness of the *Merchant's Tale* to the Melibean tract has been frequently remarked, and it was preponderantly this derivative relationship that misled Tatlock into concluding that the two were composed at about the same time' (231).

37 Pratt has argued convincingly that Robert Holkot's commentary on the Book of Wisdom, *Super Sapientiam Salomonis*, is an important source for passages on dreams in the *Nun's Priest's Tale* ('Some Latin Sources'). This suggests that Holkot's commentary may also have been a source for Chaucer's *Tale of Melibee*, which alludes extensively to the Book of Wisdom. On Holkot's wide influence, see Coleman, *Medieval Memories* 263–5 and Minnis, 'Discussions of "authorial role."'

38 Interestingly, May's perspective on whether marriage is heavenly or hellish is left unanswered: 'I dar nat to yow telle / Or wheither hire thoughte it paradys or helle' (IV.1963–4).

39 Chaucer alludes explicitly both to Claudian (IV.2232) and, exactly two hundred lines earlier, to the *Rose* (IV.2032). In the *House of Fame*, Chaucer groups Claudian with Virgil and Dante as one of those who know about hell (449–50); Claudian also appears as a statue on a pillar of sulphur, the author who 'bar up al the fame of hell' (1510). The significance of the *De raptu Proserpinae* has been discussed by Pratt, 'Chaucer's Claudian'; Donovan, 'Image of Pluto' and 'Chaucer's January and May'; Wentersdorf, 'Theme and Structure'; and Otten, 'Proserpine.' On the *Roman de la rose* and the *Merchant's Tale*, see Economou, 'January's Sin'; Wallace, 'Chaucer and the European Rose'; Diekstra, 'Chaucer and the *Romance of the Rose*'; *and* Calabrese, 'May Devoid of All Delight.'

40 Neuse, 'Marriage' 116. It is important to note that the allegory of the names May and January is original to Chaucer, as is the age difference; other versions of the tale do not mention the husband's age. See Dempster, 'Merchant's Tale'; Benson and Andersson, *Literary Context*. Though one of

the versions, 'The Blind Man and his Wife' written by one Adolphus in 1315, refers to the lover as a 'juvenis' (234, line 6), it does not specify the age of the husband.

41 Neuse 122.

42 Calabrese 262–3. Economou explores in greater detail the role of the mirror in both works, linking the mirror of Januarie's imagination to the deceptive mirrors described by Nature in Jean de Meun's continuation of the *Roman de la rose* (253–4).

43 Patterson, *Subject of History* 339. Patterson gives a particularly interesting interpretation of the relation of the teller to the tale (333–44). See also Edwards, who describes how some aspects of the tale 'counter the merchant's viewpoint, even as they presumably provide material for him ... The discontinuity of character and narrative, narration and doctrine, is an essential quality of the tale' ('Narration and Doctrine' 366).

44 Though Chaucer uses the term 'fantasye' in the *Merchant's Tale* to describe imagination specifically in the context of faculty psychology, in other tales he uses the term in something closer to the modern, more general sense of the term (e.g., I.3845; III.516). For a reading of January's 'male fantasye' in terms of 'projected wish-fulfillment' (76), see Collette, *Species, Phantasms* 75–88.

45 In this context, it is interesting to note that Chaucer likens the wedding of Januarie and May to that of Mercury and Philology, recounted by Martianus Capella (IV.1732–8), for Mercury is not only representative of eloquence but is also the patron of merchants.

46 Collette describes a similar failure of vision in her analysis of the *Franklin's Tale*: she suggests that Dorigen's will fails to regulate her imagination, causing the black rocks to be sometimes visible, sometimes not. 'Augustine['s] first point – that the will can fail to regulate the images that crowd the mind – parallels Dorigen's *derke fantasye*, her inability to deal with the visual experience, the sight of the rocks. ... Rather than accepting the rocks as part of creation and "looking" past them, she can deal with them only at the physical level, unable to move the *phantasms* of the rocks out of the imaginative section of the brain to the estimative, unable to deal rationally with them' ('Seeing and Believing' 405–6).

47 'Lumen oculi naturale radiositate sua visui conferre' (I.46); on infection or injury to the eye, see I.54; 'In distinctione visibilium rationem imperceptibiliter operari. Nullum enim visibile cognoscitur sine distinctione intentionum visibilium vel sine collatione aut relatione ad universalia cognitorum prius a sensibilius abstracta, qui fieri non possunt absque ratiocinatione' (I.57). John Pecham, *Perspectiva communis*, ed. Lindberg 128–9; 134–5; 136–7.

48 Peter Brown contextualizes Januarie's blindness more generally: 'when January becomes physically blind, it is but the external manifestation of internal disorders ('Optical Theme' 241).

49 Dempster, 'Merchant's Tale' 343–53. Two late fifteenth-century versions of the tale substitute Jove for God, but this is because the tale is collected within a nominally pre-Christian work, the Latin *Aesop* (Dempster 354–6).

50 Otten 284. Note that May is connected with Rebekke (IV.1704) and Ester (IV.1744).

51 Rosenberg, 'Cherry-Tree Carol.' Wurtele explores the significance of Chaucer's 'determin[ation] to place May in juxtaposition with Mary' ('Ironical Resonances' 75). See also Bleeth, 'Joseph's Doubting.'

52 In the language of Melibee, May is 'the fer cause and the ny cause' (VII.1395). Chaucer's exploration of the infinite regression of causality in the *House of Fame* suggests that, in the *Merchant's Tale*, final determination of a primary cause is impossible.

53 Otten 287.

54 Brewer remarks 'If the *Shipman's Tale* is closest to the fabliau-type, the *Merchant's Tale* is by general agreement furthest away ... It does not fit into any simple category ...' ('Fabliaux' 309). Arrathoon notes that 'the perception that *The Merchant's Tale* differs radically from the traditional fabliaux ... depends ... on the fact that treatment of marital strife in *The Merchant's Tale* is taken beyond the domestic sphere and given mythic significance' ('For craft is al' 264); she provides a detailed discussion of genre and the *Merchant's Tale* (241–328).

55 These flames affect Damyan as well (IV.1777, 1783, 1875–6), reinforcing the infernal aspect of the garden all three characters inhabit at the tale's climax.

56 Although Bernard of Clairvaux's identification of the bride with the Church is perhaps better known, Alanus de Insulis interprets the bride of the Song of Songs as the Virgin Mary in his *Elucidatio in Cantica Canticorum* (*PL* 120: 51–110); on Alanus's commentary, see Astell, *Song of Songs* 42–72.

57 Ovid, *Metamorphoses* I.622–721.

58 Brown points out that May's apparently scientific argument is actually specious: 'the science is misapplied since it is used to account for an event which was not an optical illusion at all.' Brown, 'Optical Theme' 234.

59 'Sed termini visi a nobis per quorum intentiones aliarum rerum memoriae nostrae non occurrunt, inintelligibiles apud nos dicuntur.' Dumbleton, *Summa* book 1, ch. 2 (1: 3–4).

60 Miller notes that 'competence in Christian allusion and symbolism – even in doctrine – was, it is clear, far from universal. Chaucer's pilgrims themselves exhibit a wide range of literacy which probably reflects that of his

own audience, capable of misapplying or of being ignorant of conventional doctrinal and allegorical materials' ('Allegory' 335).

61 Holley makes this point very emphatically: 'Chaucer's vision is simultaneously conceptual (showing how one could conceive things to be) and scientific (showing things *as the eye sees*) ... Each time the energy of a text strikes its reader, its contexts and its images come to life by *vertu* of the observing eye. The text is regenerative' (*Chaucer's Measuring Eye* 143).

9: Division and Darkness

1 'Prout oculis pictura imaginabatur, animalium celebratur concilium.' *De planctu Naturae*, prosa 1, trans. Sheridan 86.

2 Ibid., metrum 2, trans. Sheridan 106.

3 'the light so in my face / Bigan to smyte, so persing euer in one / On euere part, where that I gan gone, / That I ne myght nothing, as I would, / About me considre and bihold / The wondre hestres, for brightnes of the sonne' (Lydgate, 'Temple of Glass,' lines 24–9; ed. Norton-Smith 67.

4 Christine de Pizan, *Chemin de long estude* lines 2384–9, 2414; ed. Püschel. Subsequent citations are by line number in the text; translations are my own.

5 'Sompnium prioris de Sallono ad regem Francie super materia scismatis' in Honoré Bonet [Bouvet], *L'Apparicion Maistre Jehan de Meun et le Somnium super materia Scismatis*, ed. Arnold; cited in the text by page number.

6 Christine de Pizan, *Mutacion de Fortune*, lines 404–5 (identified as Nature at 469); ed. Solente.

7 Christine de Pizan, *Cité des dames*, ed. Curnow, trans. Richards; cited in the text by chapter number. See the illustration from one of the manuscripts of the *Cité des dames* executed under Christine's supervision (British Library, Ms. Harley 4431, fol. 290), reproduced in Richards's translation opposite p. 3. On the left, the illumination shows Christine in her room along with Raison, Rectitude, and Justice; on the right, it depicts Christine and Raison labouring side by side.

8 William Langland, *Piers Plowman: The B Version*, ed. Kane and Donaldson 535 (passus 15, lines 23–34).

9 Their names are, in order: 'Pax,' 'Concordia,' 'Guerra,' 'Oppinio,' 'Misericordie,' 'Sine Humanitate,' 'Inquisitionis vallis,' 'Ceca Ignorantia,' and 'Dulceloquium.' See Honoré Bouvet, *Somnium super materia Scismatis* 70–1.

10 Jean Gerson, 'Le Traictié d'une vision faite contre *Le Ronmant de la rose* par le chancelier de Paris,' ed. Hicks; cited in the text by line number.

11 Christine de Pizan, *The Book of Deeds of Arms and of Chivalry*, trans. Sumner Willard, ed. Charity Cannon Willard 13 (1.1). An edition of Willard's text is forthcoming.

12 'Adont, surprise de somnie en mon lit couchiee, m'apparut en dormant par semblance une creature sicomme en la fourme d'un tres sollempnel homme d'abit, de chiere et de maintien d'un pesant ancien saige auctorisee juge.' Bibliothèque Nationale, Ms. f. fr. 603, fol. 49, v.1; quoted in Coopland, trans., *The Tree of Battles of Honoré Bonet* 24n47; trans. Willard, *The Book of Deeds of Arms and of Chivalry* 143 (3.1).

13 Honoré Bouvet, *L'Apparicion Maistre Jehan de Meun*, ed. Arnold; cited in the text by line number.

14 Honoré Bouvet, *The Tree of Battles* 79.

15 Macrobius, ed. Willis 10 (I.3.10); trans. Stahl 90.

16 John Gower, *Confessio Amantis*, ed. Macaulay, Prologue 833, 851–5.

17 Christine de Pizan, *Lavision-Christine*, ed. Towner 108.9–10 (I.29).

18 Dinshaw, *Getting Medieval* 1, 158.

19 See Dinshaw, 'Chaucer's Queer Touches' and *Getting Medieval* 150–2.

20 Watson, 'Desire for the Past' 94–5.

21 Ibid. 60 and passim.

22 Bynum, 'Why all the Fuss' 29; see chapter 2, p. 22.

23 That is not to say that devotional literature is devoid of claims to authority: see Nicholas Watson, *Richard Rolle and the Invention of Authority*. The claim of authority within devotional narrative is, however, made indirectly and with what can only be described as anguish, as Jenifer Sutherland shows in her analysis of inexpressibility topoi in the writings of Walter of Wimborne and Margery Kempe.

24 On whether 'discourses,' in the Foucauldian sense of the term, can be said to be present in medieval culture, see Akbari, 'Orientation and Nation' 102–3, 124.

BIBLIOGRAPHY

Abbreviations

CCCM Corpus Christianorum, Continuatio Mediaevalis (Turnhout: Brepols, 1966–)

PL Patrologia Cursus Completus, Series Latina, ed. J.-P. Migne (221 vols. Paris, 1841–64)

PMLA *Proceedings of the Modern Language Association of America*

Ancient and Medieval Works

Alanus de Insulis. *Anticlaudianus*. Ed. Robert Bossuat. Paris: Vrin, 1955. Trans. James J. Sheridan. *Anticlaudianus or the Good and Perfect Man*. Toronto: Pontifical Institute of Mediaeval Studies, 1973.

– 'De planctu Naturae.' In *The Anglo-Latin Satirical Poets and Epigrammatists of the Twelfth Century*. Ed. Thomas Wright. 2 vols. London, 1882. 2: 429–522. Trans. James J. Sheridan. *The Plaint of Nature*. Toronto: Pontifical Institute of Mediaeval Studies, 1980.

– 'Elucidatio in Cantica Canticorum.' *PL* 210, cols 51–110.

– 'Summa de arte praedicatoria.' *PL* 210, cols 109–98.

Albertus Magnus. *De anima*. Ed. Clemens Stroick. Opera Omnia 7, part 1 (1968). Gen. ed. Bernhard Geyer. Münster, 1951–.

– *De homine* (part 2 of the *Summa de creaturis*). Ed. A. Borgnet. Opera Omnia 35 (1896). 38 vols. Paris, 1890–9.

– *De sensu et sensato*. Ed. A. Borgnet. Opera Omnia 11–12 (1891). 38 vols. Paris, 1890–9.

Alhazen [Ibn al-Haytham]. *Opticae Thesaurus. Alhazeni arabis libri septem* ... 1572.

The Sources of Science 94. Introduction by David C. Lindberg. New York: Johnson Reprint Corporation, 1972.

– *Perspectiva.* In *The Optics of Ibn al-Haytham, Books I–III on direct vision.* Trans. with introduction by A.I. Sabra. 2 vols. Studies of the Warburg Institute 40. London: Warburg Institute, 1989.

Andreas Capellanus. *The Art of Courtly Love.* Trans. John Jay Parry. 1941. Records of Civilization. New York: Columbia UP, 1990.

– *On Love.* Ed. and trans. P.G. Walsh. Duckworth Classical, Medieval, and Renaissance Editions. London: Duckworth, 1982.

Aristotle. *The 'Art' of Rhetoric.* Trans. John Henry Freese. Loeb Classical Library. Cambridge: Harvard UP, 1975.

– *De anima.* In *Aristotle: On the Soul, Parva Naturalia, On Breath.* Trans. W.S. Hett. Loeb Classical Library 288. 1957. Cambridge: Harvard UP, 1986.

– *De generatione animalium.* In *Aristotle: Generation of Animals.* Trans. A.L. Peck. Loeb Classical Library 366. 1942. Cambridge: Harvard UP, 1963.

– *De sensu et sensato.* In *Aristotle: On the Soul, Parva Naturalia, On Breath.* Trans. W.S. Hett. Loeb Classical Library 288. 1957. Cambridge: Harvard UP, 1986.

– *Poetics.* Trans. W. Hamilton Fyfe. Loeb Classical Library 199. Cambridge: Harvard UP, 1975.

Augustine. *De Genesi ad litteram.* In *Bibliothèque augustinienne: Oeuvres de saint Augustin,* vols. 48–9. Ed. Joseph Zycha. Trans. P. Agaësse and A. Solignac. 2 vols. Paris: Desclée de Brouwer, 1972. Trans. John Hammond Taylor. *The Literal Meaning of Genesis.* 2 vols. New York and Ramsay, NJ: Newman P, 1982.

– *De Trinitate.* In *Sancti Aurelii Augustini: De Trinitate libri XV.* Ed. W.J. Mountain and Fr. Glorie. 2 vols. Corpus Christianorum, Series Latina 50 and 50A. Turnhout: Brepols, 1968. Trans. Stephen McKenna. *The Trinity.* The Fathers of the Church 45. Washington, D.C.: Catholic U of America, 1963.

Avicenna [Ibn-Sina]. *Danishnama.* In *Le Livre de science.* Trans. Mohammad Achena and Henri Massé. 2 vols. Paris: 1955–8.

Bacon, Roger. *De multiplicatione specierum.* In *Roger Bacon's Philosophy of Nature: A Critical Edition, with English Translation, Introduction, and Notes of* De multiplicatione specierum *and* De speculis comburentibus. Ed. and trans. David C. Lindberg. Oxford: Clarendon P, 1983.

– 'De signis.' In K.M. Fredborg, Lauge Nielsen, and Jan Pinborg, 'An Unedited Part of Roger Bacon's *Opus maius: De signis.*' *Traditio* 34 (1978): 75–136.

– *Perspectiva* [*Opus Maius,* book 5]. In *Roger Bacon and the Origins of Perspectiva in the Middle Ages: A Critical Edition and English Translation of Bacon's* Perspectiva *with Introduction and Notes.* Ed. and trans. David C. Lindberg. Oxford: Clarendon P, 1996.

Bartholomaeus Anglicus. *De rerum Proprietatibus.* Frankfurt, 1601. Rpt. Frankfurt: Minerva, 1964.

– *On the Properties of the Soul and Body.* Ed. R. James Long. Toronto Medieval Latin Texts 9. Toronto: Pontifical Institute of Mediaeval Studies, 1979.

Bartholomaeus Anglicus. *On the Properties of Things: John Trevisa's Translation of Bartholomaeus Anglicus,* De proprietatibus rerum: *A Critical Text.* Ed. M.C. Seymour et al. 3 vols. Oxford: Clarendon P, 1975–8.

Benvenutus Grassus. *De oculis eorumque egritudinibus et curis.* Ed. and trans. Casey A. Wood. Stanford: Stanford UP, 1929.

– *De oculis eorumque egritudinibus et curis.* In *The Wonderful Art of the Eye: A Critical Edition of the Middle English Translation of His* De Probatissima Arte Oculorum. Ed. and trans. Laurence M. Eldredge. Medieval Texts and Studies 19. East Lansing: Michigan State UP, 1996.

Bernard of Clairvaux. 'De consideratione.' In *Bernard of Clairvaux: Selected Works.* Ed. and trans. G.R. Evans. New York: Paulist P, 1987.

Bernardus Silvestris [attrib.]. *The Commentary on Martianus Capella's* De nuptiis Philologiae et Mercurii *Attributed to Bernardus Silvestris.* Ed. Haijo Jan Westra. Studies and Texts 80. Toronto: Pontifical Institute of Mediaeval Studies, 1986.

– [attrib.]. *The Commentary on the First Six Books of the* Aeneid *of Vergil Commonly Attributed to Bernardus Silvestris.* Ed. Julian Ward Jones and Elizabeth Frances Jones. Lincoln: U of Nebraska P, 1977. Trans. Earl G. Schreiber and Thomas E. Maresca. *Commentary on the First Six Books of Virgil's* Aeneid. Lincoln: U of Nebraska P, 1979.

– *Cosmographia.* Ed. Peter Dronke. Leiden: E.J. Brill, 1978. Trans. Winthrop Wetherbee. *The Cosmographia of Bernardus Silvestris.* 1973. Records of Western Civilization. New York: Columbia UP, 1990.

Boethius. *De consolatione Philosophiae.* In *Boethius: The Theological Tractates.* Trans. S.J. Tester. Loeb Classical Library 74. 1973. Cambridge: Harvard UP, 1978.

[Bouvet] Bonet, Honoré. *L'Apparicion Maistre Jehan de Meun et le Somnium super materia Scismatis de Honoré Bonet.* Ed. Ivor Arnold. Publications de la faculté des lettres de l'Université de Strasbourg 28. Paris, 1926.

– *The Tree of Battles of Honoré Bonet.* Trans. G.W. Coopland. Liverpool: Liverpool UP, 1949.

Brunetto Latini. *Li Livres dou Tresor de Brunetto Latini.* Ed. Francis J. Carmody. University of California Publications in Modern Philology 22. Berkeley: U of California P, 1948.

– *Il Tesoretto (The Little Treasure).* Ed. and trans. Julia Bolton Holloway. Garland Library of Medieval Literature 2, Series A. New York: Garland, 1981.

Calcidius. [See Plato, *Timaeus.*]

Chaucer, Geoffrey. *The Complete Poetry and Prose of Geoffrey Chaucer.* Ed. John H. Fisher. New York: Holt, Rinehart and Winston, 1977.

– *The Riverside Chaucer.* Ed. Larry D. Benson. 3rd ed. Boston: Houghton Mifflin, 1987.

Chrétien de Troyes. *Le Chevalier au lion (Yvain).* Ed. Mario Roques. 1952. Paris: Champion, 1980.

– 'Yvain.' In *Chrétien de Troyes: Arthurian Romances.* Trans. W.W. Comfort. 1914. New York: Dutton/Everyman, 1970.

Christine de Pizan. *The Book of Deeds of Arms and of Chivalry.* Trans. Sumner Willard. Ed. Charity Cannon Willard. University Park: Penn State UP, 1999.

– *La città delle dame* [French text with Italian translation]. Ed. Patrizia Caraffi and Earl Jeffrey Richards. Biblioteca medievale 2. Milan: Luni, 1997. See also '*Le Livre de la Cité des Dames* of Christine de Pizan: A Critical Edition.' Ed. Maureen Cheney Curnow. 2 vols. PhD diss., Vanderbilt University, 1975. Trans. Earl Jeffrey Richards. *The Book of the City of Ladies.* New York: Persea, 1982.

– *Epître d'Othéa à Hector.* Ed. Gabriella Parussa. Textes littéraires français 517. Geneva: Droz, 1999. See also 'Classical Mythology in the Works of Christine de Pizan, with an Edition of *L'Epistre d'Othéa* from the MS. Harley 4431.' Ed. Halina D. Loukopoulos. PhD diss., Wayne State University, 1977. Trans. Jane Chance. *Christine de Pizan's Letter of Othea to Hector.* Focus Library of Medieval Women. Newburyport, MA: Focus, 1990.

– *Lavision-Christine.* Ed. Mary Louis Towner. Washington, DC: Catholic U of America P, 1932. Rpt. New York: AMS, 1969.

– *Le Livre de la mutacion de Fortune.* Ed. Suzanne Solente. 4 vols. Paris: Picard, 1959–66.

– *Le Livre du chemin de long estude.* Ed. Robert Püschel. 1881. Rpt. Geneva: Slatkine, 1974.

[Pseudo-]Cicero. *Rhetorica ad Herennium.* Trans. Harry Caplan. Loeb Classical Library 28. Cambridge: Harvard UP, 1981.

Dante Alighieri. *De vulgari eloquentia.* Ed. Pier Vicenzo Mengaldo. Vulgares eloquentes 3. Padua: Antenore, 1968.

– *La Commedia secondo l'antica vulgata.* Ed. Giorgio Petrocchi. 4 vols. Società Dantesca Italiana. Milan: Mondadori, 1966–68. Trans. Charles S. Singleton. *The Divine Comedy.* 3 vols. Bollingen Series 80. Princeton: Princeton UP, 1970–5.

– *Il Convivio.* Ed. Maria Simonelli. Bologna: Riccardo Pàtron, 1966. Trans. Christopher Ryan. *The Banquet.* Stanford French and Italian Studies 61. Saratoga: Anma Libri, 1989.

– *Epistolam X ad Canem Grandem della Scala* 8. In *Le opere di Dante: Testo critico della Società Dantesca Italiana.* Ed. Ermengildo Pistelli. 2nd ed. Florence,

1960. Trans. Robert S. Haller. *Literary Criticism of Dante Alighieri.* Regents Critics Series. Lincoln: U of Nebraska P, 1973.

– [attrib.]. *Il Fiore.* Ed. Gianfranco Contini. Milan: Montadori, 1984.

– *Vita nuova.* Ed. Dominico de Robertis. Milan and Naples: Riccardo Riccardi, 1980. Trans. with introduction by Mark Musa. *Dante's 'Vita Nuova.'* 2nd ed. Bloomington: Indiana UP, 1973.

Le Débat sur la Roman de la Rose. Ed. with introduction by Eric Hicks. Bibliothèque du XVe siècle 43. Paris: Champion, 1977.

Dominicus Gundissalinus. *De divisione philosophiae.* Ed. Ludwig Baur. Munster, 1903.

Dumbleton, John. *Summa logicae et philosophiae naturalis.* In 'Summa logicae et philosophiae naturalis according to MS Vat. Lat. 6750.' Ed. James A. Weisheipl. 2 vols. DPhil. diss., University of Oxford, 1955.

Eneas: A Twelfth-Century Romance. Ed. John A. Yunck. New York: Columbia UP, 1974.

Galen, *De placitis.* In *On the Doctrines of Hippocrates and Plato.* Ed. and trans. Phillip De Lacy. Corpus medicorum graecorum 4.1.2. 3 vols. Berlin: Akademie-Verlag, 1978–84.

Gerson, Jean. 'Le traictié d'une vision faite contre *Le Ronmant de la rose* par le Chancelier de Paris [1402].' In *Le Débat sur le* Roman de la Rose. Ed. Eric Hicks. Bibliothèque du XVe siècle 43. Paris: Champion, 1977. 59–87.

Gower, John. *Confessio Amantis.* Ed. G.C. Macaulay in *The Works of John Gower.* 4 vols. Oxford, 1901.

Grosseteste, Robert. *Le Chateau d'Amour.* Ed. J. Murray. Paris: Champion, 1918.

– *Le Chateau d'Amour.* In *The Middle English Translations of Robert Grosseteste's* Chateau d'Amour. Ed. Kari Sajavaara. Mémoires de la Société Néophilologique de Helsinki 32. Helsinki: Société Néophilologique de Helsinki, 1967.

– *De luce.* In *Die philosophischen Werke des Robert Grosseteste, Bischofs von Lincoln.* Ed. Ludwig Baur. Beiträge zur Geschichte der Philosophie des Mittelalters 9. Münster: Aschendorff, 1912. Trans. Clare C. Riedl. *On Light.* Milwaukee: Marquette UP, 1942.

– *Hexaëmeron.* Ed. Richard C. Dales and Servius Gieben. Auctores Britannici Medii Aevi. Oxford: British Academy/Oxford UP, 1982. Trans. C.F.J. Martin. *On the Six Days of Creation.* Auctores Britannici Medii Aevi 6.2. Oxford: British Academy/Oxford UP, 1996.

Guido Cavalcanti. *Le Rime.* Ed. Guido Favati. Documenti di Filologia 1. Milan and Naples: Ricciardi, 1957.

– *The Poetry of Guido Cavalcanti.* Ed. and trans. Lowry Nelson, Jr. Garland Library of Medieval Literature 18, Series A. New York: Garland, 1986.

Guillaume de Lorris and Jean de Meun. *Le Roman de la rose.* Ed. Félix Lecoy. 3

vols. Paris: Champion, 1965–70. Trans. Charles Dahlberg. *The Romance of the Rose.* 1971. Hanover, NH: UP of New England, 1983.

Hunayn ibn-Ishaq. *De oculis.* In *Le Livre des questions sur l'oeil de Honaïn ibn Ishaq.* Trans. Paul Sbath and Max Meyerhof. Mémoires de l'Institut d'Egypte 36. Cairo: Institut Français d'Archéologie Orientale, 1938.

Isidore of Seville. *Etymologiarum sive Originum libri XX.* Ed. W.M. Lindsay. 2 vols. Oxford: Clarendon P, 1911.

– *The Medical Writings.* Trans. William D. Sharpe. Transactions of the American Philosophical Society 54.2. Philadelphia, 1964.

Jean de Meun. *La Vie et les epistres Pierres Abaelart et Heloys sa fame.* Ed. with introduction by Eric Hicks. 2 vols. Nouvelle bibliothèque du moyen âge 16. Paris and Geneva: Champion-Slatkine, 1991.

– '*La Vie et les epistres Pierres Abaelart et Heloys sa fame,* A Translation by Jean de Meun, and an Old French Translation of Three Related Texts: A Critical Edition of MS 920 (Bibliothèque Nationale).' Ed. Elisabeth Schultz. PhD diss., University of Washington, 1969.

– *Li Abregemens noble homme Vegesce Flave René des Establissemenz apartenanz a chevalerie.* Ed. Leena Löfstedt. Annales Academiae Scientiarum Fennicae, Series B, no. 200. Helsinki: Suomalainen Tiedeakatemia, 1977.

– *L'art de chevalerie: traduction du* De re militari *de Végèce par Jean de Meun,* ed. Ulysse Robert. Société des Anciens Textes Français, 1897. New York: Johnson Reprint, 1965.

– *Li Livres de Confort de Philosophie.* In 'Boethius' *De Consolatione* by Jean de Meun.' Ed. L.V. Dedeck-Héry. *Mediaeval Studies* 14 (1952): 165–275.

– 'Li Testament de maistre Jehan de Meung.' In *Le Roman de la Rose par Guillaume de Lorris et Jehan de Meung.* Ed. M. Méon. 4 vols. Paris, 1814.

– 'Li Testament de maistre Jehan de Meung.' In *Le Testament maistre Jehan de Meun. Un caso letterario.* Ed. Silvia Buzzetti Gallarati. Scritti e scrittori 4. Alessandria: Edizioni dell'orso, 1989.

John of Damascus. *On the Divine Images: Three Apologies Against Those Who Attack the Divine Images.* Trans. David Anderson. Crestwood, NY: St Vladimir's Seminary P, 1980.

Langland, William. *Piers Plowman: The B Version. Will's Visions of Piers Plowman, Do-Well, Do-Better, and Do-Best.* Ed. George Kane and E. Talbot Donaldson. Rev. ed. London: Athlone/Berkeley: U of California P, 1975.

Lucretius. *De rerum natura IV.* Ed. and trans. with notes by John Godwin. Warminster: Aris and Phillips, 1986.

Lydgate, John. *The Serpent of Division.* Ed. H.C. MacCracken. Oxford, 1919.

– 'The Temple of Glass.' In *John Lydgate: Poems.* Ed. John Norton-Smith, 67–112. Oxford: Clarendon, 1966.

Macrobius. *Commentarii in Somnium Scipionis.* Ed. James Willis. Leipzig:
Teubner, 1963. Trans. William Harris Stahl. *Commentary on the Dream of Scipio.*
Records of Western Civilization. 1952. New York: Columbia UP, 1990.

– *Saturnalia.* Ed. James Willis. Leipzig: Teubner, 1963. Trans. Henri Bornecque
and François Richard. *Les Saturnales.* 2 vols. Paris: Garnier Frères, 1937.

Martianus Capella. *De nuptiis Philologiae et Mercurii.* Ed. Adolf Dick. 1925.
Stuttgart: B.G. Teubner, 1969. Trans. William Harris Stahl and Richard
Johnson with E.L. Burge. 'On the Marriage of Philology and Mercury.' In
Martianus Capella and the Seven Liberal Arts. Records of Civilization. New York:
Columbia UP, 1977.

Ovid. *Ars amatoria.* In *The Art of Love and Other Poems.* Trans. J.H. Mozley. Loeb
Classical Library 232. 1929. 2nd ed. Cambridge: Harvard UP, 1979.

– *Ars amatoria.* Anonymous Old French translation in *L'Art d'Amours (The Art of
Love).* Trans. Lawrence B. Blonquist. Garland Library of Medieval Literature
32, series A. New York: Garland, 1987.

– *Metamorphoses.* Trans. Frank Justus Miller. 2nd ed. 2 vols. Loeb Classical
Library 42–3. Cambridge: Harvard UP, 1984.

Pecham, John. *Perspectiva communis.* In *John Pecham and the Science of Optics:*
Perspectiva communis. Ed. and trans. David C. Lindberg. Madison: U of
Wisconsin P, 1970.

Plato. *Timaeus.* In *Plato's Cosmology: The* Timaeus *of Plato.* Trans. Francis
Macdonald Cornford. London: Kegan Paul, Trench, Trubner, 1937.

– *Timaeus* (Latin translation of Calcidius). *Timaeus a Calcidio translatus
commentarioque instructus.* Ed. J.H. Waszink. Plato Latinus 4. Corpus
Platonicum Medii Aevi. London: Warburg Institute/Leiden: Brill, 1975.

Pseudo-Dionysius. 'The Celestial Hierarchy.' In *Pseudo-Dionysius: The Complete
Works.* Trans. Colm Luibheid. New York and Mahwah: Paulist P, 1987.

Ptolemy. *Optica.* In *L'Optique de Claude Ptolémée dans la version latine d'après
l'arabe de l'émir Eugène de Sicile.* Ed. Albert Lejeune. Louvain: Publications
Universitaires de Louvain, 1956.

Vegetius. *De re militari* [see Jean de Meun, *L'art de chevalerie*].

Vincent of Beauvais. *Speculum Quadruplex sive Speculum Maius.* 1624. Graz:
Akademische Druck- und Verlagsanstalt, 1964.

William of Conches. [*Dragmaticon*] *Dialogus de substantiis physicis ante annos
ducentos confectus, à Vuilhelmo Aneponymo philosopho.* Ed. Guilielmi Gratarolus.
Strasbourg, 1567. Rpt. Frankfurt: Minerva, 1967.

– *Dragmaticon Philosophiae.* Ed. Italo Ronca. CCCM 152. Turnhout: Brepols,
1997. Trans. Italo Ronca and Matthew Curr. *A Dialogue on Natural Philosophy
(Dragmaticon Philosophiae).* Notre Dame Studies in Medieval Culture 2. Notre
Dame: U of Notre Dame P, 1997.

– 'Glosses on Macrobius' (excerpts). In Helen Rodnite [Lemay]. 'The Doctrine of the Trinity in Guillaume de Conches' *Glosses on Macrobius.*' PhD diss., Columbia University, 1972.

[William of Conches] Guillaume de Conches. *Glosae super Platonem.* Ed. Edouard Jeauneau. Paris: Vrin, 1965.

Witelo. *Perspectiva.* In *Witelonis Perspectivae liber primus: Book I of Witelo's* Perspectiva. Ed. and trans. Sabetai Unguru. Studia Copernicana 15. Wroclaw: Ossolineum, 1977; *Witelonis Perspectivae liber secundus et liber tertius: Books II and III of Witelo's* Perspectiva. Ed. and trans. Sabetai Unguru. Studia Copernicana. Wroclaw: Ossolineum, 1991; *Witelonis Perspectivae liber quintus: Book V of Witelo's* Perspectiva. Ed. and trans. A. Mark Smith. Studia Copernicana 23. Wroclaw: Ossolineum, 1983.

Wodeham, Adam. *Lectura secunda in librum primum Sententiarum.* Ed. Rega Wood. 3 vols. St Bonaventure, NY: St Bonaventure UP, 1990.

Secondary Sources

Adams, Marilyn McCord. 'Ockham on Will, Nature, and Morality.' In *The Cambridge Companion to Ockham.* Ed. Paul Vincent Spade. Cambridge: Cambridge UP, 1999. 245–72.

Aers, David. 'The *Parliament of Fowls*: Authority, the Knower and the Known.' *Chaucer Review* 16 (1981): 1–17.

Ahern, John. 'Troping the Fig: *Inferno* XV 66.' *Lectura Dantis* 6 (1990): 80–91.

Aiken, Pauline. 'Arcite's Illness and Vincent of Beauvais.' *PMLA* 51 (1936): 361–9.

– 'The Summoner's Malady.' *Studies in Philology* 33 (1936): 40–4.

– 'Vincent of Beauvais and Chaucer's Knowledge of Alchemy.' *Studies in Philology* 41 (1944): 371–89.

– 'Vincent of Beauvais and Dame Pertolote's Knowledge of Medicine.' *Speculum* 10 (1935): 281–7.

– 'Vincent of Beauvais and the Green Yeoman's Lecture on Demonology.' *Studies in Philology* 35 (1938): 1–9.

– 'Vincent of Beauvais and the "Houres" of Chaucer's Physician.' *Studies in Philology* 53 (1956): 22–4.

Akbari, Suzanne Conklin. 'Orientation and Nation in the *Canterbury Tales.*' In *Chaucer's Cultural Geography.* Ed. Kathryn L. Lynch. London: Routledge, 2002. 102–34.

Arrathoon, Leigh A., ed. *Chaucer and the Craft of Fiction.* Rochester, MI: Solaris Press, 1986.

– '"For craft is al, whoso that do it kan": The Genre of the *Merchant's Tale.*' Arrathoon, *Craft*, 241–328.

Ashworth, E.J., and P.V. Spade, 'Logic in Late Medieval Oxford.' Catto and Evans 35–64.

Astell, Ann W. *The Song of Songs in the Middle Ages.* Ithaca: Cornell UP, 1990.

Baeumker, Clemens. *Witelo, ein Philosoph und Naturforscher des XIII. Jahrhunderts.* 1908. Beiträge zur Geschichte und Theologie des Mittelalters 3.2. Münster: Aschendorff, 1991.

Baldassano, Lawrence. 'Dante's Hardened Heart: The Cocytus Cantos.' *Lectura dantis newberryana* (1990): 3–20.

Baltrusaitis, Jurgis. *Essai sur une légende scientifique. Le miroir: révélations, science-fiction et fallacies.* Paris: Elmayan/le seuil, 1978.

Baranski, Zygmunt G., and Patrick Boyde, eds. *The* Fiore *in Context: Dante, France, Tuscany.* Notre Dame: U of Notre Dame P, 1997.

Barber, Joseph. 'A Statistical Analysis of the *Fiore*.' *Lectura Dantis* 6 (1990): 100–22.

Barney, Stephen J. *Allegories of History, Allegories of Love.* Hamden, CT: Archon Books, 1979.

Barolini, Teodolinda. 'Arachne, Argus, and St. John: Transgressive Art in Dante and Ovid.' *Medievalia* 13 (1987): 207–26.

– *Dante's Poets: Textuality and Truth in the 'Comedy.'* Princeton: Princeton UP, 1984.

– *The Undivine Comedy: Detheologizing Dante.* Princeton: Princeton UP, 1992.

Baswell, Christopher. 'Men in the *Roman d'Eneas*: The Construction of Empire.' In *Medieval Masculinities: Regarding Men in the Middle Ages.* Ed. Clare A. Lees. Medieval Cultures 7. Minneapolis: U of Minnesota P, 1994. 149–68.

Baumgartner, Emmanuèle. 'De Lucrèce à Heloïse: remarques sur deux *exemples* du *Roman de la Rose* de Jean de Meun.' *Romania* 95 (1974): 433–42.

Bellezza, Francis S. 'The Mind's Eye in Expert Memorizers' Descriptions of Remembering.' *Metaphor and Symbolic Activity* 7 (1992): 119–33.

Benjamin, Walter. *Ursprung des deutschen Trauerspiels* (1928). In *Gesammelte Schriften*, Vol. 1. Frankfurt: Suhrkamp, 1980. Trans. John Osborne. *The Origin of German Tragic Drama.*. London: NLB, 1977.

Bennett, J.A.W. 'Some Second Thoughts on the *Parlement of Foules*.' In *Chaucerian Problems and Perspectives: Essays Presented to Paul E. Beichner.* Ed. Edward Vasta and Zacharias P. Thundy. Notre Dame: U of Notre Dame P, 1979. 132–46.

Benson, Larry D., and Theodore M. Andersson. *The Literary Context of Chaucer's Fabliaux: Texts and Translations.* Indianapolis: Bobbs-Merrill, 1971.

Bernardo, Aldo S., and Anthony L. Pellegrini, eds. *Petrarch, Boccaccio: Studies in the Italian Trecento in Honor of Charles S. Singleton.* Binghamton: Medieval and Renaissance Texts and Studies, 1983.

Binkley, Peter, ed. *Pre-Modern Encyclopedic Texts, Proceedings of the Second COMERS*

Congress, Groningen, 1–4 July 1996. Brill's Studies in Intellectual History 79. Leiden: E.J. Brill, 1997.

Bischoff, Bernhard. 'Die europäische Verbreitung der Werke Isidors von Sevilla.' In *Isidoriana.* Ed. Manuel C. Diaz y Diaz. Leon, 1961. 317–44.

Bleeth, Kenneth. 'Joseph's Doubting of Mary and the Conclusion of the *Merchant's Tale.*' *Chaucer Review* 21 (1986): 58–66.

Bloch, R. Howard. *Etymologies and Genealogies: A Literary Anthropology of the French Middle Ages.* Chicago: U of Chicago P, 1983.

Bloomfield, Morton. 'A Grammatical Approach to Personification Allegory.' *Modern Philology* 60 (1963): 161–71.

Blumenfeld-Kosinski, Renate. '*Overt* and *Covert*: Amorous and Interpretive Strategies in the *Roman de la Rose.*' *Romania* 3 (1990): 432–53.

Blumenfeld[-Kosinski], Renate. 'Remarques sur *songe/mensonge.*' *Romania* 101 (1980): 385–90.

Boase, Roger. *The Origin and Meaning of Courtly Love: A Critical Study of European Scholarship.* Manchester: Manchester UP, 1977.

Bornstein, Diane. 'Chaucer's *Tale of Melibee* as an Example of the *Style Clergial.*' *Chaucer Review* 12 (1978): 236–54.

Boucher, Holly Wallace. 'Nominalism: The Difference for Chaucer and Boccaccio.' *Chaucer Review* 20 (1986): 213–20.

Boyde, Patrick. *Perception and Passion in Dante's* Comedy. Cambridge: Cambridge UP, 1993.

Brandeis, Irma. *The Ladder of Vision: A Study of Dante's Comedy.* Garden City, NY: Doubleday, 1962.

Brewer, D.S. 'The Fabliaux.' Rowland, *Companion*, 296–325.

Brind'Amour, Lucie and Eugene Vance, eds. *Archéologie du signe.* Papers in Mediaeval Studies 3. Toronto: Pontifical Institute of Mediaeval Studies, 1982.

Brook, Leslie C. 'Comment évaluer une traduction du treizième siècle? Quelques considérations sur la traduction des lettres d'Abélard et d'Héloïse faite par Jean de Meun.' In *The Spirit of the Court: Selected Proceedings of the Fourth Congress of the International Courtly Literature Society.* Ed. Glyn S. Burgess and Robert A. Taylor. Cambridge: D.S. Brewer, 1985. 62–8.

– 'The Continuator's Monologue: Godefroy de Lagny and Jean de Meun.' *French Studies* 45 (1991): 1–16.

Brown, Peter. 'An Optical Theme in the *Merchant's Tale.*' *Studies in the Age of Chaucer: The Yearbook of the New Chaucer Society* (1985): 231–43.

– 'Chaucer's Visual World: A Study of His Poetry and the Medieval Optical Tradition.' DPhil. diss., University of York, U.K., 1982.

Brownlee, Kevin. 'Jean de Meun and the Limits of Romance: Genius as Rewriter of Guillaume de Lorris.' In *Romance: Generic Transformation from Chrétien de Troyes to Cervantes.* Hanover, NH: UP of New England, 1985. 114–34.

– 'Orpheus' Song Re-sung: Jean de Meun's Reworking of *Metamorphosis, x.*' *Romance Philology* 36 (1982): 201–9.

– 'The Problem of Faux Semblant: Language, History, and Truth in the *Roman de la Rose.*' In *The New Medievalism.* Ed. Marina S. Brownlee, Kevin Brownlee, and Stephen G. Nichols, 253–71. Baltimore: Johns Hopkins UP, 1991.

– 'Reflections in the *Miroër aus Amoreus:* The Inscribed Reader in Jean de Meun's *Roman de la Rose.*' In *Mimesis: From Mirror to Method, Augustine to Descartes.* Ed. John D. Lyons and Stephen G. Nichols, Jr. Hanover, NH: UP of New England, 1982. 60–70.

Brownlee, Kevin, and Sylvia Huot, eds. *Rethinking the* Romance of the Rose: *Text, Image, Reception.* Philadelphia: U of Pennsylvania P, 1992.

Bryan, W.F., and Germaine Dempster, eds. *Sources and Analogues of Chaucer's* Canterbury Tales. 1941. New York: Humanities P, 1958.

Buck-Morss, Susan. *The Dialectics of Seeing: Walter Benjamin and the Arcades Project.* Cambridge: MIT Press, 1990.

Bundy, Murray Wright. *The Theory of Imagination in Classical and Mediaeval Thought.* University of Illinois Studies in Language and Literature 12. Urbana: U of Illinois P, 1927.

Burgwinkle, William. 'Knighting the Classical Hero: Homo/hetero Affectivity in *Eneas.*' *Exemplaria* 5 (1993): 1–43.

Burnett, Charles. 'The Superiority of Taste.' *Journal of the Warburg and Courtauld Institutes* 54 (1991): 230–8.

Burnett, Charles, and Danielle Jacquart, eds. *Constantine the African and 'Ali ibn al'Abbas al-Magusi: The* Pantegni *and Related Texts.* Studies in Ancient Medicine 10. Leiden: E.J. Brill, 1994.

Burnley, J. David. *Chaucer's Language and the Philosopher's Tradition.* Chaucer Studies 2. Cambridge: D.S. Brewer, 1979.

– 'Some Terminology of Perception in the *Book of the Duchess.*' *English Language Notes* 23 (1986): 15–22.

Bynum, Caroline Walker. 'Faith Imagining the Self: Somatomorphic Soul and Resurrection Body in Dante's *Divine Comedy.*' In *Faithful Imagining: Essays in Honor of Richard R. Niebuhr.* Ed. Sang Hyun Lee, Wayne Proudfoot, and Albert Blackwell. Atlanta: Scholars P, 1995. 81–104.

– 'Material Continuity, Personal Survival and the Resurrection of the Body: A Scholastic Discussion in its Medieval and Modern Contexts.' In *Fragmentation and Redemption: Essays on Gender and the Human Body in Medieval Religion.* New York: Zone, 1991. 239–97.

– 'Why All this Fuss about the Body? A Medievalist's Perspective.' *Critical Inquiry* 22 (1985): 1–33.

Calabrese, Michael A. 'May Devoid of All Delight: January, the *Merchant's Tale*, and the *Romance of the Rose*.' *Studies in Philology* 87 (1990): 261–84.

Camille, Michael. *The Gothic Idol: Ideology and Image-Making in Medieval Art*. New York: Cambridge UP, 1989.

Cantarino, Vincente. 'Dante and Islam: Theory of Light in the *Paradiso*.' *Kentucky Romance Quarterly* 15 (1968): 3–35.

Carruthers, Mary. *The Book of Memory: A Study of Memory in Medieval Culture*. Cambridge Studies in Medieval Literature 10. Cambridge: Cambridge UP, 1990.

– *The Craft of Thought: Meditation, Rhetoric, and Images, 400–1200*. Cambridge Studies in Medieval Literature 34. Cambridge: Cambridge UP, 1998.

– 'Italy, *Ars Memorativa*, and Fame.' *Studies in the Age of Chaucer: Proceedings* 2 (1986): 179–88.

– 'The Poet as Master Builder: Composition and Locational Memory in the Middle Ages.' *New Literary History* 24 (1993): 881–904.

Cassell, Anthony K. 'Santa Lucia as Patroness of Sight: Hagiography, Iconography, and Dante.' *Dante Studies* 109 (1991): 71–88.

Catto, J.I., and Ralph Evans, eds. *The History of the University of Oxford. Vol. II: Late Medieval Oxford*. 1992. Oxford: Clarendon, 1995.

Chenu, M.-D. *Nature, Man, and Society in the Twelfth Century: Essays on New Theological Perspectives on the Latin West*. [Orig. *La théologie au douzième siècle* (1957).] Trans. Jerome Taylor and Lester K. Little. Chicago: U of Chicago P, 1968.

Cherniss, Michael D. 'Chaucer's Last Dream Vision: The *Prologue* to the *Legend of Good Women*.' *Chaucer Review* 20 (1986): 183–99.

Chiarenza, Marguerite Mills. 'The Imageless Vision and Dante's *Paradiso*.' *Dante Studies* 90 (1972): 77–91.

Cioffari, Vincenzo. 'Dante's Use of Lapidaries: A Source Study.' *Dante Studies* 109 (1991): 149–62.

Coleman, Janet. *Ancient and Medieval Memories: Studies in the Reconstruction of the Past*. Cambridge: Cambridge UP, 1992.

Colish, Marcia. *Medieval Foundations of the Western Intellectual Tradition, 400–1400*. New Haven: Yale UP, 1997.

– *The Stoic Tradition from Antiquity to the Early Middle Ages*. 2 vols. Leiden: E.J. Brill, 1985.

Collette, Carolyn P. 'Heeding the Counsel of Prudence: A Context for the *Melibee*.' *Chaucer Review* 29 (1995): 416–33.

– 'Seeing and Believing in the *Franklin's Tale*.' *Chaucer Review* 26 (1992): 395–410.

- *Species, Phantasms, and Images: Vision and Medieval Psychology in* The Canterbury Tales. Ann Arbor: U of Michigan P, 2001.

Contini, Gianfranco. 'Un nodo della cultura medievale: la serie *Roman de la rose, Fiore, Divine commedia.*' In *Un'idea di Dante: Saggi danteschi.* Turin: Einaudi, 1976. 245–83.

Copeland, Rita, *Rhetoric, Hermeneutics, and Translation in the Middle Ages: Academic Traditions and Vernacular Texts.* Cambridge Studies in Medieval Literature 11. Cambridge: Cambridge UP, 1991.

Copeland, Rita and Stephen Melville. 'Allegory and Allegoresis, Rhetoric and Hermeneutics.' *Exemplaria* 3 (1991): 159–87.

Corti, Maria. *La felicità mentale: Nuove prospettive per Cavalcanti e Dante.* Turin: Einaudi, 1983.

Courcelle, Pierre. *La Consolatione de Philosophie dans la tradition littéraire.* Paris, 1967.

Courtenay, William J. *Adam Wodeham: An Introduction to his Life and Writings.* Leiden: Brill, 1978.

- 'The Dialectic of Divine Omnipotence in the Age of Chaucer: A Reconsideration.' Keiper et al., 111–21.

- *Schools and Scholars in Fourteenth-Century England.* Princeton: Princeton UP, 1987.

Crombie, A.C. 'Grosseteste's Position in the History of Science.' *Science, Optics, and Music in Medieval and Early Modern Thought.* London and Ronceverte: Hambleton P, 1990. 115–38.

- *Robert Grosseteste and the Origins of Experimental Science, 1100–1700.* Oxford: Clarendon P, 1953.

Dahlberg, Charles. 'First Person and Personification in the *Roman de la Rose.* Amant and Dangier.' *Mediaevalia* 3 (1977): 37–58.

- 'Love and the *Roman de la Rose.*' *Speculum* 44 (1969): 568–84.

- 'Macrobius and the Unity of the *Roman de la Rose.*' *Studies in Philology* 58 (1961): 573–82.

d'Alverny, Marie-Thérèse. 'Les traductions d'Avicenne (Moyen Age et Renaissance).' *Avicenna nella storia della cultura medioevale.* Problemi attuali di scienza e di cultura 40. Rome: Accademia Nazionale dei Lincei, 1957. 71–87.

- 'Notes sur les traductions médiévales d'Avicenna.' *Archives d'histoire doctrinale et littéraire du moyen âge* 19 (1952): 337–58.

- 'Translations and Translators.' *Renaissance and Renewal in the Twelfth Century.* Ed. R.L. Benson and G. Constable. Cambridge: Harvard UP, 1982. 421–62.

De Lacy, Phillip. 'The Third Part of the Soul.' *Le opere psicologiche di Galeno. Atti del terzo colloquio Galenico internazionale Pavia, 10–12 Settembre 1986.* Ed. Paola Manuli and Mario Vegetti. Elenchos 13. Naples: Bibliopolis, 1988. 43–63.

Delany, Sheila. 'The Late Medieval Attack on Analogical Thought: Undoing
Substantial Connection.' *Mosaic* 5 (1972): 33–52. Rpt. in *Chaos and Form:
History and Literature, Ideas and Relationships*. Ed. Kenneth McRobbie. Mosaic
Essay Series. Winnipeg: U of Manitoba P, 1972. 37–58.

– '"Phantom" and the *House of Fame*.' *Chaucer Review* 2 (1971): 67–74.

Delasanta, Rodney. 'Chaucer and Strode.' *Chaucer Review* 26 (1991): 205–18.

– 'Nominalism and the *Clerk's Tale* Revisited.' *Chaucer Review* 31 (1997): 209–31.

de Libera, Alain. *Albert la Grand et la philosophie*. Paris: Vrin, 1990.

de Man, Paul. *Allegories of Reading: Figural Language in Rousseau, Nietzsche, Rilke,
and Proust*. New Haven: Yale UP, 1979.

– 'The Rhetoric of Temporality.' In *Blindness and Insight: Essays in the Rhetoric of
Contemporary Criticism*. Rev. ed. Theory and History of Literature 7. Minneapolis: U of Minnesota P, 1983. 187–228.

Demats, Paule. 'D'*Amoenitas* à *Deduit*: André le Chapelain et Guillaume de
Lorris.' In *Mélanges de langue et de littérature du moyen âge et de la Renaissance
offerts à Jean Frappier*. Publications Romanes et Françaises 112. Geneva: Droz,
1970. 1: 217–33.

Dembowski, Peter F. 'Learned Latin Treatises in French: Inspiration, Plagia-
rism, and Translation.' *Viator* 17 (1986): 225–69.

Dempster, Germaine. 'The Merchant's Tale.' Bryan and Dempster 341–56.

Devons, Samuel. 'Optics Through the Eyes of the Medieval Churchmen.' In
Science and Technology in Medieval Society. Ed. Pamela O. Long. Annals of the
New York Academy of Sciences 441. New York: New York Academy of Sci-
ences, 1985. 205–24.

Dewan, Lawrence. 'St. Albert, the Sensibles, and Spiritual Being.' Weisheipl,
Albertus 291–300.

Diekstra, F.N.M. 'Chaucer and the *Romance of the Rose*.' *English Studies* 69 (1988):
12–26.

Dinshaw, Carolyn. 'Chaucer's Queer Touches/A Queer Touches Chaucer.'
Exemplaria 7 (1995): 76–92.

– *Getting Medieval: Sexualities and Communities, Pre- and Postmodern*. Durham:
Duke UP, 1999.

Donovan, Mortimer J. 'Chaucer's January and May: Counterparts in Claudian.'
In *Chaucerian Problems and Perspectives: Essays Presented to Paul E. Beichner*. Ed.
Edward Vasta and Zacharias P. Thundy. Notre Dame: U of Notre Dame P,
1979. 59–69.

– 'The Image of Pluto and Proserpine in the *Merchant's Tale*.' *Philological
Quarterly* 36 (1957): 49–60.

Douie, Decima. 'John XXII and the Beatific Vision.' *Dominican Studies* 3.2
(1950): 154–74.

Dronke, Peter. 'Andreas Capellanus.' In *Sources of Inspiration: Studies in Literary Transformations, 400–1500.* Storia e Letteratura 196. Rome: Edizioni di Storia e Letteratura, 1997. 101–16.

– 'Boethius, Alanus, and Dante.' In *The Medieval Poet and his World.* Storia e Letteratura 164. Rome: Edizioni di Storia e Letteratura, 1984. 431–8.

– *Fabula: Explorations into the Uses of Myth in Medieval Platonism.* Mittellateinische Studien und Texte 9. Leiden: Brill, 1974.

– *The Medieval Lyric.* London: Hutchinson, 1968.

– 'Symbolism and Structure in *Paradiso* 30.' *Romance Philology* 43 (1989): 29–48.

Dronke, Peter, ed. *A History of Twelfth-Century Western Philosophy.* Cambridge: Cambridge UP, 1988.

Dronke, Peter, and Jill Mann. 'Chaucer and the Medieval Latin Poets.' In *Geoffrey Chaucer: The Writer and His Background.* Ed. Derek Brewer. Cambridge: D.S. Brewer, 1974. 155–83.

Duhem, Pierre. *Le Système du monde: histoire des doctrines cosmologiques de Platon à Copernic.* 5 vols. Paris, 1913–17.

Durling, Robert M. 'Deceit and Digestion in the Belly of Hell.' In *Allegory and Representation: Selected Papers from the English Institute, 1979–80.* Ed. Stephen J. Greenblatt. Baltimore: Johns Hopkins UP, 1981. 61–93.

Durling, Robert M., and Ronald L. Martinez. *Time and the Crystal: Studies in Dante's Rime Petrose.* Berkeley: U of California P, 1990.

Eastwood, Bruce S. *Astronomy and Optics from Pliny to Descartes: Texts, Diagrams, and Conceptual Structures.* London: Variorum, 1989.

Eastwood, Bruce S. *The Elements of Vision: The Micro-Cosmology of Galenic Visual Theory According to Hunayn ibn Ishaq.* Transactions of the American Philosophical Society 72.5. Philadelphia, 1982. Rpt. in Eastwood, *Astronomy and Optics,* 1–59.

Eastwood, Bruce S. 'Medieval Empiricism: The Case of Grosseteste's Optics.' *Speculum* 43 (1968): 306–21. Rpt. in Eastwood, *Astronomy and Optics,* 306–21.

Eberle, Patricia J. 'The Lovers' Glass: Nature's Discourse on Optics and the Optical Design of the *Romance of the Rose.*' *University of Toronto Quarterly* 46 (1977): 241–62.

Economou, George D. *The Goddess Natura in Medieval Literature.* Cambridge: Harvard UP, 1972.

– 'January's Sin Against Nature: The *Merchant's Tale* and the *Romance of the Rose.*' *Comparative Literature* 17 (1965): 251–7.

– 'Saying Spirit in Terms of Matter: The Epic Embrace in Medieval Poetic Imagination.' *Lectura Dantis* 11 (1992): 72–9.

– 'The Two Venuses and Courtly Love.' In *In Pursuit of Perfection.* Ed. Joan M.

Ferrante, George D. Economou, and Frederick Goldin. Port Washington, NY: Kennikat, 1975. 17–50.

Edgerton, Samuel Y., Jr. *The Heritage of Giotto's Geometry: Art and Science on the Eve of the Scientific Revolution*. Ithaca: Cornell UP, 1991.

– *The Renaissance Rediscovery of Linear Perspective*. New York: Basic Books, 1975.

Edwards, Robert R. *The Dream of Chaucer: Representation and Reflection in the Early Narratives*. Durham: Duke UP, 1989.

– 'Narration and Doctrine in the *Merchant's Tale*.' *Speculum* 66 (1991): 342–67.

Eldredge, Laurence M. 'The Anatomy of the Eye in the Thirteenth Century: The Transmission of Theory and the Extent of Practical Knowledge.' *Micrologus* 5 (1997): 145–60.

– 'The Textual Tradition of Benvenutus Grassus' *De arte probatissima oculorum*.' *Studi medievali*, 3rd series 34 (1993): 95–138.

Elford, Dorothy. 'Developments in the Natural Philosophy of William of Conches: A Study of the *Dragmaticon* and a Consideration of its Relationship to the *Philosophia*.' PhD diss., University of Cambridge, 1983.

– 'William of Conches.' Dronke, *Twelfth-Century* 308–27.

Ellis, Steve. 'Chaucer, Dante, and Damnation.' *Chaucer Review* 22 (1988): 282–94.

– 'The Death of the *Book of the Duchess*.' *Chaucer Review* 29 (1995): 249–58.

Entzminger, Robert L. 'The Pattern of Time in *The Parlement of Fouls*.' *Journal of Medieval and Renaissance Studies* 5 (1975): 1–11.

Fenellosa, Ernest. *The Chinese Written Character as a Medium for Poetry*. Ed. Ezra Pound. 1936. San Francisco: City Lights, 1983.

Ferrante, Joan M. '"Cortes Amor" in Medieval Texts.' *Speculum* 55 (1980): 686–95.

– *The Political Vision of the* Divine Comedy. Princeton: Princeton UP, 1984.

– *Woman as Image in Medieval Literature, From the Twelfth Century to Dante*. New York: Columbia UP, 1975.

– 'Words and Images in the *Paradiso*: Reflections of the Divine.' Bernardo and Pellegrini 115–32.

Ferster, Judith. 'Reading Nature: The Phenomenology of Reading in the *Parliament of Fowls*.' *Mediaevalia* 3 (1977): 189–213.

Flamant, Jacques. *Macrobe et le Néo-Platonisme Latin à la fin du IVe siècle*. Études préliminaires aux religions orientales dans l'empire Romain 58. Leiden: E.J. Brill, 1977.

Fleming, John V. *The* Roman de la Rose: *A Study in Iconography and Allegory*. Princeton: Princeton UP, 1969.

Fleming, John V. '"Smoky Reyn": From Jean de Meun to Geoffrey Chaucer.' Arrathoon, *Craft* 1–21.

Fletcher, Angus. *Allegory: Theory of a Symbolic Mode.* Ithaca: Cornell UP, 1964.

Fletcher, J.M. 'Developments in the Faculty of Arts, 1370–1520.' Catto and Evans 315–45.

Fourrier, Anthime, ed. *L'Humanisme médiévale dans les littératures romanes du XIIe au XIVe siècle.* Actes et Colloques 3. Paris: Klincksieck, 1964.

Frangenberg, Thomas. 'Perspectivist Aristotelianism: Three Case-studies of Cinquecento Visual Theory.' *Journal of the Warburg and Courtauld Institutes* 54 (1991): 137–58.

Frappier, Jean. 'Variations sur le thème du miroir, de Bernard de Ventadour à Maurice Scève.' *Cahiers de l'Association internationale des études françaises* 11 (1959): 134–58.

Freccero, John. 'An Introduction to the *Paradiso.*' In *Dante: The Poetics of Conversion.* Ed. Rachel Jacoff. Cambridge: Harvard UP, 1986. 209–20.

Fyler, John M. *Chaucer and Ovid.* New Haven: Yale UP, 1979.

– 'Irony and the Age of Gold in the *Book of the Duchess.*' *Speculum* 52 (1977): 314–28.

Gallarati, Silvia Buzzetti. *Le Testament maistre Jehan de Meun. Un caso letterario.* Scritti e scrittori 4. Alessandria: Edizioni dell'orso, 1989.

– '"Mots sous les mots": una firma per il *Testament.*' *Medioevo Romanzo* 15 (1990): 259–76.

Gaunt, Simon. 'From Epic to Romance: Gender and Sexuality in the *Roman d'Eneas.*' *Romanic Review* 83 (1992): 1–27.

Gersh, Stephen. *Middle Platonism and Neoplatonism: The Latin Tradition.* 2 vols. Publications in Medieval Studies 33.1–2. Notre Dame: U of Notre Dame P, 1986.

Gibson, Margaret T., ed. *Boethius: His Life, Thought, and Influence.* Oxford: Basil Blackwell, 1981.

Goldin, Frederick. *The Mirror of Narcissus in the Courtly Love Lyric.* Ithaca: Cornell UP, 1967.

Grant, Edward, ed. *A Source Book in Medieval Science.* Cambridge: Harvard UP, 1974.

Gregory, Tullio. *Anima mundi: La filosofia de Guglielmo di Conches e la scuola di Chartres.* Florence: G.C. Sansoni, 1955.

– 'The Platonic Inheritance.' Dronke, *Twelfth-Century* 54–80.

Grennen, Joseph E. 'The Calculating Reeve and His *Camera Obscura.*' *Journal of Medieval and Renaissance Studies* 14 (1984): 245–59.

– 'Chaucer and Chalcidius: The Platonic Origins of the *Hous of Fame.*' *Viator* 15 (1984): 237–62.

Gunn, Alan M.F. *The Mirror of Love: A Reinterpretation of 'The Romance of the Rose.'* Lubbock: Texas Tech P, 1952.

Hackett, Jeremiah, ed. *Roger Bacon and the Sciences: Commemorative Essays.* Studien und Texte zur Geistesgeschichte des Mittelalters 57. Leiden: Brill, 1997.

Haefili-Till, Dominique. *Der 'Liber de oculis' des Constantinus Africanus: Übersetzung und Kommentar.* Zürcher Medizingeschichtliche Abhandlungen n.s. 121. Zürich: Juris, 1977.

Hagen, Susan K. *Allegorical Remembrance: A Study of* The Pilgrimage of the Life of Man *as a Medieval Treatise on Seeing and Remembering.* Athens: U of Georgia P, 1990.

Hajdu, Helga. *Das mnemotechnische Schrifttum des Mittelalters.* Vienna: Franz Leo, 1936.

Hanning, Robert W. 'Chaucer's First Ovid: Metamorphosis and Poetic Tradition in *The Book of the Duchess* and *The House of Fame.*' Arrathoon, *Craft* 121–63.

– 'Poetic Emblems in Medieval Narrative Texts.' In *Vernacular Poetics in the Middle Ages.* Ed. Lois Ebin. Studies in Medieval Culture 16. Kalamazoo: Western Michigan University/Medieval Institute, 1984. 1–32.

Hansen, Elaine Tuttle. *Chaucer and the Fictions of Gender.* Berkeley: U of California P, 1992.

Hardman, Phillipa. 'The *Book of the Duchess* as a Memorial Monument.' *Chaucer Review* 28 (1994): 205–15.

Harley, Marta Powell. 'Narcissus, Hermaphroditus, and Attis: Ovidian Lovers at the Fontaine d'Amors in Guillaume de Lorris' *Roman de la rose.*' *PMLA* 101 (1986): 324–37.

Harrison, Robert Pogue. 'The Bare Essential: The Landscape of the *Fiore.*' Brownlee and Huot 290–303.

– *The Body of Beatrice.* Baltimore: Johns Hopkins UP, 1988.

Hart, Thomas Elwood. 'Medieval Structuralism: "Dulcarnoun" and the Five-Book Design of Chaucer's *Troilus.*' *Chaucer Review* 16 (1981): 129–70.

Hartung, Albert H. 'A Study of the Textual Affiliations of Chaucer's *Melibeus* Considered in Relation to the French Source.' PhD diss., Lehigh University, 1957.

Harvey, E. Ruth. *The Inward Wits: Psychological Theory in the Middle Ages and the Renaissance.* Warburg Institute Surveys 6. London: Warburg Institute, 1975.

Hewitt, Kathleen. '"Ther It Was First": Dream Poetics in the *Parliament of Fowls.*' *Chaucer Review* 24 (1989): 20–8.

Hill, Thomas D. 'Narcissus, Pygmalion, and the Castration of Saturn: Two Mythographical Themes in the *Roman de la Rose.*' *Studies in Philology* 71 (1974): 404–26.

Hillman, Larry. 'Another Look Into the Mirror Perilous: The Role of the Crystals in the *Roman de la Rose.*' *Romania* 101 (1980): 225–38.

Hoenen, Maarten J.F.M., and Lodi Nauta, eds. *Boethius in the Middle Ages: Latin and Vernacular Traditions of the* Consolatio Philosophiae. Studien und Texte zur Geistesgeschichte des Mittelalters 58. Leiden: E.J. Brill, 1997.

Hollander, Robert. *Dante's Epistle to Cangrande.* Recentiores. Ann Arbor: U of Michigan P, 1993.

– *'Purgatorio* XIX: Dante's Siren/Harpy.' Bernardo and Pellegrini 77–88.

Holley, Linda Tarte. *Chaucer's Measuring Eye.* Houston: Rice UP, 1990.

– 'Medieval Optics and the Framed Narrative in Chaucer's *Troilus and Criseyde.*' *Chaucer Review* 21 (1986): 26–44.

Honig, Edwin. *Dark Conceit: The Making of Allegory.* 1959. New York: Oxford UP, 1966.

Hult, David F. 'Closed Quotations: The Speaking Voice in the *Roman de la Rose.*' *Yale French Studies* 67 (1984): 248–69.

– 'Gaston Paris and the Invention of Courtly Love.' In *Medievalism and the Modernist Temper.* Ed. R. Howard Bloch and Stephen G. Nichols. Baltimore: Johns Hopkins UP, 1996. 192–224.

– 'Language and Dismemberment: Abelard, Origen, and the *Romance of the Rose.*' Brownlee and Huot 101–30.

– *Self-Fulfilling Prophecies: Readership and Authority in the First* Roman de la Rose. Cambridge: Cambridge UP, 1986.

Huot, Sylvia. 'Authors, Scribes, Remanieurs: A Note on the Textual History of the *Romance of the Rose.*' Brownlee and Huot 203–33.

– 'Medieval Readers of the *Roman de la Rose*: The Evidence of Marginal Notations.' *Romance Philology* 43 (1990): 400–20.

– 'Notice sur les fragments poétiques dans un manuscrit du *Roman de la Rose.*' *Romania* 109 (1988): 119–21.

– *The* Romance of the Rose *and Its Medieval Readers: Interpretation, Reception, Manuscript Transmission.* Cambridge Studies in Medieval Literature 16. Cambridge: Cambridge UP, 1993.

Irvine, Martin. 'Grammatical Theory and the *House of Fame.*' *Speculum* 60 (1985): 850–76.

Jackson, W.T.H. 'Allegory and Allegorization.' In *The Challenge of the Medieval Text.* Ed. Joan M. Ferrante and Robert W. Hanning, 157–71. 1964. New York: Columbia UP, 1985.

Jacquart, Danielle. 'A l'aube de la renaissance médicale des Xie-XIIe siècles: L'"Isagoge Johannitii" et son traducteur.' *Bibliothèque de l'Ecole des Chartes* 144 (1986): 209–40; rpt. in *La science médicale occidentale entre deux renaissances (XIIe s.–XVe s.).* Aldershot: Variorum, 1997.

– *La médecine médiévale dans le cadre Parisien, XIVe-XVe siècle.* Paris: Fayard, 1998.

Jacquart, Danielle, and Françoise Micheau, *La médecine arabe et l'occident médiévale*. Paris: Maisonneuve et Larose, 1990.

Jacquart, Danielle, and Claude Thomasset. *Sexuality and Medicine in the Middle Ages*. [Orig. *Sexualité et savoir médical au Moyen Age* (1985).] Trans. Matthew Adamson. Princeton: Princeton UP, 1988.

Jauss, Hans Robert. 'La Transformation de la forme allégorique entre 1180 et 1240: d'Alain de Lille à Guillaume de Lorris.' Fourrier, *L'Humanisme* 107–64.

Jay, Martin. *Downcast Eyes: The Denigration of Vision in Twentieth-century French Thought*. Berkeley: U of California P, 1993.

Jeauneau, Edouard. *'Lectio philosophorum': Recherches sur l'Ecole de Chartres*. Amsterdam: Adolf Hakkert, 1973.

– 'L'Usage de la notion d'*integumentum* à travers les gloses de Guillaume de Conches.' *Archives d'histoire doctrinale et littéraire du moyen âge* 24 (1957): 35–100. Rpt. Jeauneau, *Lectio philosophorum* 125–92.

– 'Macrobe source du platonisme chartrain.' *Studi medievali*, 3rd series, 1 (1960): 3–24. Rpt. Jeauneau, *Lectio philosophorum* 279–300.

Johnson, Lynn Staley. 'Inverse Counsel: Contexts for the *Melibee.' Studies in Philology* 87 (1990): 137–55.

Jonsson, Einar Màr. 'Le sens du titre *Speculum* aux XIIe et XIIIe siècles et son utilisation par Vincent de Beauvais.' Paulmier-Foucart et al. 11–32.

Jung, Marc-René. *Etudes sur la poème allégorique en France au moyen âge*. Berne, 1971.

Katzenellenbogen, Adolf. *Allegories of the Virtues and Vices in Medieval Art from Early Christian Times to the Thirteenth Century*. 1939. Medieval Academy Reprints for Teaching. Toronto: U of Toronto P, 1989.

Kay, Richard. 'The Spare Ribs of Dante's Michael Scot.' *Dante Studies* 103 (1985): 1–14.

Keiper, Hugo, Christoph Bode, and Richard J. Utz, eds. *Nominalism and Literary Discourse: New Perspectives*. Ed. Critical Studies 10. Amsterdam and Atlanta: Rodopi, 1997.

Kelley, Theresa M. *Reinventing Allegory*. Cambridge Studies in Romanticism 22. Cambridge: Cambridge UP, 1997.

Kelly, Douglas. '"Li chastiaus ... Qu'Amors prist puis par ses esforz": The Conclusion of Guillaume de Lorris' *Rose*.' In *A Medieval French Miscellany: Papers of the 1970 Kansas Conference on Medieval French Literature*. Ed. Norris J. Lacy. University of Kansas Humanistic Studies 42. Lawrence: U of Kansas Publications, 1972. 61–78.

– *Medieval Imagination: Rhetoric and the Poetry of Courtly Love*. Madison: U of Wisconsin P, 1978.

– '*Translatio Studii*: Translation, Adaptation, and Allegory in Medieval French Literature.' *Philological Quarterly* 57 (1978): 287–310.

Kirkham, Victoria. 'A Canon of Women in Dante's *Commedia.*' *Annali d'Italianistica* 7 (1987): 16–41.

Klassen, Norman. *Chaucer on Love, Knowledge, and Sight.* Cambridge: D.S. Brewer, 1995.

Klibansky, Raymond. *The Continuity of the Platonic Tradition during the Middle Ages.* London: Warburg Institute, 1939.

Klinck, Roswitha. *Die lateinische Etymologie des Mittelalters.* Medium Aevum Philologische Studien 17. Munich: Wilhelm Fink, 1970.

Klubertanz, George. *The Discursive Power: Sources and Doctrine of the 'Vis cogitativa' According to St. Thomas Aquinas.* St Louis: Modern Schoolman, 1952.

Knapp, Daniel. 'The Relyk of a Saint: A Gloss on Chaucer's Pilgrimage.' *English Literary History* 39 (1972): 1–26.

Knoespel, Kenneth. *Narcissus and the Invention of Personal History.* New York: Garland, 1985.

Köhler, Erich. 'Narcisse, la Fontaine d'Amour et Guillaume de Lorris.' Fourrier, *L'Humanisme* 147–64.

Kolve, V.A. *Chaucer and the Imagery of Narrative: The First Five* Canterbury Tales. Stanford: Stanford UP, 1984.

Kruger, Steven F. *Dreaming in the Middle Ages.* Cambridge Studies in Medieval Literature 14. Cambridge: Cambridge UP, 1992.

Kuhn, Thomas S. *The Structure of Scientific Revolutions.* 1962. 2nd ed. Chicago: U of Chicago P, 1970.

Langlois, Ernest. *Origines et sources de* Roman de la rose. Paris. Bibliothèque des Ecoles Françaises d'Athenes et de Rome, 1891.

[Lemay], Helen Rodnite. 'The Doctrine of the Trinity in Guillaume de Conches' *Glosses on Macrobius.*' PhD diss., Columbia University, 1972.

Lemay, Richard. *Abu Ma'shar and Latin Aristotelianism in the Twelfth Century: The Recovery of Aristotle's Natural Philosophy Through Arabic Astrology.* Beirut, 1962.

– 'Roger Bacon's Attitude Toward the Latin Translations and Translators of the Twelfth and Thirteenth Centuries.' Hackett, *Roger Bacon* 25–47.

Lewis, C.S. *The Allegory of Love: A Study in Medieval Tradition.* 1936. New York: Oxford UP/Galaxy, 1958.

Leyerle, John. 'Chaucer's Windy Eagle.' *University of Toronto Quarterly* 40 (1971): 247–65.

– 'The Rose-Wheel Design and Dante's *Paradiso.*' *University of Toronto Quarterly* 46 (1977): 280–308.

Lidaka, Juris G. 'Bartholomaeus Anglicus in the Thirteenth Century.' Binkley 393–406.

– 'John Trevisa and the English and Continental Traditions of Bartholomaeus Anglicus' *De proprietatibus rerum.*' *Essays in Medieval Studies* 5 (1988): 71–92.

Lieser, Ludwig. *Vinzenz von Beauvais als Kompilator und Philosoph. Eine*

Untersuchung seiner Seelenlehre im Speculum Maius. Forschungen zur Geschichte der Philosophie und der Pädagogik 3.1. Leipzig: Felix Meiner, 1928.

Lindberg, David C. 'Alhazen's Theory of Vision and its Reception in the West.' *Isis* 58 (1967): 321–41. Rpt. in Lindberg, *Studies* 321–41.

– 'Alkindi's Critique of Euclid's Theory of Vision.' *Isis* 62 (1971): 469–89. Rpt. in Lindberg, *Studies* 469–89.

– *Roger Bacon and the Origins of Perspectiva* [see Bacon, *Perspectiva*].

– *Roger Bacon's Philosophy of Nature* [see Bacon, *De multiplicatione specierum*].

– 'Bacon, Witelo, and Pecham: The Problem of Influence.' *Actes du XIIe Congrès international d'histoire des sciences, Paris 1968*. Vol. 3A. Paris, 1971. 103–7. Rpt. in Lindberg, *Studies* 103–7.

– Introduction. *Opticae Thesaurus. Alhazeni arabis libri septem* ... 1572. The Sources of Science 94. New York: Johnson Reprint Corporation, 1972.

– 'Roger Bacon on Light, Vision, and the Universal Emanation of Force.' In Hackett, *Roger Bacon* 243–75.

– *Studies in the History of Medieval Optics*. London: Variorum, 1983.

– *Theories of Vision from al-Kindi to Kepler*. Madison: U of Wisconsin P, 1976.

Louth, Andrew. *Denys the Areopagite*. Outstanding Christian Thinkers. Wilton, CT: Morehouse-Barlow, 1989.

– *Discerning the Mystery: An Essay on the Nature of Theology*. Oxford: Clarendon P, 1983.

– *The Origins of the Christian Mystical Tradition: From Plato to Denys*. Oxford: Clarendon P, 1981.

Lowes, J.L. 'The Prologue to the *Legend of Good Women* as Related to the French *Marguerite* Poems, and the *Filostrato*.' *PMLA* 19 (1904): 595–683.

Luria, Maxwell. *A Reader's Guide to the* Roman de la Rose. Hamden, CT: Archon Books, 1982.

Luscombe, David. 'John of Salisbury in Recent Scholarship.' *The World of John of Salisbury*. Ed. Michael Wilks. Studies in Church History, subsidia 3. Oxford: Basil Blackwell, 1984. 21–37.

Lynch, Kathryn L. 'The *Book of the Duchess* as a Philosophical Vision: The Argument of Form.' *Genre* 21 (1988): 279–305.

– 'The *Parliament of Fowls* and Late Medieval Voluntarism.' *Chaucer Review* 25 (1990): 1–16, 85–95.

Magee, John. *Boethius on Signification and Mind*. Philosophia Antiqua 52. Leiden: E.J. Brill, 1989.

Maierù, Alfonso. 'Le ms. Oxford canonici misc. 219 et la *Logica* de Strode.' In *English Logic in Italy in the Fourteenth and Fifteenth Centuries*. Ed. Alfonso Maierù. Naples: Bibliopolis, 1982. 87–110.

Malato, Enrico. *Dante e Guido Cavalcanti. Il Dissidio per la* Vita Nuova *e il 'disdegno' di Guido.* Quaderni di 'Filologia e critica' 11. Rome: Salerno, 1997.

Maloney, Thomas. 'Roger Bacon on the *Significatum* of Words.' Brind'Amour and Vance 187–211.

Matthews, Lloyd J. 'The Date of Chaucer's *Melibee* and the Stages of the Tale's Incorporation in the *Canterbury Tales.*' *Chaucer Review* 20 (1986): 221–34.

Mazzeo, Joseph A. 'Light, Love and Beauty in the *Paradiso.*' *Romance Philology* 11 (1957): 1–17.

– *Medieval Cultural Tradition in Dante's 'Commedia.'* Ithaca: Cornell UP, 1960. Rpt. New York: Greenwood, 1992.

– *Structure and Thought in the 'Paradiso.'* Ithaca: Cornell UP, 1958.

Mazzotta, Giuseppe. *Dante's Vision and the Circle of Knowledge.* Princeton: Princeton UP, 1993.

McEvoy, James. *The Philosophy of Robert Grosseteste.* Oxford: Clarendon P, 1982.

McInerny, Ralph, ed. with essays. *Aquinas Against the Averroists: On There Being Only One Intellect.* Purdue University Series in the History of Philosophy. West Lafayette, IN: Purdue UP, 1993.

Michaud-Quantin, Pierre. 'Les champs semantiques de *species.* Tradition Latine et traductions du Grec.' In *Etudes sur le vocabulaire philosophique du Moyen Age.* Rome: Ateneo, 1970. 113–50.

Milbank, John. 'Sacred Triads: Augustine and the Indo: European Soul.' *Modern Theology* 13 (1997): 451–74.

Miller, James L. 'The Three Mirrors of Dante's *Paradiso.*' *University of Toronto Quarterly* 46 (1977): 263–79.

Miller, Robert P. 'Allegory in the *Canterbury Tales.*' Rowland, *Companion* 326–51.

Minnis, A.J. '"Authorial Intention" and "Literal Sense" in the Exegetical Theories of Richard Fitzralph and John Wyclif.' *Proceedings of the Royal Irish Academy* 75 (1975): 1–31.

– *Chaucer and Pagan Antiquity.* Cambridge: D.S. Brewer, 1982.

– 'Discussions of "authorial role" and "literary form" in late medieval scriptural exegesis.' *Beiträge zur Geschichte der deutschen Sprache und Literatur* 99 (1977): 37–65.

– 'Looking for a Sign: The Quest for Nominalism in Chaucer and Langland.' In *Essays on Ricardian Literature in Honour of J.A. Burrow.* Ed. A.J. Minnis, Charlotte C. Morse, and Thorlac Turville-Petre. Oxford: Clarendon, 1997. 142–78.

– *Medieval Theory of Authorship.* 1984. 2nd ed. Aldershot: Wildwood, 1988.

Minnis, A.J., ed. *Chaucer's* Boece *and the Medieval Tradition of Boethius.* Chaucer Studies 18. Cambridge: D.S. Brewer, 1993.

– *The Medieval Boethius.* Cambridge: D.S. Brewer, 1987.

Minnis, A.J., and A.B. Scott with David Wallace, eds. *Late Medieval Theory and Criticism, c.1100–c.1375. The Commentary Tradition.* Rev. ed. Oxford: Clarendon P, 1988.

Molland, George. 'John Dumbleton and the Status of Geometrical Optics.' *Actes du XIIIe Congrès International d'Histoire des Sciences* 3/4 (1974): 1–6. Rpt. *Mathematics and the Medieval Ancestry of Physics.* Aldershot: Variorum, 1995.

Murray, H.J.R. *A History of Chess.* Oxford, 1913.

Murrin, Michael. *The Veil of Allegory: Some Notes Toward a Theory of Allegorical Rhetoric in the English Renaissance.* Chicago: U of Chicago P, 1969.

Muscatine, Charles. 'The Emergence of Psychological Allegory in Old French Romance.' *PMLA* 68 (1953): 1160–82.

Nardi, Bruno. *Dante e la cultura medievale.* 2nd ed. Bari: Laterza, 1949.

– 'Noterella polemica sull'averroismo di Guido Cavalcanti.' *Rassegna di filosofia* 3 (1954): 47–71.

– *Saggi di filosofia dantesca.* 2nd ed. Florence: Nuova Italia, 1967.

Neaman, Judith S. 'Brain Physiology and Poetics in *The Book of the Duchess.*' *Res publica litterarum* 3 (1980): 101–13.

Neuse, Richard. 'Marriage and the Question of Allegory in the *Merchant's Tale.*' *Chaucer Review* 24 (1989): 114–31.

Newman, Francis X. 'St. Augustine's Three Visions and the Structure of the *Commedia.*' *Modern Language Notes* 82 (1967): 56–78.

Newman, Francis X., ed. *The Meaning of Courtly Love.* Albany: State University of New York P, 1972.

Newton, Francis. 'Constantine the African and Monte Cassino: New Elements and the Text of the *Isagoge.*' Burnett and Jacquart 16–47.

Nolan, Edward. *Now Through a Glass Darkly: Specular Images of Being and Knowing from Virgil to Chaucer.* Ann Arbor: U of Michigan P, 1990.

North, J.D. *Chaucer's Universe.* 1988. Rev. ed. Oxford: Clarendon Press, 1990.

– 'Natural Philosophy in Late Medieval Oxford.' Catto and Evans 65–102.

O'Daly, Gerard. *Augustine's Philosophy of Mind.* London: Duckworth, 1987.

– *The Poetry of Boethius.* Chapel Hill: U of North Carolina P, 1991.

O'Neill, Ynez Violé. 'William of Conches and the Cerebral Membranes.' *Clio Medica* 2 (1967): 13–21.

– 'William of Conches' Descriptions of the Brain.' *Clio Medica* 3 (1968): 203–23.

O'Rourke, Fran. *Pseudo-Dionysius and the Metaphysics of Aquinas.* Leiden: Brill, 1992.

Otten, Charlotte F. 'Proserpine: *Liberatrix suae gentis.*' *Chaucer Review* 5 (1971): 277–87.

Owen, Charles A., Jr. 'The *Tale of Melibee*.' *Chaucer Review* 7 (1973): 267–80.

Palomo, Dolores. 'What Chaucer Really Did to *Le Livre de Mellibee*.' *Philological Quarterly* 53 (1974): 304–20.

Park, David. *The Fire Within the Eye: A Historical Essay on the Nature and Meaning of Light*. Princeton: Princeton UP, 1997.

Parronchi, Alessandro. 'La perspettiva dantesca.' *Studi danteschi* 36 (1959): 5–103. Rpt. in *Studi su la dolce prospettiva*. Milan: Aldo Martello, 1964. 3–90.

Patterson, Lee. *Chaucer and the Subject of History*. Madison: U of Wisconsin P, 1991.

– '"What Man Artow?"': Authorial Self-Definition in *The Tale of Sir Thopas* and *The Tale of Melibee*.' *Studies in the Age of Chaucer* 11 (1989): 117–75.

Paulmier-Foucart, Monique. 'L'Atelier Vincent de Beauvais: Recherches sur l'état des connaissances au Moyen Age d'après une encyclopédie du XIIIe siècle.' *Le Moyen Age* 80 (1979): 87–99.

Paulmier-Foucart, Monique, Serge Lusignan, and Alain Nadeau, eds. *Vincent de Beauvais: Intentions et réceptions d'une oeuvre encyclopédique au Moyen-Age. Actes du XIVe Colloque de l'Institut d'études médiévales, organisé conjointement par l'Atelier Vincent de Beauvais (A.R.Te.M., Université de Nancy II) et l'Institut d'études médiévales (Université de Montréal) 27–30 avril 1988*. Cahiers d'études médiévales, Cahier spécial 4. Montréal: Bellarmin/Paris: Vrin, 1990.

Paxson, James J. *The Poetics of Personification*. Cambridge: Cambridge UP, 1994.

Peck, Russell A. 'Chaucer and the Nominalist Questions.' *Speculum* 53 (1978): 745–60.

– 'Theme and Number in Chaucer's *Book of the Duchess*.' In *Silent Poetry: Essays in Numerological Analysis*. Ed. Alastair Fowler. London: Routledge and Kegan Paul, 1970. 73–115.

Peden, Alison M. 'Macrobius and Medieval Dream Literature.' *Medium Aevum* 54 (1985): 59–73.

Pelikan, Jaroslav. *Eternal Feminines: Three Theological Allegories in Dante's Paradiso*. New Brunswick: Rutgers UP, 1990.

Penn, Stephen. 'Literary Nominalism and Medieval Sign Theory: Problems and Perspectives.' Keiper et al. 157–89.

Pertile, Lino. 'A Desire for Paradise and a Paradise of Desire.' In *Dante: Contemporary Perspectives*. Ed. Amilcare A. Iannucci. Toronto: U of Toronto P, 1997. 148–66.

– '*Paradiso*: A Drama of Desire.' In *Word and Drama in the* Divine Comedy. Ed. John Barnes and Jennifer Petrie. Dublin: Foundation of Italian Studies/Irish Academic P, 1993. 26–60.

Phillips, Heather. 'John Wyclif and the Optics of the Eucharist.' In *From Ockham to Wyclif*. Ed. Anne Hudson and Michael Wilks. Studies in Church History, subsidia 5. Oxford: Basil Blackwell, 1987. 245–58.

Pickens, Rupert T. '*Somnium* and Interpretation in Guillaume de Lorris.'
    Symposium 28 (1974): 175–86.
Poirion, Daniel. 'De la signification selon Jean de Meun.' Brind'Amour and
    Vance 165–85.
– 'Les mots et les choses selon Jean de Meun.' *L'Information littéraire* 26 (1974):
    7–11.
– 'Narcisse et Pygmalion dans *Le Roman de la rose*.' In *Essays in Honor of Louis
    Francis Solano*. Ed. Raymond J. Cormier and Urban T. Holmes. University
    of North Carolina Studies in the Romance Languages and Literatures 92.
    Chapel Hill: U of North Carolina P, 1970. 153–65.
Pratt, Robert A. 'Chaucer's Claudian.' *Speculum* 22 (1947): 419–29.
– 'Some Latin Sources of the Nonnes Preest on Dreams.' *Speculum* 52 (1977):
    538–70.
Prior, Sandra Pierson. '*Routhe* and *Hert-Huntyng* in the *Book of the Duchess*.'
    Journal of English and Germanic Philology 85 (1986): 3–19.
Quilligan, Maureen. *The Language of Allegory: Defining the Genre*. Ithaca and
    London: Cornell UP, 1979.
– 'Words and Sex: The Language of Allegory in the *De planctu naturae*, the
    Roman de la Rose, and Book III of the *Faerie Queene*.' *Allegorica* 2 (1977):
    195–216.
Redle, M.J. 'Beatific Vision.' In *New Catholic Encyclopedia*. 19 vols. Washington,
    D.C.: Catholic U of America P/New York: McGraw-Hill, 1967–79. 2: 186–93.
Regalado, Nancy Freeman. '"Des Contraires Choses": La fonction poétique de
    la citation de des *exempla* dans le 'Roman de la Rose' de Jean de Meun.'
    Littérature 41 (1981): 62–81.
Rendall, Thomas. '*Gawain* and the Game of Chess.' *Chaucer Review* 27 (1992):
    186–99.
Reydellet, Marc. 'La diffusion des *Origines* d'Isidore de Seville au Haut Moyen
    Age.' *Mélanges d'archéologie et d'histoire de l'École Française de Rome* 78 (1966):
    383–437.
Richards, Earl Jeffrey. 'Dante's *Commedia* and Its Vernacular Narrative Context.'
    PhD diss., Princeton University, 1978.
– 'Reflections on Oiseuse's Mirror: Iconographic Tradition, Luxuria, and the
    Roman de la Rose.' *Zeitschrift für romanische Philologie* 98 (1982): 296–311.
Ricklin, Thomas. 'Vue et vision chez Guillaume de Conches et Guillaume
    de Saint-Thierry: Le récit d'une controverse.' *Micrologus* 5 (1997): 19–41.
Rigg, A.G. *A History of Anglo-Latin Literature, 1066–1422*. Cambridge: Cam-
    bridge UP, 1992.
Robertson, D.W. *A Preface to Chaucer*. Princeton: Princeton UP, 1962.
– 'The Doctrine of Charity in Medieval Literary Gardens: A Topical Approach
    Through Symbolism and Allegory.' *Speculum* 26 (1951): 24–49.

Ronca, Italo. 'The Influence of the *Pantegni* on William of Conches' *Dragmaticon.*' Burnett and Jacquart 266–85.

Ronchi, Vasco. *Storia della luce.* Bologna, 1939. Expanded French translation, J. Taton, *Histoire de la lumière.* Paris: Armand Colin, 1956. Expanded English translation, V. Barocas, *The Nature of Light: An Historical Survey.* London: Heinemann, 1970.

Roney, Lois. *Chaucer's Knight's Tale and Theories of Scholastic Psychology.* Tampa: U of South Florida P, 1990.

Rosenberg, Bruce A. 'The "Cherry-Tree Carol" and the *Merchant's Tale.*' *Chaucer Review* 5 (1971): 264–76.

Rossi, Albert L. '"Miro gurge" (*Par.* xxx, 68): Virgilian Language and Textual Pattern in the River of Light.' *Dante Studies* 103 (1985): 79–101.

Rowland, Beryl. 'Bishop Bradwardine, the Artificial Memory, and the *House of Fame.*' In *Chaucer at Albany.* Ed. Rossell Hope Robbins. New York: Burt Franklin, 1973. 41–62.

– 'The Chess Problem in Chaucer's *Book of the Duchess.*' *Anglia* 80 (1962): 384–89.

Rowland, Beryl, ed. *Companion to Chaucer Studies.* 2nd ed. Oxford: Oxford UP, 1979.

Russell, Gül. 'The Anatomy of the Eye in 'Ali ibn al-'Abbas al-Magusi: A Textbook Case.' Burnett and Jacquart 247–65.

Russell, J. Stephen. *The English Dream Vision: Anatomy of a Form.* Columbus: Ohio State UP, 1988.

Rutledge, Monica. 'Dante, the Body, and Light.' *Dante Studies* 113 (1995): 151–65.

Sambursky, Samuel. *Physics of the Stoics.* New York: Macmillan, 1959.

Sandler, Lucy Freeman. 'Face to Face with God: A Pictorial Image of the Beatific Vision.' In *England in the Fourteenth Century: Proceedings of the 1985 Harlaxton Symposium.* Ed. W.M. Ormrod. Woodbridge, Suffolk: Boydell, 1986. 224–35.

Schless, Howard H. *Chaucer and Dante: A Revaluation.* Norman, OK: Pilgrim, 1984.

Severs, J. Burke. 'The Tale of Melibeus.' Bryan and Dempster 560–614.

Seymour, Michael C. 'Some Medieval English Owners of the *De proprietatibus rerum.*' *Bodleian Library Record* 9 (1974): 156–65.

– 'Some Medieval French Readers of the *De proprietatibus rerum.*' *Scriptorium* 28 (1974): 100–3.

Shoaf, R.A. *Dante, Chaucer, and the Currency of the Word: Money, Images, and Reference in Late Medieval Poetry.* Norman, OK: Pilgrim, 1983.

Siegel, Rudolph E. *Galen on Psychology, Psychopathology, and Function and Diseases of the Nervous System. An Analysis of his Doctrines, Observations, and Experiments.* Galen's System of Physiology and Medicine 3. Basel: S. Karger, 1973.

– *Galen on Sense Perception: His Doctrines, Observations and Experiments of Vision, Hearing, Smell, Taste, Touch and Pain, and their Historical Background.* Galen's System of Physiology and Medicine 2. Basel: S. Karger, 1970.

Singleton, Charles S. *An Essay on the* Vita Nuova. Cambridge: Harvard UP, 1949.

– *'Commedia': Elements of Structure.* Dante Studies 1. Cambridge: Harvard UP, 1954. Rpt. Baltimore: Johns Hopkins UP, 1977.

– *Journey to Beatrice.* Dante Studies 2. Cambridge: Harvard UP, 1958. Rpt. Baltimore: Johns Hopkins UP, 1967.

Siraisi, Nancy G. *Medieval and Early Renaissance Medicine: An Introduction to Knowledge and Practice.* Chicago: U of Chicago P, 1990.

Smith, A. Mark. 'Getting the Big Picture in Perspectivist Optics.' *Isis* 72 (1981): 568–89.

– 'The Psychology of Visual Perception in Ptolemy's *Optics.*' *Isis* 79 (1988): 189–207.

Spade, Paul Vincent. *The Medieval Liar: A Catalogue of the Insolubilia-Literature.* Toronto: Pontifical Institute of Mediaeval Studies, 1975.

– 'Robert Fland's *Insolubilia:* An Edition, with Comments on the Dating of Fland's Works.' *Medieval Studies* 40 (1978): 56–80.

Spearing, A.C. *The Medieval Poet as Voyeur: Looking and Listening in Medieval Love-Narratives.* Cambridge: Cambridge UP, 1993.

Spillenger, Paul W. 'Dante's Mnêmosunê: Memory and Poetry in the *Divine Comedy.*' PhD diss., Columbia University, 1992.

Stanbury, Sarah. 'The Lover's Gaze in *Troilus and Criseyde.*' In *Chaucer's* Troilus and Criseyde: *'Subgit to alle Poesye.' Essays in Criticism.* Ed. R.A. Shoaf. Medieval and Renaissance Texts and Studies 104. Binghamton, NY: Pegasus, 1992. 224–38.

– *Seeing the* Gawain-*Poet: Description and the Act of Perception.* Philadelphia: U of Pennsylvania P, 1991.

Steadman, John M. 'Chaucer's Eagle: A Contemplative Symbol.' *PMLA* 75 (1960): 153–9.

Steinmetz, David C. 'Late Medieval Nominalism and the *Clerk's Tale.*' *Chaucer Review* 12 (1977): 38–54.

Steneck, Nicholas [H.] 'Albert on the Psychology of Sense Perception.' Weisheipl, *Albertus* 263–90.

– 'The Problem of the Internal Senses in the Fourteenth Century.' PhD diss., University of Wisconsin, Madison, 1970.

Stock, Brian. *Augustine the Reader: Meditation, Self-Knowledge, and the Ethics of Interpretation.* Cambridge: Belknap/Harvard UP, 1996.

– *Myth and Science in the Twelfth Century: A Study of Bernard Silvester.* Princeton: Princeton UP, 1972.

Strohm, Paul. 'The Allegory of the *Tale of Melibee.*' *Chaucer Review* 2 (1967): 32–42.

– 'Guillaume as Narrator and Lover in the *Roman de la Rose.*' *Romanic Review* 59 (1968): 3–9.

Sturges, Robert S. *Medieval Interpretation: Models of Reading and Literary Narrative, 1100–1500.* Carbondale and Edwardsville: Southern Illinois UP, 1991.

Sturm-Maddox, Sarah. 'The *Rime Petrose* and the Purgatorial Palinode.' *Studies in Philology* 84 (1987): 119–33.

Sudhoff, Walther. 'Die Lehre von den Hirnventrikeln in textlicher und graphischer Tradition des Altertums und Mittelalters.' *Sudhoffs Archiv für Geschichte der Medizin und der Naturwissenschaften* 7 (1914): 149–205.

Sutherland, Jenifer. 'The Inexpressible Self: Biblical Autobiography in the Poetry of Walter of Wimborne and *The Book of Margery Kempe.*' PhD diss., University of Toronto, 2002.

Sylla, Edith D. 'The Oxford Calculators and Mathematical Physics: John Dumbleton's *Summa Logicae et Philosophiae Naturalis*, Parts II and III.' In *Physics, Cosmology, and Astronomy, 1300–1700.* Ed. Sabetai Unguru. Dordrecht: Kluwer, 1991. 129–61.

– *The Oxford Calculators and the Mathematics of Motion, 1320–1350: Physics and Measurement by Latitudes.* New York: Garland, 1991.

Tachau, Katherine H. '"Et maxime visus, cuius species venit ad stellas et ad quem species stellarum veniunt": *Perspectiva* et *Astrologia* in Late Medieval Thought.' *Micrologus* 5 (1997): 201–24.

– *Vision and Certitude in the Age of Ockham: Optics, Epistemology, and the Foundations of Sematics, 1250–1345.* Leiden: Brill, 1988.

– 'What Senses and Intellect Do: Argument and Judgment in Late Medieval Theories of Knowledge.' In *Argumentationstheorie: Scholastische Forschungen zu den logischen und semantischen Regeln korrekten Folgerns.* Ed. Klaus Jacobi. Studien und Texte zur Geistesgeschichte des Mittelalters 38. Leiden: Brill, 1993. 653–68.

Taylor, Karla. *Chaucer Reads the* Divine Comedy. Stanford: Stanford UP, 1989.

Tempier, Margherita de Bonfils. 'Genesi di un'allegoria.' *Dante Studies* 105 (1987): 79–94.

– 'Il dantesco "amoroso uso di Sapienza": sue radici platoniche.' *Stanford Italian Review* 7 (1987): 5–27.

– 'Le due ineffabilitadi del *Convivio.*' *Dante Studies* 108 (1990): 67–78.

– 'Ragion e intelletto nel *Convivio.*' In *Italiana: Selected Papers of the Proceedings of the Third Annual Conference of the AATI.* River Forest, IL: Rosary College, 1988. 77–86.

Teskey, Gordon. *Allegory and Violence.* Ithaca: Cornell UP, 1996.

Thonnard, François-Joseph. 'La notion de lumière en philosophie august-
inenne.' *Recherches Augustiniennes* 2 (1962): 125–75.

Tinkle, Theresa. *Medieval Venuses and Cupids: Sexuality, Hermeneutics, and English
Poetry.* Stanford: Stanford UP, 1996.

Torti, Anna. *The Glass of Form: Mirroring Structures from Chaucer to Skelton.* Cambridge: D.S. Brewer, 1991.

Toynbee, Paget. 'Some Unacknowledged Obligations of Dante to Albertus
Magnus.' *Romania* 24 (1895): 399–412.

Travis, Peter W. 'Chaucer's Heliotropes and the Poetics of Metaphor.' *Speculum*
72 (1997): 399–427.

– 'Chaucer's Trivial Fox Chase and the Peasants' Revolt of 1381.' *Journal of
Medieval and Renaissance Studies* 18 (1988): 195–220.

Trottman, Christian. *La vision béatifique: Des disputes scholastiques à sa définition
par Benoît XII.* Bibliothèque des Écoles Françaises d'Athènes et de Rome 289.
Rome: École Française de Rome, 1995.

Trout, John M. *The Voyage of Prudence: The World View of Alan of Lille.* Washington, D.C.: UP of America, 1979.

Tuve, Rosemond. *Allegorical Imagery: Some Mediaeval Books and Their Renaissance
Posterity.* Princeton: Princeton UP, 1966.

Twomey, Michael W. 'Towards a Reception History of Western Medieval Encyclopedias in England before 1500.' Binkley 329–62.

Tyler, Stephen A. 'The Vision Quest in the West, or What the Mind's Eye Sees.'
Journal of Anthropological Research 40 (1984): 23–39.

Uitti, Karl D. '"Cele [qui] doit estre Rose clamee" (*Rose*, vv. 40–44): Guillaume's Intentionality.' Brownlee and Huot 39–64.

– 'From *Clerc* to *Poète*: The Relevance of the *Romance of the Rose* to Machaut's
World.' In *Machaut's World: Science and Art in the Fourteenth Century.* Ed.
Madeleine Pelner Cosman and Bruce Chandler. Annals of the New York
Academy of Sciences 314. New York: New York Academy of Sciences, 1978.
209–16.

Utz, Richard J. '"As Writ Myn Auctor Called Lollius": Divine and Authorial
Omnipotence in Chaucer's *Troilus and Criseyde.*' Keiper et al. 123–44.

Utz, Richard J., and William H. Watts. 'Nominalist Perspectives on Chaucer's
Poetry: a Bibliographical Essay.' *Mediaevalia et Humanistica* 20 (1994): 147–
73.

Vance, Eugene. *'Mervelous Signals': Poetics and Sign Theory in the Middle Ages.*
Lincoln: U of Nebraska P, 1986.

van den Boogaard, Nico. 'Le *Roman de la rose* de Guillaume de Lorris et l'art de
mémoire.' In *Jeux de mémoire.* Ed. Bruno Roy and Paul Zumthor. Montréal:
PU de Montréal/Paris: Vrin, 1985. 85–90.

Verbeke, Gérard. *L'Évolution de la doctrine du* pneuma *du Stoicisme à S. Augustin: étude philosophique.* Paris: Éditions de l'Institut Supérieure de Philosophie, 1945.

Verhuyck, Paul. 'Guillaume de Lorris *ou* la multiplication des cadres.' *Neophilologus* 58 (1974): 283–93.

Vescovini, Graziella Federici. 'La "perspectiva" nell'enciclopedia del sapere medievale.' *Vivarium* 6 (1968): 35–45.

– *Studi sulla prospettiva medievale.* 1965. Rev. ed. Turin: G. Giappichelli, 1987.

– 'Vision et Réalité dans la perspective au XIVe siècle.' *Micrologus* 5 (1997): 161–80.

Vitz, Evelyn Birge. 'Inside/Outside: Guillaume's *Roman de la Rose* and Medieval Selfhood.' In *Medieval Narrative and Modern Narratology: Subjects and Objects of Desire.* New York: New York UP, 1989. 64–95.

Vollmann, Benedikt Konrad. 'La vitalità delle enciclopedie di scienza naturale: Isidoro di Siviglia, Tommaso di Cantimpré, e le Redazioni del cosiddetto Tommaso III.' In *L'Enciclopedismo medievale.* Ed. Michelangelo Picone. Ravenna: Longo, 1994. 135–45.

Voorbij, J.B. 'The *Speculum Historiale*: Some Aspects of its Genesis and Manuscript Tradition.' In *Vincent of Beauvais and Alexander the Great: Studies on the Speculum Maius and its Translation into Medieval Vernaculars.* Ed. W.J. Aerts, E.R. Smits, and J.B. Voorbij. Mediaevalia Groningana 7. Groningen: Egbert Forsten, 1986. 11–55.

Wack, Mary Frances. 'From Mental Faculties to Magical Philters: The Entry of Magic into Academic Medical Writing on Lovesickness, 13th–17th Centuries.' In *Eros and Anteros: The Medical Traditions of Love in Renaissance Culture.* Ed. Donald Beecher and Massimo Ciavolella. Montreal: McGill/Queens UP, 1989. 9–31.

– *Lovesickness in the Middle Ages: The* Viaticum *and Its Commentaries.* Philadelphia: U of Pennsylvania P, 1990.

Walker, Denis. 'Narrative Inconclusiveness and Consolatory Dialectic in the *Book of the Duchess.*' *Chaucer Review* 18 (1983): 1–17.

Wallace, David. 'Chaucer and the European Rose.' *Studies in the Age of Chaucer: Proceedings* 1 (1984): 61–7.

– *Chaucerian Polity: Absolutist Lineages and Associational Forms in England and Italy.* Stanford: Stanford UP, 1997.

Walters, Lori. 'Appendix: Author Portraits and Textual Demarcation in Manuscripts of the *Romance of the Rose.*' Brownlee and Huot 359–73.

– 'Illuminating the *Rose*: Gui de Mori and the Illustration of MS 101 of the Municipal Library, Tournai.' Brownlee and Huot 167–200.

Waterhouse, Ruth and Gwen Griffiths. '"Sweete Wordes of Non-sense": The

Deconstruction of the Moral *Melibee.*' *Chaucer Review* 23 (1989): 338–61; 24 (1989): 53–63.

Watson, Nicholas. 'Desire for the Past.' *Studies in the Age of Chaucer* 21 (1999): 59–97.

– *Richard Rolle and the Invention of Authority.* Cambridge: Cambridge UP, 1991.

Weisheipl, James A. 'Early Fourteenth-Century Physics and the Merton "School" with Special Reference to Dumbleton and Heytesbury.' DPhil. diss., University of Oxford, 1956.

– 'Ockham and Some Mertonians.' *Mediaeval Studies* 30 (1968): 168–213.

– 'The Place of John Dumbleton in the Merton School.' *Isis* 50 (1959): 439–54.

Weisheipl, James A., ed. *Albertus Magnus and the Sciences: Commemorative Essays.* Studies and Texts 49. Toronto: Pontifical Institute of Mediaeval Studies, 1980.

Wentersdorf, Karl P. 'Theme and Structure in the *Merchant's Tale*: The Function of the Pluto Episode.' *PMLA* 80 (1965): 522–7.

Wetherbee, Winthrop. 'The Literal and the Allegorical: Jean de Meun and the "de Planctu Naturae."' *Mediaeval Studies* 33 (1971): 264–91.

Whitman, Jon. *Allegory: The Dynamics of an Ancient and Medieval Technique.* Cambridge: Harvard UP, 1987.

✓ Williams, Rowan. *Christian Spirituality: A Theological History from the New Testament to Luther and St. John of the Cross.* Atlanta: John Knox P, 1979.

Wilson, Curtis. *William Heytesbury: Medieval Logic and the Rise of Mathematical Physics.* Madison: U of Madison P, 1960.

Windeatt, B.A., ed. and trans. *Chaucer's Dream Poetry: Sources and Analogues.* Chaucer Studies 7. Cambridge: D.S. Brewer, 1982.

Wolfson, Harry A. 'The Internal Senses in Latin, Arabic, and Hebrew Philosophical Texts.' *Harvard Theological Review* 28 (1935): 69–133.

Wood, Rega. 'Adam Wodeham on Sensory Illusions, with an Edition of "Lectura secunda," Prologus, Quaestio 3.' *Traditio* 38 (1982): 213–52.

Wurtele, Douglas. 'Ironical Resonances in the *Merchant's Tale.*' *Chaucer Review* 13 (1978): 66–79.

Yager, Susan. '"A Whit Thyng in hir Ye": Perception and Error in the *Reeve's Tale.*' *Chaucer Review* 28 (1994): 393–404.

Yates, Frances A. *The Art of Memory.* Chicago: U of Chicago P, 1966.

Zumthor, Paul. 'Narrative and Anti-Narrative: *Le Roman de la Rose.*' *Yale French Studies* 51 (1974): 185–204.

INDEX

Abelard, Peter, 44, 79, 101–2, 110–11
abstraction, 13–14, 104–7, 211–15,
 217–21, 238–9
Adam, 144–5, 158, 229
Adams, Marilyn, 301n30
Adelard of Bath, 32–3, 35
Adonis, 84, 88
Aelred of Rievaulx, 110
Aers, David, 178, 289nn1–2, 295n52
afterimage, 28
Ahern, John, 150, 283n20
Aiken, Pauline, 39
Akbari, Suzanne, 305n24
Alanus de Insulis, *Anticlaudianus*, 15,
 18–19, 66, 109, 216, 246n23,
 249n69, 250n70, 257n15, 258n22,
 262n67, 264nn96–7, 100, 266n23,
 270n91, 283n27, 284n35, 286n54,
 289n71, 303n56; *De planctu*
 Naturae, 7, 10, 15, 19, 50, 54, 62,
 72, 75, 79–81, 85–6, 88–90, 96, 99,
 105, 107–8, 110, 151, 235–7;
 Summa de arte praedicatoria, 52
Albertus Magnus, 35–6, 39–40, 90–1,
 98, 114–18, 122, 129–32, 134–5,
 141, 147, 152–3, 158–9, 164, 166–
 7, 255n76, 268nn58–9, 272n6,

273nn8, 11, 274nn12–13, 15–16,
 277nn34–5, 40, 278n48, 282n14,
 284nn32, 37, 285nn38, 41,
 286nn47, 51, 54
Alceste, 181–4, 226, 231–2
Alcione, 187
Alhazen (Ibn al-Haytham), 35–7, 80,
 89–93, 129, 253nn53, 61,
 266nn36–7, 267nn38–9, 42–3,
 273n7, 278n46, 286n48
al-Kindi. *See* Kindi, al-
allegoresis, 11–12, 14–17, 57, 188
allegory (defined), 7–13, and passim;
 allegory of poets, 11; allegory of
 theologians, 11; horizontal allegory, 12–15, 19, 88, 105; imposed
 allegory, 11–12; structural allegory,
 14, 46–55, 76, 79, 83, 88, 91, 109–
 10, 189, 197; vertical allegory, 12–
 14, 19, 88, 104, 214, 228
Amant, 52–4, 66–70, 72–6, 82–4, 90,
 98, 100, 102–4, 106, 108–9, 112
ambiguity, 9, 106, 118, 177, 187, 195,
 203, 209
Amis, 66–7, 69, 73, 75, 101, 116–17
Amore. *See* Amors
Amors, 51, 64, 66–9, 71–3, 75–6, 79,

83–5, 87, 98, 106, 116, 119, 122, 124, 127, 132, 146, 177, 180–4, 212, 215, 226, 231–2

analogy, 6, 22, 28, 44, 48, 68, 129, 133, 141, 152, 160–1, 211, 229, 242; analogical, 20, 23, 211, 241; analogous, 22, 146, 149, 164, 193–4, 196, 206

Andreas Capellanus, 71, 80, 99, 116, 263nn78–9, 273n10

anger, 134–5. *See also* wrath

antiocularcentrism, 21

apophasis, 5

Aquinas, Thomas, 114–15, 164, 167, 243, 272n7. *See also* Thomism

aranea. *See* retina

Argus, 163, 231–2

Aristotle, 4, 9–10, 25–6, 28–9, 35, 39–40, 66, 80, 129, 148, 154, 245n4, 247nn32–5, 251nn13, 20, 262n65, 277n42, 285n40; Aristotelian, 18, 35–6, 100, 131, 147, 188

Arrathoon, Leigh, 303n54

Ashworth, E.J., 296n70, 297n73

Astell, Ann, 303n56

atomism, 24, 129, 131, 147

Augustine, 3–5, 10, 26–8, 30, 32, 39, 41, 48, 54–5, 114, 129, 243, 245n1, 246nn10, 21, 247n37, 255n84, 258n26, 296n60, 302n46; Augustinian, 10, 39; Augustinianism, 114

authority, 45, 89, 96, 99, 111, 184–6, 209, 229, 237–9, 243; authoritative, 108

Averroës (Ibn Rushd), 35–6, 40, 129, 272n6, 275n21

Avicenna (Ibn Sina), 32, 35–6, 40–1, 129, 147, 158, 277n42, 286n47

Babel, 6–7, 44, 53, 234, 242

Bacon, Roger, 18, 24, 35, 37, 43, 90–4, 104, 111, 115–16, 129–31, 141, 164, 207, 234, 236, 249n68, 254nn63, 65, 69, 267nn40–4, 46–8, 268nn50–1, 53–4, 270n81, 278nn44, 46, 286n48

Baeumker, Clemens, 280n6

Baldassano, Lawrence, 287n61

Baltrušaitis, Jurgis, 246n23

Baranski, Zygmunt, 280n1

Barber, Joseph, 280n1

Barney, Stephen, 46, 48, 257n8, 263n91

Barolini, Teodolinda, 114, 272nn1, 4, 280n4, 286n46, 288n64, 289n72

Bartholomaeus Anglicus, 38–9, 41, 255nn73, 75, 84, 294n45

Baswell, Christopher, 261n57

Baumgartner, Emmanuèle, 269n72

beam (visual beam), 25–7, 35, 67, 122, 129, 136, 147, 149–51, 154, 164–5, 167–70, 173, 178–80, 191. *See also* extramission

beatific vision, 4, 88

Beatrice, 117–27, 137–46, 148, 155–6, 160–70, 172, 175

Bel Acueil, 48, 53–4, 60, 66–7, 69–75, 81, 86, 98, 106, 108

Bellezza, Francis, 256n87

Benjamin, Walter, 8, 246n25

Bennett, J.A.W., 295n51

Benson, Larry, 301n40

Benvenutus Grassus, 31–2, 252nn31–3

Bernard of Clairvaux, 155, 167, 180–3, 281n8, 290n10, 291nn13–14, 303n56

Bernardus Silvestris, 58, 62, 65, 82, 104, 249nn65, 67, 259n36, 262n66, 269n69, 270n83; *Cosmographia*, 15, 17, 107, 109

Bertran de Born, 151–2

Biauté, 64, 106

Bischoff, Bernhard, 255n73

Blanche, Duchess of Lancaster, 178, 186–7, 194

Bleeth, Kenneth, 303n51

blindness, 186, 218, 223–32

Bloch, R. Howard, 11, 13, 248n42, 249n60

Bloomfield, Morton, 247n39

Blumenfeld-Kosinski, Renate, 257nn11, 13

Boase, Roger, 256n1

Boccaccio, Giovanni, 21, 179

body, 18, 20, 22, 24, 36, 41, 43, 59, 105, 112–13, 120, 123, 131, 136, 140–3, 149, 159–61, 163–4, 166, 186–7, 194, 212, 216, 218, 223, 229–30, 235–6

Boethius, 3, 26, 28–30, 34, 42, 57, 62, 110–11, 135–6, 146, 191, 212, 217–19, 235–7, 240, 251n20, 256n85, 270n106, 279n57, 290n8, 295n51

Bonaventure of Siena, 243, 281n10

Book of the Twenty-four Philosophers, 155

Bornstein, Diane, 300n25

Boucher, Holly, 296n66

Bouvet, Honoré, 237–43, 304nn5, 9, 305nn13–14

Boyde, Patrick, 11, 248n41, 273n7

Bradwardine, Thomas, 199–200, 256n86

brain 27, 30, 33, 35, 41, 57, 131–2, 135

Brandeis, Irma, 272n7

Brewer, D.S., 303n54

Brook, Leslie, 80, 265n7, 269n73

Brown, Peter, 248n41, 303nn48, 58

Brownlee, Kevin, 46, 82, 84, 256n6, 265nn8, 12

Brunetto Latini, 150–1, 155–6, 283nn22–4

Buck-Morss, Susan, 246n25

Bundy, Murray, 255nn81, 83

Burgwinkle, William, 261n57

Buridan, Jean, 198

Burnett, Charles, 245n9

Burnley, J. David, 290n5, 293n44, 300n27

Butler, Judith, 22

Bynum, Caroline, 22, 242, 250nn4–5, 282n19, 284n28, 305n22

Cadmus, 86

Calabrese, Michael, 224, 301n39, 302n42

Calcidius, 3, 25–6, 34, 58, 60–1, 253n46, 260nn51–2, 293n43

Camille, Michael, 87–8, 266nn18, 22, 24

Cantarino, Vincente, 272n7, 280n5

carbuncle, 79, 97–8

Carruthers, Mary, 42, 245n7, 246n19, 247n31, 255n81, 256nn86–7, 261n60, 288nn65, 67, 294n48, 298n90

cataphasis, 5–6

category, 24, 100, 110, 145, 184, 187–8, 201, 204, 242

certitude, 181; certainty, 183, 195, 200–1, 208–9, 220–2, 234, 236

Ceyx, 187

chaff, 196

chain, 13, 19, 28, 144, 235

Chaucer, Geoffrey, 20, 39, 44, 248n41; *Book of the Duchess*, 178–9, 183, 185–95, 201–4, 209, 212–13, 228, 238; *Canterbury Tales* (including individual tales), 20, 106, 178–210 passim, 211–33; *House of Fame*,

179–80, 185–6, 195–7, 201, 203–
10, 212, 215–16, 227, 232, 236;
Legend of Good Women, 179–86,
226–7, 231–2, 235; *Parlement
of Fowls*, 178, 185–6, 195–204,
213–15, 219–21, 235; *Troilus and
Criseyde*, 184–5, 189, 192, 201, 217
Chenu, M.-D., 247n40
Cherniss, Michael, 291n15
chess, 187, 193–4
Chiarenza, Marguerite, 288n66
Chrétien de Troyes, 21, 45, 257n13,
258n23, 263n84
Christine de Pizan, 108, 236–43,
270n95, 287n57, 304nn4, 6–7,
305nn11, 17
Ciacco, 148
[pseudo-]Cicero, *Rhetorica ad
Herennium*, 6, 42, 246n20, 256n86,
261n60
Cioffari, Vincenzo, 273n8
Claudian, 224, 301n39
Col, Pierre, 108
Coleman, Janet, 245n7, 288n65,
294n48
Colish, Marcia, 27, 251nn11, 17, 19,
21, 301nn30, 37
Collette, Carolyn, 289n3, 290n4,
295n54, 300n22, 302nn44, 46
colour, 40, 43, 56, 60–1, 63, 65–6,
118, 129
common sense, 4, 19, 41–2, 116, 123,
176, 225
complexion, 30
conceit, 155. *See also* vehicle
conception, 17–18, 26, 229–30
cone (visual cone), 25, 35–6, 58–9,
164
Constantinus Africanus, 30–1, 34,
39–41, 57, 251n26, 253n50

Contini, Gianfranco, 280n1, 283n24
Copeland, Rita, 7–8, 11–15, 185,
247n30, 248nn43, 45–6, 49, 51,
57–8, 259n34, 271n106, 291nn17,
23–4
corn, 196. *See also* seed
Cortoisie, 64
Courcelle, Pierre, 279n56
Courtenay, William, 199, 221,
254n71, 297n72, 301n32
courtly love, 45, 66, 71, 76, 79–80, 83
Criseyde, 183–4, 201, 217
Crombie, A.C., 90, 253n56, 266n33
crystal, 47, 55–7, 60, 65–6, 75, 79, 94,
96–8, 107
crystalline humour. *See* lens
Cupid, 214–15

Dahlberg, Charles, 45, 61, 70, 256n4,
257n18, 262n76, 264n98, 264n1,
265nn4, 15
daisy, 179–83, 227, 232, 235
d'Alverny, Marie-Thérèse, 252n35,
278n42
Dangiers, 48, 66–7, 70–1, 75, 106,
112
Dante, 11–12, 20–1, 179, 246n21,
248n47, 290n9, 296n66, 301n39;
Commedia, 114–37 passim, 138–77,
189, 203–4, 223, 229–30, 235–6,
243; *Convivio*, 114–37 passim, 146–
8, 163, 170, 173; *De vulgari
eloquentia*, 110; *Fiore* (attrib.), 151;
Vita nuova, 114–37, 146
David, 167
deduit, 61–4, 69, 79, 82, 84, 94, 98,
107; Deduit (personification), 51–
2, 62, 64, 79, 97, 100, 106–7, 162,
223
de Deguileville, Guillaume, 8, 10

De Lacy, Phillip, 251n24

Delany, Sheila, 11, 211, 248n42, 298nn, 91, 3

Delasanta, Rodney, 199, 296n66, 297nn71, 75

de Libera, Alain, 272n6

de Man, Paul, 8, 246n25

Demats, Paule, 263n78

Dembowski, Peter, 110, 271nn104, 112

Democritus, 122, 131

Dempster, Germaine, 301n40, 303n49

Descartes, René, 37

Deschamps, Eustache, 184

Despars, Jacques, 32

detuitio, 60–6, 69, 98, 107. *See also* refraction

Devons, Samuel, 262n65

Dewan, Lawrence, 253n54

Dialectic (personification), 9

diaphanous, 18, 25, 44, 119–20, 123, 129–30, 132, 136–40, 143, 147–8, 154, 156, 158–9, 163, 165, 167, 169

dichotomy, 14, 161, 185, 188–93, 214

Dido, 205–6

Diekstra, F.N.M., 301n39

Dinshaw, Carolyn, 242–3, 305nn18–19

Dis, 149

division, 6–7, 41, 113, 240–3

Dominicans, 173

Dominicus Gundissalinus, 17, 249n62

Donovan, Mortimer, 301n39

double vision, 26–8, 32, 227–30

Douie, Decima, 287n58

dream, 10, 19, 48–51, 54–5, 60, 68, 75–6, 89, 95–6, 109, 113, 161, 176, 179–82, 185–8, 193, 195–7, 201–2, 204, 210–12, 235, 240–1

Dronke, Peter, 57, 247n40, 249n59, 252n39, 259n34, 262n77, 271n110, 282n11, 288n63, 289n71, 299n11

Duhem, Pierre, 251n18

Dumbleton, John, 38, 199–201, 208, 213, 220–1, 231, 254n72, 296nn64–5, 70, 297n76, 298nn7–8, 301n31, 303n59

Durling, Robert, 272nn1, 7, 273n8, 279n56, 280n3, 281n6, 282nn11, 17, 19, 286n51, 287n59

eagle, 157, 167, 174, 201–9, 215

Eastwood, Bruce, 253n59, 259n32

Eberle, Patricia, 90–1, 98, 266nn26, 30–3, 35, 268nn52, 61

Echo, 68–9, 83–4

eclipse, 28, 35, 59, 139

Economou, George, 96, 268n56, 281n9, 299n10, 301n39, 302n42

Eden, 6, 111, 144, 162, 229. *See also* paradise

Edgerton, Samuel, 250n9

Edwards, Robert, 185, 291n22, 292n31, 302n43

ekphrasis, 7, 209

Eldredge, Laurence, 252n33

Elford, Dorothy, 32, 34, 249n64, 252nn36, 39, 253n49, 260n48

Elie, Pierre, 16

Ellis, Steve, 292n30, 297n80

emblem, 174

encyclopedia, 16, 38–41; encyclopedic, 15, 147

Eneas, 62; Eneas (character), 205

ennarratio poetarum, 11–12, 17

Entzminger, Robert, 202, 297n79

envy, 158–9

Epicureans, 24, 33, 37

Eriugena, John Scotus, 26

etymology, 13, 16, 58, 61–2
Euclid, 30, 35–6, 80, 90, 148
euphemism, 103, 105
exchange, 195, 212, 214, 233, 243
exegesis, 11–12, 17, 51, 57, 161, 185
experiment, 139–43, 159; experimental, 139, 141, 164
extramission, 23–7, 29–33, 35–6, 38–9, 41, 67, 93, 116, 122, 129, 132, 136, 147–8, 153–5, 164, 168–71, 178–9, 212, 235
eye (anatomy), 30–3, 35–6, 38–40, 57, 68, 94, 135–6, 152, 156, 158–9. *See also* pupil, iris, retina, lens, vitreous humour, tunic
eye of the mind, 28, 133–4, 152; eye of intelligence, 29, 42, 127, 148

fabliau, 15
faculty (mental), 4, 40–2, 93, 103, 116, 123, 175–8, 192–3, 224–31
Fair Welcome. *See* Bel Acueil
Fame (personification), 207–9, 215–16
fantasy, 95–6, 176, 222, 224–31. See also *phantasia*
Faral, Edmond, 46
Farinata, 149
fear, 134–6
Fenellosa, Ernest, 247n36
Ferrante, Joan, 69–70, 256n1, 262nn74–5, 264n102, 272n1, 281nn8, 11, 286n50, 299n10
Ferster, Judith, 197–8, 214–15, 295n57, 299nn13–14
fertile, 15, 83, 85, 108; fertility, 16, 79, 85, 96, 98–9, 101, 151, 162, 184. *See also* reproduction
figurative language, 7, 79, 103–5, 107, 109, 116, 127, 134, 147, 188, 204, 227–9

Filosofia. *See* Philosophy
fire (visual fire), 25, 27, 147–8, 153–4, 164, 168–9, 173, 178
Flamant, Jacques, 252n43
Fleming, John, 56, 69, 107, 259n28, 262n73, 263n89, 264n1, 270n92, 291n19
Fletcher, Angus, 8, 246nn24, 26
Fletcher, J.M., 296n69
Florence, 150
Folco of Marseilles, 165
form, 17–20, 24–30, 32–3, 36, 41, 58–9, 61, 75, 87, 93, 115–16, 120–3, 127, 129–32, 135–6, 140, 143–5, 151, 154–61, 167–8, 171, 174–5, 179, 189, 191, 225, 229–30, 235. *See also* species
Fortune (personification), 76, 150–1, 155, 193, 212, 215, 222
Franchise, 65
Franciscans, 173
Frangenberg, Thomas, 285n44
Frappier, Jean, 56, 259n28, 265n13
Freccero, John, 175, 288n66
Froissart, Jean, 184, 186
Fyler, John, 292n30, 299n15

Galatea, 86, 88
Galen, 24, 30–3, 36, 41, 132, 147, 251nn19, 24–5, 27, 252n41, 255n80, 259n32, 278n46; galenic, 31, 33–5, 38, 41, 57, 158
garden, 51–2, 55, 62–3, 68–70, 72, 76, 79, 96–7, 99, 107, 162, 201, 214, 223–6, 229–30
gate, 147, 149, 157, 197–8, 213
Gaunt, Simon, 261n57
gaze, 22, 28, 62, 87, 116–18, 122, 127, 136, 149–54, 161, 165, 167, 170, 172–4, 179–81, 189, 204, 235

genealogy, 13, 195

generation. *See* seed

Genesis, 65, 234, 241–2

Genius (personification), 79, 85–6, 88–9, 96–8, 101, 107–8, 110–12, 239

genre, 11, 14–15, 20, 100, 109–10, 188, 202, 223–4, 232, 237–8

geometry, 58, 89, 169; geometrical, 90, 117–18

Gersh, Stephen, 252n43

Gerson, Jean, 110, 238–9, 271n107, 304n10

Gibson, Margaret, 251n22

gloss, 11, 103, 109, 196; glossator, 11–12; glosses, 32, 34, 57–9, 62, 67, 206

gluttony, 148

Goldin, Frederick, 262n71

Gower, John, 239, 241, 305n16

Gregory, Tullio, 58, 252n38, 259n37

Grennen, Joseph, 205, 289n3, 297n85

griffin, 163

Grosseteste, Robert, 26, 35–7, 39, 43, 90–1, 93, 122, 129, 141, 167, 206–7, 234, 241, 253nn56–9, 256n89, 266nn33, 36, 277n34, 281n6

Gui de Mori, 46, 82–3, 99–100

Guido Cavalcanti, 120–5, 170, 275nn18–21, 276nn23, 27

Guillaume de Lorris and Jean de Meun, *Roman de la rose*, 14–15, 18–21, 44–113, 138, 146, 150–1, 155, 157, 162, 170–4, 183–4, 186, 194, 212, 214, 223–4, 235–6, 239, 241, 285n43, 287n59

Guillaume de Saint-Amour, 82

Gunn, Alan, 45, 88, 256n4, 264n1, 266nn16, 25

habitus, 130–1. *See also* species

Haefili-Till, Dominique, 251n28

Hagen, Susan, 10, 248n41

Hajdu, Helga, 256n86

Haly Abbas, 31

Hanning, Robert, 288n63, 292n28

Hansen, Elaine, 292n26

Hardman, Phillipa, 292n27

Harley, Marta, 62, 75–6, 261n58, 264nn92, 101, 276nn24–5, 281n11, 284n29

Hart, Thomas, 189, 293n40

Hartung, Albert, 300n23

Harvey, E. Ruth, 245n6, 255nn78, 82, 276n30, 294n45

hearing, 178, 203–4. *See also* sound

heat, 28, 35, 59, 122, 161, 166–7

Heloïse, 79, 101–2, 110

hesitation, 186, 195, 200–1, 208, 221

Hewitt, Kathleen, 299n11

Heytesbury, William, 199–200

Hicks, Eric, 110

hierarchy, 13

Hill, Thomas, 87, 266nn19–21

Hillman, Larry, 258n27

Hippocrates, 251n24

Hoenen, Maarten, 251n22

Holkot, Robert, 188, 301n37

Hollander, Robert, 280n4, 288n69

Holley, Linda, 248n41, 289n3, 290n4, 304n61

Homer, 148

homoeroticism, 62, 86, 108; homoerotic, 72, 78

homosexuality, 54, 72

Honig, Edwin, 8, 247n29

Hult, David, 46, 56, 63, 82, 99–104, 256nn1, 5, 257n16, 258n24, 259n29, 261n61, 264nn103–4,

265n9, 268n63, 269nn64–5, 70, 75, 78, 80, 270nn82, 89

humours, 30, 200. *See also* melancholy

Hunain ibn Ishaq, 31, 57, 89, 255n80, 259n32

Huot, Sylvia, 82–3, 102–3, 256n3, 265n10, 269nn76–7

iconography, 7, 9

illumination, 9, 43, 161, 165, 206, 209, 234, 236–7

imagination, 4, 19, 35, 40–2, 90, 116, 123–4, 158, 175–7, 191–3, 224–31

impression, 25, 28–9, 42, 131, 158, 176–7, 191, 224–30

insomnium, 50–1, 60, 79, 89, 96, 176

integumentum, 17, 57–9, 62, 75, 102, 104, 109, 112, 235

intellect, 120–7, 142–3, 148–9, 156, 176, 195, 231–2. *See also* reason

intention, 11, 24, 113, 184–5, 213, 225. *See also* species

intermediary, 115, 118–19, 130, 163, 165, 169, 207, 238. *See also* mediation

intromission, 21, 23–6, 29, 32–4, 36, 41, 116, 122, 129–32, 136, 147–8, 153–5, 158, 162, 168–70, 191

Irigaray, Luce, 21

iris, 135, 156–7

Irvine, Martin, 206, 298n87

Isidore of Seville, 15–16, 38–9, 61, 216, 249n59, 255nn73–4, 260n54, 299n18

Jackson, W.T.H., 265n2

Jacquart, Danielle, 32, 249nn60, 66, 251n26, 252nn30, 34, 260n54

Jalousie, 53

Januarie, 186, 211, 222–32

Jauss, Hans Robert, 109

Jay, Martin, 21, 245n5, 250n1

Jean de Meun, 15, 18–20, 44, 115, 129, 138, 171, 184, 214, 235–6, 240, 284n30, 291n19, 302n42; *Li Livres de Confort de Philosophie*, 110; *Roman de la rose, see under* Guillaume de Lorris; *Testament*, 111–13, 150–1

Jeauneau, Edouard, 57–8, 252n44, 259nn33, 35, 37–42, 260nn43, 49–50, 53, 261n59

Jesus Christ, 8, 15, 144–6, 161, 163, 181, 226–7, 230, 232

John (apostle), 167

John of Damascus, 6

John of Gaunt, 194

John of Salisbury, 33

Johnson, Lynn, 299n21, 300n23

John the Baptist, 161

Joinece, 64–5, 68, 212, 224

Jonsson, Einar, 255n77

judgment, 4, 42, 134, 192, 224–32. *See also* reason

Julian of Norwich, 242–3

Jung, Marc-René, 263n78

Jupiter, 100–1

Juvenal, 57

Katzenellenbogen, Adolf, 246n22, 299nn18–19

Kay, Richard, 282n16

Kelley, Theresa, 246n25

Kelly, Douglas, 13, 47, 83, 111, 257n10, 265n11, 271nn108–9

Kempe, Margery, 242–3

Kindi, al-, 35–6, 129

Kirkham, Victoria, 281n11

Klassen, Norman, 248n41, 290n4

Klibansky, Raymond, 260n48

Klinck, Roswitha, 249nn60–2, 259n39

Klubertanz, George, 255n81

Knapp, Daniel, 298n93

Knoespel, Kenneth, 56–7, 89, 91, 97, 106, 259nn30–1, 262n70, 268n57, 270n88

Köhler, Erich, 56, 259n28, 265n2

Kolve, V.A., 298n90

Kruger, Steven, 188, 257n12, 292n34

Kuhn, Thomas, 256n90

ladder, 5, 165, 174, 235

lamp, 152, 164

Langland, William, 14, 211, 238, 247n28, 304n8

Langlois, Ernest, 257n18

Largeice, 64–5

latitudes, 199–201, 208, 220–1

Laziness. See Oiseuse

Leah, 161, 174

Lecoy, Félix, 80, 89–90

Leesce, 64

Lejeune, Rita, 46

Lemay, Helen, 259n34, 260nn42, 45, 262n68

Lemay, Richard, 253n60

lens (of the eye), 36, 41, 57, 94

lenses (glass), 92, 141

Lewis, C.S., 56, 108, 259n28, 265n2, 266n16, 270nn87, 93, 96, 280n5, 298n88

Lidaka, Juris, 255n73

Lieser, Ludwig, 255n76

light, 13, 18, 23, 25–9, 35–8, 40, 43, 59, 93, 97–8, 114, 116, 118–24, 127, 129, 132, 136–49, 152–70, 178–9, 191–2, 195, 203–7, 225, 236–7, 241

Limbo, 148

Lindberg, David, 23, 37, 90, 93, 250n2, 251n14, 252n36, 253nn48–9, 52–3, 55–6, 60–1, 254nn63–5, 67, 259n32, 266nn34, 36, 267nn38–44, 46–9, 268nn50–1, 273n7, 277n40, 278nn42, 44–6, 280n6

literal, 11, 14–15, 104, 107, 109, 127–8, 136–7, 148, 221, 223, 227–8, 236, 241

Louth, Andrew, 5, 246nn10–12, 14, 16–17

Love (personification). See Amors

Lowes, J.L., 290n6

Lucretius, 17, 24, 249n63, 284n34

Lucy, Saint, 143, 146, 157, 203

luminescence, 153

Luria, Maxwell, 263n90

Lydgate, John, 236, 241, 243, 304n3

Lynch, Kathryn, 178, 198, 200, 212–13, 289n2, 295nn58–9, 296nn60–2, 298n6

Machaut, Guillaume de, 184, 186–7, 292nn29, 31

Macrobius, 26, 28, 33–4, 49–50, 57–9, 62, 67, 96, 195–7, 204, 240, 251n18, 252nn41–4, 257nn14, 17–18, 258n19, 268n55, 292n32, 295nn51, 53, 305n15

macrocosm, 3, 35, 45

Magee, John, 251n20, 256n85

magnet, 198–201

Maierù, Alfonso, 296n70

Male Bouche, 53, 69–70, 72

Maloney, Thomas, 104, 270n81, 271n111

Mark (gospel of), 85

Mars, 94, 120

Martianus Capella, 9, 34, 58, 62, 104,
239, 247n31, 259n36, 269n69,
270n83, 302n45
Mary (Virgin Mary), 43, 113, 143,
157–8, 167, 227–30, 239
Matelda, 154, 159, 161
Matthew (gospel of), 9
Matthews, Lloyd, 300n23, 301n36
May, 211, 222–31
Mazzeo, Joseph, 272n7, 289n70
Mazzotta, Giuseppe, 272n1, 287n58,
288n67, 289n70
McEvoy, James, 253n58
McInerny, Ralph, 275n21
mediation, 7, 14, 19, 42, 44, 119, 144,
189, 235; mediator, 6–7, 21, 43–4,
144, 146, 235; medium, 18, 20, 23,
37–8, 40, 43–4, 50, 62, 79, 93, 116,
119–20, 124, 129–30, 132–3, 137,
140–1, 143, 147–8, 154, 156, 158,
162–3, 165–70, 175, 234, 237;
mediating, 23, 25; mediate, 23,
115, 132, 179, 203, 209, 234, 236.
See also intermediary, mezzo
melancholy, 187, 192
Melibee, 216–23, 226, 231
Melville, Stephen, 8, 12, 14
memory, 4, 6, 9, 26, 41–2, 61, 116,
120, 123–4, 126, 159, 174–5, 192–
3, 209–10, 213, 238; memorial,
186, 193–4, 231–2
Mercury, 163, 231–2, 239
metamorphosis, 187, 190, 203, 215
metaphor, 9–10, 13, 15–17, 28–9, 49,
86, 91, 103–4, 120, 122, 131–5, 138,
155–6, 171, 178, 184, 186, 191, 198,
206, 229, 234–6, 243; metaphorical,
111, 115, 133, 140, 142
'metaphysics of light,' 25, 28, 36, 90,
141

metonymy, 13, 212, 224
mezzo, 117–19, 124, 130, 132–3, 159,
162–3, 165, 167–9. See also mediation
Michaud-Quantin, Pierre, 251n10
microcosm, 3, 35, 45, 168
Milbank, John, 5, 246n13
Miller, James, 280n3
Miller, Robert, 303n60
Minnis, A.J., 12, 246n16, 248nn44–5,
47, 50, 249n62, 251n22, 296n67,
297n77, 298n4, 301n37
mirror, 7, 10, 13–14, 18, 28, 35, 40,
42–4, 47–52, 54–5, 61, 66–8, 75,
79, 87–9, 91–2, 94–6, 98–9, 105–6,
109, 116–17, 121, 123, 130–1, 135,
138–43, 146, 157, 159–68, 171–4,
194, 216, 224–9, 242
mnemonic, 6, 9, 42
Mohammed, 151–2
Molland, George, 255n72
moon, 28, 35, 40, 138–40, 142–3,
146, 149, 161–3, 165, 172, 175
Morpheus, 187
Moses, 4
motion, 38, 41. See also Oxford
Calculators
'multiplication of species,' 37, 39,
93–4, 96, 111, 115–16, 129, 140,
164, 169, 179, 206–8, 236
multiplicity, 5, 79, 84, 91, 209, 215,
235; multiplied, 36, 141, 153, 160;
multiplication, 47, 66–7, 91, 111,
140, 144, 164, 167, 169, 206–7,
234, 241, 243
Murrin, Michael, 7
Musa, Mark, 119, 273n9, 274n14,
275nn17, 22
Muscatine, Charles, 105, 270n86
Myrrha, 88

myth, 57–8, 62, 79, 86, 90, 99, 109, 173, 224, 239; mythological, 100; mythic, 110

Narcissus, 47–8, 50–3, 55–6, 62–3, 65, 67–9, 72, 74–6, 79, 83–8, 91, 94, 96–9, 107, 110, 170–4; narcissistic, 26, 56, 62, 83, 85–6, 105–6, 109, 146; narcissism, 45, 98
Nardi, Bruno, 115, 272nn1, 7, 273n8, 275n21, 280n3
Nature (personification), 7, 46, 81, 88–96, 98–9, 101, 105, 107–8, 138, 146, 172, 198, 201–3, 212–15, 219, 235–9
Neaman, Judith, 294n47
Neoplatonism, 3–5, 33–4, 58; Neoplatonic, 5, 10, 13, 18–19, 27–9, 33–6, 39, 49, 59, 107, 132, 135; Neoplatonists, 25–7
Neuse, Richard, 224, 301n40, 302n41
Newman, Francis, 256n1, 272n3
Newton, Francis, 251n26
Nolan, Edward, 246n23, 248n41, 250n2, 287n60
nominalism, 38, 44, 101–2, 199, 211, 213; nominalist, 11, 13, 20, 38, 44, 103–4, 199, 211–3
North, J.D., 293n41, 301n31
number, 58, 127, 135, 189, 202; numerical, 174, 189, 202

Ockham, William of, 20, 37–8, 44, 199, 213, 221, 234, 254nn70, 72, 293n43, 301n30
O'Daly, Gerard, 251n15, 279n57
Oiseuse, 7, 48, 51–2, 65–6, 107–9, 157, 212
Olivi, Peter, 37–8, 234, 254n69
O'Neill, Ynez, 253n50

opinion, 182, 238
optic nerve, 30–1, 35, 41, 57, 132, 135, 159
Origen, 22, 101
origin, 16
O'Rourke, Fran, 272n5, 286n49
Orpheus, 62, 215
Otten, Charlotte, 227, 301n39, 303nn50, 53
Ovid, 71–2, 74, 80, 111, 179, 185, 187, 263nn83–4, 87–8, 264n103, 290n9, 292n31, 303n57
Ovide moralisé, 12
Owen, Charles, 299n22
Oxford Calculators, 38, 199–200, 220–1
Oyarses, 107

Palomo, Dolores, 300n25
Pandarus, 192
Paphus, 88
parable, 8, 15, 85
'paradigm shift,' 43
paradise, 113, 138, 145, 147, 159, 161–2, 223–5. See also Eden
paradox, 9, 116, 190, 192–3, 241
Park, David, 250n
Parronchi, Alessandro, 141, 273n7
Patterson, Lee, 220, 222, 224, 300nn23, 29, 301n35, 302n43
Paul, Saint, 4, 10, 44, 49, 88, 113
Paulmier-Foucart, Monique, 255n73
Paxson, James, 211, 247n39, 298nn1, 3
peacock, 163
Pecham, John, 37, 42, 90, 141, 164, 206–7, 225, 254nn64–5, 278n44, 285n39, 297n86, 302n47
Peck, Russell, 189, 293nn41, 43
Peden, Alison, 292n32, 295n51

Pelikan, Jaroslav, 276n31, 282n12
Penn, Stephen, 296n67, 298nn3–4
personification, 8, 10, 14, 69–71, 84, 89–90, 103–9, 116, 119, 146, 186, 211–22, 237–41
perspectiva, 32, 35, 39, 42, 89–90, 92–3, 167–9, 207; perspectivist, 23, 32, 37–9, 42, 115, 129, 141, 147–8, 156, 162, 164–5, 167, 179, 206–7, 225
Pertile, Lino, 114, 272n2
Petrarch, 179
phantasia, 41, 89, 193. See also imagination, fantasy
phantasm, 24. See also species
Phillips, Heather, 251n10
Philosophy (personification), 28, 116, 126–7, 133, 135–6, 138, 143, 146, 212, 217–19, 235–6, 240
Phlegyas, 148
Phronesis, 18
Pickens, Rupert, 257n18, 258nn20, 24
Plato, 3, 21, 25–6, 34, 57–8, 62, 116, 148, 152, 164, 205, 245n3, 251n12, 259n40, 293n43. See also Calcidius
Platonism, 4–6, 189; Platonist, 10, 58, 153; Platonic, 26, 30, 33, 35, 67, 115, 122, 147, 149, 164, 168, 173, 178–9, 205, 209, 212–13
Pleasaunce, 215
Pliny the Elder, 32
Plotinus, 4–5, 26, 28, 34
plow, 85–6, 111
Pluto, 223–7, 231
pneuma, 25, 27, 29–31, 33, 57
Poirion, Daniel, 46, 72, 75, 101–2, 256n5, 263n86, 264n93, 269n71
polysemy, 8, 190, 192; polysemous, 62, 88, 104, 145, 188–9

Porphyry, 26, 34, 49
Praevidentia, 163, 216
Pratt, Robert, 301nn37, 39
Prior, Sandra, 293n37
Priscian, 17, 34, 206
Proclus, 4–5, 26, 246n14
Proserpine, 161, 226
Prudence (personification), 7, 186, 211, 216–23, 228, 238
pseudo-Dionysius, 4–7, 10, 13, 26, 36, 114, 164, 234, 243, 246n14, 246n16, 248n55, 251n14
Ptolemy, 34–6, 80, 90, 252n42; Ptolemaic, 34
punning, 8, 64, 188; puns, 88
pupil, 33, 35, 41, 130–1, 135–6, 140–1, 167
Pygmalion, 79, 83–4, 86–8, 96–9, 109
Pythagoreans, 25; Pythagorean, 27, 58, 189, 209

Quilligan, Maureen, 8, 12, 14, 75, 103, 188, 211, 246n24, 247n28, 248nn48, 56, 264n95, 265n3, 269n79, 293n36, 298n1
Quintilian, 7, 247n39, 294n48

Rachel, 146, 161, 174
Rahab, 169
rainbow, 35, 40, 43, 60–1, 65–6, 82, 92, 148, 160, 162, 164, 173
ray (visual ray), 30, 33–4, 37, 59, 61, 115, 117–19, 123, 130–1, 136–8, 142, 146, 148, 153, 157, 163–7, 169–70; radiation, 114
realism, 20, 44, 211, 213; realist, 103, 199, 213
reason, 3–4, 10, 14, 29, 40–2, 134, 148, 198, 221, 237; Reason (personification), 3, 51, 66, 79, 81–2,

84, 90, 100, 102–4, 108–9, 112, 146, 151, 170, 215, 219, 224–32, 236–9

Redle, M.J., 245n8, 287n58

reflection, 34, 38–40, 42–3, 47–8, 55, 59–63, 65, 88, 123, 139–42, 146, 153, 162; reflected, 69, 75, 148, 154, 160, 162–3, 165, 170–4

refraction, 34, 39, 43, 47, 59–63, 65, 79, 82, 84, 98, 141, 162–3, 173; refracted, 69, 148, 241

Regalado, Nancy, 100–1, 269nn66–8

Rendall, Thomas, 294n49

reproduction, 23, 28, 101, 111, 138; reproductive, 89, 104; reproduce, 112. *See also* fertility

retina, 36, 94

revelation, 4, 9, 237

Reydellet, Marc, 255n73

Reynaud de Louen, 218–20

Rhea, 104

rhetoric, 7, 9, 11, 15

Rhetorica ad Herennium. See [pseudo-] Cicero, *Rhetorica ad Herennium*

Richards, Earl, 108–9, 258n21, 270nn97–8, 283n24

Richards, I.A., 279n51

Richece, 64–5, 214

Richesse. See Richece

Ricklin, Thomas, 34, 252n36, 253nn49, 51, 260n44

riddle, 9–10

Robertson, D.W., 56, 258n23, 259n28, 264n1; Robertsonian, 90

Ronca, Italo, 252nn37, 39–40, 44, 253n50

Ronchi, Vasco, 250n9, 273n7

Roney, Lois, 289n3

rose, 45–113 passim; 138, 146, 162, 169–70, 174, 235–6, 239

Rosenberg, Bruce, 303n51

Rossi, Albert, 287n55, 289n71

Rowland, Beryl, 209–10, 294n50, 297n81, 298nn90, 92

Russell, Gül, 251nn27, 29

Russell, J. Stephen, 187–8, 292nn33, 35, 296n66

Rutledge, Monica, 273n7, 285n45

Salerno, 31

Sandler, Lucy, 287n58

Sapientia. *See* Wisdom

Saturn, 79, 90, 100–1, 103–4, 109, 111, 166

schismatics, 151–2

Schless, Howard, 297n80

science, 196, 202, 216, 237

Scipio Africanus, 195–7, 201

seal, 29, 42

seed, 15–17, 84–5, 98, 144, 151, 184 (insemination). *See also* corn

self-contemplation, 26

self-knowledge, 14, 19–20

Seneca, 32–3, 35

sense perception, 6, 26, 29–30, 40, 44, 50, 93, 203, 209, 227, 231, 234; senses, 25–6, 40, 60, 66, 82, 121, 205, 209, 218–19, 234; sensory, 122. *See also* common sense

sensus communis. See common sense

Severs, J. Burke, 300nn24–5

sexuality, 78, 87

Seymour, Michael, 255nn73, 75

Shakespeare, 8

Shoaf, R.A., 287nn60, 62

Siegel, Rudolph, 251nn24–5

sieve, 216

sign, 43, 181; signification, 44, 105, 227; signifying, 42; signify, 43, 79, 104–5, 107, 211, 213; signifier, 203

simulacrum, 24, 37, 63, 116, 119, 124, 171, 204. *See also* species
Singleton, Charles, 149, 151, 156, 159, 272nn1, 7, 281n11, 288n67
Sinon, 172
Siraisi, Nancy, 251n23
Smith, Mark, 23, 250nn7, 10
snail, 153
sodomy, 149–51
Solomon, 222–3, 226–9
somnium, 49–51, 59–60, 67, 79, 89, 96, 176, 240–1
Sordello, 157
soul, 5–6, 20, 34, 40, 49, 59, 67, 121, 126, 128, 134, 138, 140–6, 148, 156, 160–1, 164–6, 168–9, 170–1, 174, 177–8, 219, 224
sound, 179, 201–10. *See also* hearing
Spade, Paul, 296n70, 297n74
Spearing, A.C., 250n2
species, 18, 20, 24, 36–44, 94, 96, 111, 115–16, 119, 123–4, 129–31, 138, 140–2, 144–6, 148, 155–8, 167–9, 175, 191, 202, 206–8, 213, 225, 229–30, 234, 237. *See also* 'multiplication of species'
speculum. *See* mirror
Spenser, 14
Spillenger, Paul, 288n67
Stanbury, Sarah, 250n3, 289n3, 295n54
Statius, 160–1
Steadman, John, 297n81
Steinmetz, David, 296n66
Steneck, Nicholas, 253n54, 255nn78, 82
sterile, 83–4, 87, 108; sterility, 84–5, 98, 150–1, 171–2
Stock, Brian, 249n67, 251n16
Stoics, 25, 27, 29–30, 191

Strode, Ralph, 199–200, 208, 221, 296n70, 297n75
Strohm, Paul, 46, 256n5, 300n25
Sturges, Robert, 11, 13, 248nn42, 54
Sturm-Maddox, Sarah, 282n11
subjectivity, 21–2; subject, 23–5, 27, 30, 33, 36, 42–3, 114–16, 118, 122, 129, 132, 149, 153, 164, 189, 202–3, 209, 234–6, 243
Sudhoff, Walther, 255n79
sun, 25, 28, 35, 40, 56–7, 59–60, 65–6, 82, 97–8, 120–1, 126–7, 137–40, 142–3, 148–9, 154–5, 161, 163–4, 166, 173, 179–82, 204
Sutherland, Jenifer, 305n23
Swyneshed, Richard, 199–200
Sylla, Edith, 254n72, 296nn63, 68, 297n76, 301n31
symbol, 106, 127, 146, 204; symbolism, 10, 109; symbolic, 55
symmetry, 14, 47–8; symmetrical, 53, 55, 98, 235
synecdoche, 144, 154

Tachau, Katherine, 23, 38, 250nn6, 8, 251n14, 253n62, 254nn68–71, 255n82
tears, 28, 150, 172
Tempier, Margherita, 285n43
terminist logic, 210, 212–13
theology, 4–5, 32, 36, 114–15, 221, 243; theological, 15, 114–15, 164; Theology (personification), 146
Thomism, 114; Thomist, 115, 199. *See also* Aquinas
Thonnard, François-Joseph, 251n15
time, 202
Tinkle, Theresa, 299n10
Torti, Anna, 248n41

touch, 29, 42, 122, 149; tactile, 113
tower, 48, 53–5, 60, 86–7, 194
Toynbee, Paget, 115, 273n8
translation, 7, 10–11, 13, 15, 29–32,
    35–6, 45, 57, 79, 92, 101, 109–11,
    183–6, 217–18
Travis, Peter, 290n12, 297n82
Trinity, 26, 41, 43, 98, 127, 173, 189;
    trinitarian, 97
Trottman, Christian, 245n8
Trout, John, 299n16
Troy, 209
tunic (of the eye), 38, 94, 152, 156
Tuve, Rosemond, 7–8, 11–12,
    247n27, 299n17
Twomey, Michael, 255n73
Tyler, Stephen, 245n5

Uitti, Karl, 62, 257n6, 261n58,
    264n92
understanding, 26, 28, 42, 121, 126,
    158, 175. See also reason
universals, 199, 211, 213–14, 225
usurers, 151
Utz, Richard, 296n67, 297n77
uvea. See iris

Vance, Eugene, 293n40
van den Boogaard, Nico, 261n60
Vegetius, 110
vehicle, 123, 134, 147, 179. See also
    conceit
veil, 9, 16–17, 49, 51, 57–8, 62, 127,
    133, 143, 158–9, 174, 205. See also
    integumentum
Venus, 83–4, 87, 94, 100, 104, 161,
    201, 209, 213–16, 236, 239
Verbeke, Gérard, 251n11
Verhuyck, Paul, 46–8, 257nn7, 9,
    268n60

verisimilitude, 20, 23, 188, 209–10,
    236
Vescovini, Graziella, 23, 250n6,
    254n67, 278n42, 285n44
Vincent of Beauvais, 38–42, 255nn73,
    76
Virgil, 49, 66, 143–4, 147–9, 153,
    155–7, 160, 163, 169–72, 176
vision, direct, 19; reflected vision, 19;
    refracted vision, 19. See also reflec-
    tion and refraction
vitreous humour, 38, 94, 147, 156
Vitz, Evelyn, 75, 257n11, 264nn94, 99
Vollmann, Benedikt, 255n73
voluntarism, 198–203
Voorbij, J.B., 255n73
Vulcan, 94

Wack, Mary, 250n9
Walker, Denis, 191, 292n30, 293n42
Wallace, David, 217–18, 221–2,
    299nn20, 22, 300nn26, 28,
    301nn34, 39
Walters, Lori, 45, 256n3, 263n82,
    269n65
water, 47, 55, 60–1, 65, 118, 130–2,
    138, 140, 143, 147–8, 153, 156,
    158–9, 168–9, 172, 206
Waterhouse, Ruth, 300n25
Watson, Nicholas, 242–3, 305nn20–1,
    23
wax, 6, 25, 28–9, 35, 42, 157, 229–30
Weisheipl, James, 254n72, 296n64
Wentersdorf, Karl, 301n39
Wetherbee, Winthrop, 80, 89–90, 98,
    107–10, 265n5, 266n21, 268n62,
    270nn93–4, 99
White (personification), 190–5, 212–
    13, 238
Whitman, Jon, 105, 211, 247nn29,

40, 247nn29, 39–40, 270nn84, 90, 298n2

will, 26, 41, 113, 121, 158, 170, 172, 175, 195, 197–201, 220–2, 225–32

William of Conches, 3, 10, 19, 32–5, 40, 57–63, 65, 67–8, 71, 84, 99, 235, 245n2, 253nn47, 49–51, 259nn34–5, 40–2, 261n64, 262nn65, 68, 72, 263n80, 285n43, 292n32

William of Moerbeke, 141

Williams, Rowan, 4, 245n10, 246nn11, 15

Wilson, Curtis, 297n73

Windeatt, B.A., 185, 290n6, 291n21

wisdom, 166, 188, 218–19, 228; Wisdom (personification), 146, 219, 238

wit, 9, 41–2, 190–3, 197

Witelo, 37, 90, 115–16, 122, 141–2, 164, 166, 254nn65–6, 278n44, 281nn6–7, 286nn48, 52

Wodeham, Adam, 38, 44, 254n71

Wolfson, Harry, 245n6, 255n78

Wood, Rega, 254n71

wordplay, 8, 14, 188. *See also* polysemy, punning

wrath, 159. *See also* anger

Wurtele, Douglas, 303n51

Wycliff, John, 199

Yager, Susan, 289n3

Yates, Frances, 245n7, 256n86, 260n54, 261n60, 288n65, 294n48

Yowthe. *See* Joinece

Zumthor, Paul, 105, 270n85